ESSENTIAL EU
CLIMATE LAW

ESSENTIAL EU CLIMATE LAW

SECOND EDITION

EDITED BY

EDWIN WOERDMAN
Professor of Markets and Regulation and Co-Director of the Groningen Centre of Energy Law and Sustainability, University of Groningen, the Netherlands

MARTHA ROGGENKAMP
Professor of Energy Law and Director of the Groningen Centre of Energy Law and Sustainability, University of Groningen, the Netherlands

MARIJN HOLWERDA
Legal Counsel, NV Nederlandse Gasunie and Fellow of the Groningen Centre of Energy Law and Sustainability, University of Groningen, the Netherlands

Edward Elgar
PUBLISHING

Cheltenham, UK • Northampton, MA, USA

Published by
Edward Elgar Publishing Limited
The Lypiatts
15 Lansdown Road
Cheltenham
Glos GL50 2JA
UK

Edward Elgar Publishing, Inc.
William Pratt House
9 Dewey Court
Northampton
Massachusetts 01060
USA

Paperback edition 2022

A catalogue record for this book
is available from the British Library

Library of Congress Control Number: 2021943482

This book is available electronically in the **Elgar**online
Law subject collection
http://dx.doi.org/10.4337/9781788971300

ISBN 978 1 78897 129 4 (cased)
ISBN 978 1 78897 130 0 (eBook)
ISBN 978 1 78897 131 7 (paperback)

Printed and bound by CPI Group (UK) Ltd, Croydon, CR0 4YY

CONTENTS IN BRIEF

FULL CONTENTS

FIGURES

CONTRIBUTORS

Lea Diestelmeier, Assistant Professor, University of Groningen, the Netherlands. Member of the Groningen Centre of Energy Law and Sustainability (GCELS).

Joris Gazendam, Lecturer in Law, Hanze University of Applied Sciences, the Netherlands. PhD researcher, University of Groningen. Member of the Groningen Centre of Energy Law and Sustainability (GCELS).

Kars de Graaf, Professor of Public Law and Sustainability, University of Groningen, the Netherlands. Co-director of the Groningen Centre of Energy Law and Sustainability (GCELS).

Marlies Hesselman, Lecturer in International Law, University of Groningen, the Netherlands. Member of the Groningen Centre of Energy Law and Sustainability (GCELS).

Marijn Holwerda, Legal Counsel, NV Nederlandse Gasunie, the Netherlands. Formerly PhD researcher, University of Groningen. Fellow of the Groningen Centre of Energy Law and Sustainability (GCELS).

Romain Mauger, Post-doctoral researcher, University of Groningen, the Netherlands. Member of the Groningen Centre of Energy Law and Sustainability (GCELS).

Martha Roggenkamp, Professor of Energy Law, University of Groningen, the Netherlands. Director of the Groningen Centre of Energy Law and Sustainability (GCELS).

Lorenzo Squintani, Professor of Energy Law, University of Groningen, the Netherlands. Member of the Groningen Centre of Energy Law and Sustainability (GCELS).

Florian Stangl, Attorney at Law with Niederhuber & Partner Rechtsanwälte GmbH, Vienna, Austria. Formerly Master Student of Energy and Climate Law, Groningen Centre of Energy Law and Sustainability (GCELS).

Hanna Tolsma, Professor of Decision-making and Legal Protection in Environmental Law, University of Groningen, the Netherlands. Member of the Groningen Centre of Energy Law and Sustainability (GCELS).

Hans Vedder, Professor of Economic Law, University of Groningen, the Netherlands. Member of the Groningen Centre of Energy Law and Sustainability (GCELS).

Edwin Woerdman, Professor of Markets and Regulation, University of Groningen, the Netherlands. Co-director of the Groningen Centre of Energy Law and Sustainability (GCELS).

Olivia Woolley, Associate Professor in Biolaw, Durham University, United Kingdom. Formerly post-doctoral researcher, University of Groningen. Fellow of the Groningen Centre of Energy Law and Sustainability (GCELS).

PREFACE

EU climate law is one of the most dynamic and fastest growing areas of EU law. The Groningen Centre of Energy Law and Sustainability (GCELS) has therefore taken the initiative to prepare a second, revised edition of its textbook on *Essential EU Climate Law*. All authors of this book are members or fellows of GCELS at the University of Groningen, the Netherlands.

The purpose of our book is to provide an introduction to EU climate mitigation law and to review some of the main challenges affecting climate regulation in the European context. Special attention is paid to the energy sector, which is responsible for the largest share of greenhouse gas emissions in the EU. After presenting a definition of climate law, we discuss the main climate targets and instruments of the EU, such as emissions trading, the use of renewable energy sources and energy efficiency mechanisms. We also clarify their legal impact on energy network regulation, multi-level governance and human rights. The latter issues are important for understanding the broader picture of EU climate mitigation law.

The book is organised according to a specific format. Each chapter not only addresses the relevant directives and regulations, but also analyses their implementation and how it affects current policy and academic debate. This combination is what makes our introduction a comprehensive one, namely by describing, analysing and reviewing the main features and issues of EU climate mitigation law.

Content-wise, therefore, our book is unique: until this publication, an up-to-date and easily accessible volume presenting and critically reflecting upon EU climate mitigation law was missing. Process-wise, our book is unique too. Most of the chapters of the first edition have been actively reviewed by students of the LLM course on Climate Law at the University of Groningen. They provided valuable referee reports to us. The second edition was guided by a publisher survey among users of the first edition of the book, including lecturers and students across Europe. This led to various suggestions which helped us to improve the chapters. The text of the book was completely updated by covering all latest developments up to and including December 2020.

We hope that you will enjoy our exposition of the 'essence' of EU climate law.

Groningen, 15 January 2021
Edwin Woerdman, Martha Roggenkamp and Marijn Holwerda (editors)

ABBREVIATIONS

AAUs	assigned amount units
ACER	Agency for the Cooperation of Energy Regulators
AEAs	annual emission allocations
AUS CPM	Australia Carbon Pricing Mechanism
AVR	Accreditation and Verification Regulation
CBA	cost–benefit analysis
CBDR	common but differentiated responsibilities
CCS	carbon capture and storage
CDM	Clean Development Mechanism
CEN	European Committee for Standardisation (Comité Européen de Normalisation)
CENLEC	European Committee for Electrotechnical Standardisation
CERs	Certified Emission Reductions
CFCs	chlorofluorocarbons
CH_4	methane
CITL	Community Independent Transaction Log
CMP	Conference of the Parties Serving as the Meeting of Parties to the Kyoto Protocol
CO_2	carbon dioxide
COP	Conference of the Parties
CSLF	Carbon Sequestration Leadership Forum
DENA	German Energy Regulator
DSO	distribution system operator
ECCP	European Climate Change Programme
ECJ	European Court of Justice
EEC	European Economic Community
EEX	European Energy Exchange
EEZ	exclusive economic zone
EIB	European Investment Bank
EIF	European Investment Fund
EITE	Energy Intensive Trade Exposed
ENTSOE	European national transmission system operators for electricity
ENTSOG	European national transmission system operators for gas
ERUs	Emission Reduction Units
ESD	Effort Sharing Decision

ESTI	European Telecommunications Standards Institute
ETS	Emissions Trading System
EU	European Union
EU ETS	European Union Emissions Trading System
EUTL	European Union Transaction Log
EWEA	European Wind Energy Association
FIT	feed-in tariff
GCELS	Groningen Centre of Energy Law and Sustainability
GDP	gross domestic product
GHGs	greenhouse gases
GOs	Guarantees of Origin
GWP	global warming potential
HCFCs	hydrochlorofluorocarbons
HFCs	hydrofluorocarbons
ICAO	International Civil Aviation Organization
ICE	Intercontinental Exchange Futures Europe
ICT	information and communication technology
IEA	International Energy Agency
IETA	International Emissions Trading Association
IMO	International Maritime Organization
IPCC	Intergovernmental Panel on Climate Change
ITO	Independent Transmission Operator
JI	Joint Implementation
MAC	Mobile Air-Conditioning
MAD	Market Abuse Directive
MBM	market-based mechanism
MiFID	Markets in Financial Instruments Directive
MRR	Monitoring and Reporting Regulation
MRV	monitoring, reporting and verification
MSR	market stability reserve
Mt	megatonne
Mtoe	million tonnes of oil equivalent
N_2O	nitrous oxide
NAPs	National Allocation Plans
NEEAPs	National Energy Efficiency Action Plans
NF_3	nitrogen trifluoride

NIMs	National Implementation Measures
NRA	national regulatory authority
NZ ETS	New Zealand Emissions Trading Scheme
ODS	ozone-depleting substances
OFTO	offshore transmission owner
PCI	project of common interest
PFCs	perfluorocarbons
QELRO	Quantified Emission Limitation or Reduction Objective
RES	renewable energy sources
REZ	renewable energy zone
RGGI	Regional Greenhouse Gas Initiative
RMUs	removal units
SAVE	Specific Actions for Vigorous Energy Efficiency
SF_6	sulphur hexafluoride
SM-CG	Smart Meters Coordination Group
STEK	Stichting Emissiepreventie Koudetechniek
TEN	trans-European networks
TFEU	Treaty on the Functioning of the European Union
THT	tetrahydrothiophene
TSO	transmission system operator
UCPTE	Union for the Coordination of Production and Transmission of Electricity
UKDECC	UK Department of Energy and Climate Change
UNCLOS	United Nations Convention on the Law of the Sea
UNEP	United Nations Environment Programme
UNFCCC	United Nations Framework Convention on Climate Change
VAT	value-added tax
WCI	Western Climate Initiative
WMO	World Meteorological Organization
ZEP	Zero Emissions Platform

TABLE OF CASES

Court of Justice of the European Union

European Court of Human Rights

Dutch Cases

New Zealand Cases

Swiss Cases

US Cases

TABLE OF LEGISLATION

Regulations

Decisions

Resolutions

National Laws

PART I
INTRODUCTION

1
Purpose, approach and outline of the book

1.1   PURPOSE OF THE BOOK

EU climate law can be defined as all European Union legislation related to climate action. It is one of the most dynamic and fastest growing areas of EU law. This is perhaps best exemplified by the Union's recent efforts to establish a 'European Climate Law' that intends to present a legal framework for achieving climate neutrality in the EU by 2050.[1]

In general, a distinction can be made between climate mitigation law and climate adaptation law. The former focuses on the reduction ('mitigation') of greenhouse gas emissions to curb global warming, while the latter concerns the adjustment of society ('adaptation') to the consequences of global warming. Climate mitigation law thus addresses the causes of climate change, for instance by pricing carbon and by requiring a higher share of renewable energy. Climate adaptation law fights the consequences of climate change, for instance by building more flood defences.

The purpose of our book is to provide an introduction to EU climate mitigation law. We focus on mitigation law for the following four reasons. First, mitigation measures are typically defined at EU level, whereas adaptation measures are mostly constructed at (sub-)national level. Second, in the period 2021–30, mitigation requires considerably higher extra investment from the EU budget than does adaptation.[2] Third, mitigation will have long-term and global benefits, whereas adaptation mostly has short-term, local and sector-specific benefits.[3] For these reasons mitigation is typically mentioned prior to adaptation, not only on EU climate action websites[4] but also in EU legislation, such as the 2020 Taxonomy Regulation, which con-

---

[1]   Proposal for a Regulation establishing the framework for achieving climate neutrality (European Climate Law), COM/2020/80 final. See also: https://ec.europa.eu/clima/policies/eu-climate-action/law_en accessed 16 January 2021.

[2]   D. Forster et al. (2017), Climate mainstreaming in the EU Budget: preparing for the next MFF, Luxembourg: Publications Office of the European Union, pp.7–8.

[3]   A. Hof, K. de Bruin, Rob Dellink, M. den Elzen and D. van Vuuren (2010), 'Costs, benefits and interlinkages between adaptation and mitigation', in: F. Biermann, P. Pattberg and F. Zelli (eds), Global Climate Governance Beyond 2012: Architecture, Agency and Adaptation. Cambridge: Cambridge University Press, pp.235–54.

[4]   See https://ec.europa.eu/clima/policies/eu-climate-action_en accessed 10 January 2021.

tains criteria for sustainable investment.[5] But there is also a fourth and more practical reason to focus on EU climate mitigation law: a textbook that covers both mitigation and adaptation in Europe would either be excessively long or would run the risk of insufficiently treating their respective legal characteristics and issues.

Most EU climate mitigation law comes in the form of EU directives and regulations that intend to reduce greenhouse gas emissions in the fight against climate change. These do not operate in isolation from other areas of EU law, such as energy law, environmental law, competition law, human rights law or financial regulation. We will touch upon those areas of law whenever they interact with EU climate mitigation law. Moreover, law is intended to have an impact on society. We will therefore pay some attention to the implementation of EU climate law and its effects on climate policy, companies and citizens.

Our book is the first textbook solely devoted to the basics of climate mitigation law in the EU, offering a thorough introduction for students, legal professionals and others who are unfamiliar with (some parts of) EU climate regulation. Our book, however, does more than that.

First, we focus on the relationship between climate law and the energy sector, because the use of energy is an important driver of climate change. The energy sector is responsible for the largest share of greenhouse gas emissions in the EU.[6] These emissions are primarily the by-product of burning fossil fuels. Another reason for this focus is the explicit policy link made by the EU between climate and energy, for instance in the 2020 Climate and Energy Package and subsequently in the 2030 Climate and Energy Framework, adopted in 2008 and 2014 respectively.[7] These policy documents outline the key climate and energy goals of the EU. They mainly contain interlinked targets for reducing greenhouse gas emissions by *inter alia* increasing the share of renewable energy and improving energy efficiency.

Second, after discussing the main climate targets and instruments of the EU, including emissions trading as well as the promotion of renewable energy and energy efficiency, we also expand upon some related issues, such as the impact of decarbonisation on energy network operations, multi-level governance and human rights. This is important to understand the broader picture of EU climate law. Promoting the use of renewable energy sources, for instance, requires investment in network expansion and the construction of new (types of) networks and has a clear impact on network balancing. Moreover, EU climate mitigation law is shaped in a process of multi-level governance between international climate agreements, EU climate policy and climate action by Member States as well as companies, cities and citizens. Last but not least, human rights play an increasingly important role in climate law, because climate change has a negative impact on people's fundamental rights, such as the right to life, the right to health and the right to a healthy environment. We are convinced that these broader issues deserve proper attention, because EU climate regulation does not develop in isolation but evolves with and adapts to interrelated legal issues.

---

[5] Regulation (EU) 2020/852 on the establishment of a framework to facilitate sustainable investment, and amending Regulation (EU) 2019/2088 (Text with EEA relevance), PE/20/2020/INIT, OJ L 198, 22.6.2020, pp.13–43.

[6] https://ec.europa.eu/eurostat/cache/infographs/energy/bloc-4a.html accessed 14 January 2021.

[7] https://ec.europa.eu/clima/policies/strategies_en accessed 14 January 2021.

The title of our book is *Essential EU Climate Law* – and not without reason. Focusing on the 'essence' of EU climate mitigation law means that we first (a) address the relevant directives and regulations and then continue to (b) analyse their implementation and (c) review current policy and academic debate on the climate targets and instruments of the EU. This combination is what makes our introduction a comprehensive one, namely by describing, analysing and reviewing the main features and issues of the EU legal framework that aims to combat climate change. The central question of our book is therefore: what is the essence of EU climate mitigation law?

The following sub-questions are addressed:

- What are the targets and instruments of EU climate mitigation law?
- What are the relevant directives and regulations in EU climate mitigation law?
- What are the main implementation issues of EU climate mitigation law?
- How can those implementation issues be solved in a cost-effective and acceptable way?

To ensure coherence and close coordination, all authors of our book are members or fellows of the Groningen Centre of Energy Law and Sustainability (GCELS), at the University of Groningen in the Netherlands. At the time of writing we were, and most of us still are, housed in the same Faculty building, although the COVID-19 pandemic forced us to work from home for some time. We agreed to write the core chapters on EU climate mitigation law based on a tight and uniform format that contained various requirements, not only in relation to issues of style, but also in relation to an analytical approach that consistently focuses on cost-effectiveness and solidarity in EU climate mitigation law. We will now elaborate upon this common approach.

1.2   APPROACH OF THE BOOK

Our approach is to explore the relevant legislation in each selected area of EU climate mitigation law from the perspective of the following two principles:

- cost-effectiveness, and
- solidarity.

The principles of cost-effectiveness and solidarity can already be found in the so-called 'Climate and Energy Package' proposal of 2008, in which the European Commission explains:[8]

> The architecture of the proposals has been driven by two factors. First, the proposals are designed in such a way that the targets are reached in the most cost-effective way possible. Second, the effort required of particular Member States and particular industries remains

---

[8]   European Commission, 'Communication from the Commission to the European Parliament, the Council, the European Economic and Social Committee and the Committee of the Regions – 20 20 by 2020: Europe's Climate Change Opportunity' (2008) COM(2008) 30 final, 4.

balanced and proportionate, and takes their own circumstances into account. Fairness and solidarity have been at the heart of the Commission's thinking in developing the proposals.

The same principles were also highlighted in 2014 in the European Council conclusions on the EU's climate and energy policy for 2030:[9]

> The European Council endorsed a binding EU target of an at least 40% domestic reduction in greenhouse gas emissions by 2030 compared to 1990. To that end: [...] the target will be delivered collectively by the EU in the most cost-effective manner possible [and] all Member States will participate in this effort, balancing considerations of fairness and solidarity.

A wider set of principles but a comparable approach can be found in the European Commission's proposal, tabled in 2020, for a 'European Climate Law':[10]

> A fixed long-term objective is crucial to contribute to economic and societal transformation, jobs, growth, and the achievement of the United Nations Sustainable Development Goals, as well as to move in a fair and cost-effective manner towards the temperature goal of the 2015 Paris Agreement on climate change [...] taking into account the importance of promoting fairness and solidarity among Member States.

The EU aims to reach its multiple climate targets by using various instruments that ensure least-cost implementation and solidarity among its Member States. Our main hypothesis, therefore, is that EU climate mitigation law is primarily based on the principles of cost-effectiveness and solidarity. This hypothesis is tested for the different areas of EU climate regulation in each separate chapter. This approach only applies to the second part of our book, which centres on essential EU climate mitigation law itself. It is not applied in the third part, which discusses overarching issues, such as energy network management, that relate only indirectly to EU climate law. An assessment based on these two principles allows for a characterisation of the relevant area of climate law that goes beyond a mere explanation of the law as it is. As a consequence, the author of each chapter in the second part of the book also answers the following question: to what extent have the principles of cost-effectiveness and solidarity been honoured in the relevant legislation, and what balance has been struck between them?

As part of this assessment, the role of other related issues, such as the protection of domestic economic interests, is also addressed. In some parts of EU climate law, the principle of solidarity may have gained ground at the expense of the principle of cost-effectiveness. In other parts of EU climate law, we would for instance expect that particular instruments orig-

---

[9]   European Council, 'European Council (23 and 24 October 2014) Conclusions on 2030 Climate and Energy Policy Framework' (2014) SN79/14, 2.

[10]   Proposal for a Regulation of the European Parliament and of the Council establishing the framework for achieving climate neutrality and amending Regulation (EU) 2018/1999 (European Climate Law), COM/2020/80 final, Preamble (3) and Article 2(2).

inally intended to guarantee a cost-effective realisation of climate goals have been weakened to protect national industrial interests. This methodology gives the authors of the different chapters (in the second part of the book) a common writing approach to ensure consistency between the different chapters and to facilitate a critical analysis and review of the law.

1.3 CONTRIBUTION OF THE BOOK

This timely second, revised edition of our textbook offers a comprehensive introduction to EU climate law, with a special focus on the energy sector. We have updated our book based on the most recent legal developments. We have also made improvements based on a publisher survey among users of the first edition, including both lecturers and students. Alongside addressing the essential directives, regulations, decisions and court cases, our textbook analyses implementation issues, including their impact on cost-effectiveness and solidarity. The book reviews current policy and academic debate on EU targets and instruments to reduce greenhouse gas emissions. It also expands upon related legal issues, such as energy network regulation, multi-level governance and human rights. By focusing on the essence of EU climate mitigation law, it offers easily accessible insights for both students and professionals.

There are several books that deal with international and EU climate regulation or policy in one way or another. However, none of these books is solely devoted to the main features and implementation issues of EU climate mitigation law, with particular attention paid to the energy sector. There are several thematic books on climate law and policy, including very interesting books on climate change law in general, on EU Member State climate policies, on climate change adaptation law, on emissions trading, or on renewables, but none of these books provides a comprehensive introduction to essential EU climate law in relation to broader legal issues including energy law. This is what we aim to do.

Content-wise, in outlining the aforementioned subjects and issues, our book is unique. There is considerable academic literature on EU climate mitigation law dealing with very specific questions, but an up-to-date and easily accessible book that presents and critically reflects upon EU climate law was missing. As we focus on the relationship between climate mitigation and the energy sector, we will also examine specific energy-related climate mitigation developments, such as the regulation of carbon capture and storage (CCS), the promotion of renewable energy consumption and energy efficiency, as well as the decarbonisation of energy networks. Our book addresses this niche by targeting primarily students, but also others who need to familiarise themselves with EU climate law design and implementation in a comprehensive and time-efficient way.

Our book is unique in terms of process, too. Most of the chapters in the first edition were actively reviewed by students of the 2014 course on Climate Law at the University of Groningen. They provided valuable referee reports to us. This led to various questions and suggestions, on the basis of which we have been able to improve the chapters in our textbook. The second, revised edition has taken into account the most recent developments in EU climate mitigation law up to December 2020, as well as the outcomes of an EU-wide survey carried out by Edward Elgar Publishing in 2018 among users of the first edition of the book.

Both lecturers and students from across Europe provided us with a useful 'wish list' which helped us to further improve our textbook. Examples are an extra chapter on the relationship between human rights and EU climate law, a new section on the role of cities in mitigating greenhouse gas emissions, more information on the use of carbon taxation and the addition of classroom questions to each chapter (with relevant legal issues to be debated in class), among others.

1.4 OUTLINE OF THE BOOK

Our book is organised as follows. The first part introduces the subject of, as well as our approach to, EU climate law. The purpose and approach of our book have been explained above. Chapter 2, which follows, gives an overview of EU climate policy, including a discussion on the impact that international climate law had on the development of EU climate measures. That chapter therefore presents a brief outline of the international, historical and policy background to EU climate mitigation law, as further explored in the second part of the book.

The chapters in the second part of the book describe and analyse the main pillars of EU climate regulation, focusing on the principles of cost-effectiveness and solidarity. We subsequently examine the EU's greenhouse gas emissions trading scheme, abbreviated as EU ETS (Chapter 3); the regulation of emissions from sectors not covered by the EU ETS (Chapter 4); and the roles of renewable energy consumption (Chapter 5), energy efficiency (Chapter 6) and carbon capture and storage (Chapter 7) in reducing greenhouse gas emissions, as well as the regulation of fluorinated gases (Chapter 8). These chapters provide more than just a concise overview of the relevant legislation. Starting with the basics and functioning of the applicable regulation, each chapter proceeds to explore key academic and policy debates and finishes with a brief outlook for the future.

The third part of the book explores the interaction between EU climate regulation and three important and topical overarching issues. First, we examine the legal consequences of EU climate regulation for the management of energy networks (Chapter 9). Among other things, we address how decentralised energy production impacts the existing regulatory framework that was designed for centralised energy systems, and how this leads to new and innovative legal solutions. We then discuss the issue of multi-level climate governance in the context of EU climate regulation (Chapter 10). In this chapter we examine the interaction between EU climate policies, on the one hand, and climate policies at the level of EU Member States and the international level, on the other. Subsequently, we discuss the relationship between human rights and EU climate law (Chapter 11). After clarifying the human rights obligations of the EU and its Member States when taking climate action, we explain why human rights-based climate litigation is emerging.

The fourth part of the book concludes. The final chapter reflects upon the past of EU climate law and provides an outlook on its possible future (Chapter 12). In this chapter we link the findings of the different chapters and give the reader some food for thought about the steps that the EU may or perhaps should take next to further improve its climate mitigation law.

Before we start analysing the various aspects of EU climate mitigation law in the next chapters, we would like to make a comment on the science of climate change in relation to our book. In order to combat climate change, scientists want climate policies to lead to genuine reductions in greenhouse gas emissions. This must be done to limit the damage that these gases do. This damage is serious and increasingly visible, albeit surrounded by many uncertainties. There are climate sceptics and climate activists – and people that stand somewhere in between. It is not our intention to choose in favour of either of those sides. In fact, as lawyers and law and economics scholars, we lack the scientific knowledge to do so. We do observe that, according to the Intergovernmental Panel on Climate Change (IPCC), the balance of scientific evidence suggests a discernible human influence on the global climate.[11] Instead of diving into the science of climate change and instead of choosing sides between sceptics and activists, the starting point of this book is a simple but clear legal observation. The EU and its Member States have – and this is a fact – formulated all kinds of policy targets as well as various legal and regulatory instruments to reduce greenhouse gas emissions. We take it from there, and describe, analyse and review these targets and instruments in EU climate mitigation law.

1.5   ACKNOWLEDGEMENTS

This book was initiated by Marijn Holwerda, a specialist in carbon capture and storage who obtained his doctoral degree at the Faculty of Law in Groningen. At the University of Groningen, Martha Roggenkamp is a specialist in energy law and coordinates several teaching programmes such as the LLM in Energy and Climate Law. Edwin Woerdman is a specialist in emissions trading and coordinates several courses, including the master course on Climate Law. Together they form the editorial team of this book. Each chapter in this book has been refereed by two members of this editorial team. Most of the chapters of the first edition have also been reviewed by students of the 2014 course on Climate Law at the University of Groningen. The second, revised edition was improved based on a 2018 review carried out among lecturers and students throughout Europe. The chapters themselves were written by the following members or fellows of GCELS:

- Chapter 1   Edwin Woerdman, Martha Roggenkamp and Marijn Holwerda
- Chapter 2   Florian Stangl and Romain Mauger
- Chapter 3   Edwin Woerdman
- Chapter 4   Lorenzo Squintani
- Chapter 5   Olivia Woolley

---

[11]   'Human activities are estimated to have caused approximately 1.0°C of global warming above pre-industrial levels, with a likely range of 0.8°C to 1.2°C. Global warming is likely to reach 1.5°C between 2030 and 2052 if it continues to increase at the current rate. [...] Climate-related risks to health, livelihoods, food security, water supply, human security, and economic growth are projected to increase with global warming of 1.5°C and increase further with 2°C.' Quotation taken from: IPCC Special Report (2018), Global Warming of 1.5°C: Summary for Policymakers, V. Masson-Delmotte et al. (eds.), Geneva: World Meteorological Organization, p.4 and p.9.

- Chapter 6 Hans Vedder
- Chapter 7 Marijn Holwerda and Joris Gazendam
- Chapter 8 Kars de Graaf and Hanna Tolsma
- Chapter 9 Martha Roggenkamp and Lea Diestelmeier
- Chapter 10 Hans Vedder
- Chapter 11 Marlies Hesselman
- Chapter 12 Edwin Woerdman, Martha Roggenkamp and Marijn Holwerda

In the first edition, Avelien-Haan Kamminga was co-author of the chapter on carbon capture and storage (now Chapter 7), and Hannah Kruimer was co-author of the chapter on energy network management (now Chapter 9). Although they are not involved in this second edition, we do want to thank them sincerely for their hard work in preparing the first edition a few years ago.

2
EU climate policy

ABSTRACT

- EU climate policy is an aggregate of legislative measures to combat climate change and its consequences;
- The UN Framework Convention on Climate Change and the Paris Agreement have a substantial influence on the design of EU climate policy;
- EU climate policy is mainly built on targets for greenhouse gas reduction, for renewable energy consumption and for energy efficiency, which should also improve energy security;
- The EU designed various instruments, including emissions trading and various emission standards, to reach its climate targets in a way that stimulates cost-effectiveness and takes into account solidarity between its Member States;
- The primary goals of EU climate policy are to reduce greenhouse gas emissions and to secure energy supply; its secondary goals are to develop and export low-carbon technologies and to protect industrial competitiveness;
- The EU tries to lead in international climate negotiations, for instance by setting an ambitious greenhouse gas emission reduction target for 2030;
- The EU aims for a carbon-neutral economy by 2050 in which emissions and removals of greenhouse gases are balanced, thus reducing emissions to net zero.

2.1 INTRODUCTION

This chapter aims to give an overview of EU climate policy and will introduce the targets and instruments of EU climate law as well as its relevant directives and regulations.

The body of EU legislation in the field of climate protection developed gradually – from more sectoral provisions in the 1990s to a comprehensive legal framework which covers a wide range of activities such as industrial production, energy generation, transportation and housing. EU climate policy cannot be isolated from international developments. The 1997 Kyoto Protocol shaped the design of the first generation of (binding) EU climate legislation, while the 2015 Paris Agreement paved the way for more ambitious targets and policies. This chapter traces the steps taken from an early stage and describes current EU policy in the field of climate action against the background of international developments.

Section 2.2 provides the basics of 'climate' and 'policy' as well as the economic rationale behind combating global warming. Section 2.3 addresses international efforts to combat the climate crisis. Section 2.4 focuses on the most important developments in EU climate policy from its beginnings until 2020. Section 2.5 outlines the objectives and legislative actions implemented by the 2030 Climate and Energy Framework and provides a longer-term perspective by focusing on the EU Climate Policy Roadmap for 2050. Section 2.6 outlines the role of cities in combating climate change. Finally, section 2.7 concludes.

2.2 BASICS OF 'CLIMATE' AND 'POLICY'

Before talking about 'climate policy' as a set of legal tools to prevent climate change and miti-gate the effects of global warming, we should first define the concepts of:

- climate, and
- policy.

2.2.1 Climate Change Caused by Human Activities

'Climate' refers to 'conditions of the atmosphere at a particular location over a long period of time'; it is the long-term summation of atmospheric elements such as temperature, wind, precipitation and humidity.[1] Thus, climate change means a change of average daily weather on a lasting basis. A changing climate is not an abnormal development as such. The earth has gone through several climatic periods caused by natural events such as volcanic activities or changes in sun exposure.

However, the global warming now faced by humankind is to a large extent caused by human activities.[2] Due to industrialization in recent centuries, the concentration of carbon dioxide (CO_2) and other gases in the atmosphere has reached a level that leads to the so-called greenhouse effect: solar radiation is reflected from the earth's surface but cannot fully escape the atmosphere because it is re-reflected by these gas molecules. Put simply, the more of those greenhouse gases (hereafter GHGs) are emitted, the more solar radiation will 'stay' at the earth's surface and, consequently, heat it up. Although GHGs occur naturally in the atmos-phere, human activities have substantially increased their levels. The most significant GHG of anthropogenic origin is CO_2, which typically arises from the combustion of fossil fuels such as coal, petroleum and natural gas, for instance to produce industrial goods, to generate energy or to fuel cars and other vehicles. The amount of emissions is closely connected with the way we live. Emission reductions can be achieved by changing our way of transportation (for example, from individual to public transport or by developing alternatives to fuel-based cars) or by improving the insulation of our buildings to reduce energy consumption for heating. Moreover, switching from hydrocarbon-based energy production to the use of renewable energy sources (RES) can effectively decrease our CO_2 output.

---

[1]   Cf. *Encyclopædia Britannica* (online), s.v. 'climate'; available under Britannica.com/science/climate-meteorology accessed 17 June 2021.

[2]   Cf. IPCC, Special Report: Global Warming of 1.5°C – Summary for Policymakers (World Meteorological Organization, Geneva 2018) p.4 et seq; available under https://www.ipcc.ch/sr15/ accessed 3 January 2021.

Carbon dioxide is not the only gas that contributes to the greenhouse effect. Methane, nitrous oxide and fluorinated gases are even more potent than CO_2, but they are also emitted in smaller quantities. They result especially from the production or use of certain products (such as air-conditioners), agricultural activity and transport.[3] In order to provide for an effect-based comparison of the various greenhouse gases and to monitor reduction achievements, non-CO_2 GHGs are expressed in carbon dioxide 'equivalent' (CO_2e) – indicating the global warming potential of each gas.

Besides the production of GHGs, the existence of carbon sinks influences the atmospheric GHG balance. A carbon sink is anything that processes and stores CO_2 or other GHGs, so that the gases are removed from the atmosphere. One can distinguish natural from artificial reservoirs. Examples of natural sinks are plants, the soil and the ocean. Trees, for example, absorb CO_2 through the process of photosynthesis and store it. Consequently, deforestation frustrates the natural removal of carbon dioxide and contributes to climate change. Artificial carbon sinks are techniques used to store GHGs permanently. The most prominent example of this is 'carbon capture and storage' (CCS), where CO_2 gets captured from flue gases and is sequestered in deep subsoil layers.

2.2.2 Climate Policy as a Concept

The term 'policy' derives from the ancient Greek word 'ta politika' ('the political things'), which refers to issues concerning the *polis* (city or city-state).[4] It describes the aggregate of tasks necessary for the formation and maintenance of a state and its society. Whereas the terms 'polity' and 'politics' cover the formalistic and procedural aspects, 'policy' means the content-related dimension of political activity. It comprises substantial measures by a government or legislature in order to tackle a problem or provide a service. A policy is not tantamount to a certain legal act, although they are related to each other: the content of a law reflects the policy that the legislature pursues in a regulatory field and can thus be seen as its materialization in the form of binding social rules.

The EU develops policies in various fields,[5] including climate change. EU 'climate policy' encompasses measures aimed at preventing climate change, especially by reducing GHG emissions and by alleviating the consequences of global warming through adaptation strategies. Due to the high impact of energy production and consumption on the greenhouse effect, the policies on climate protection and energy supply are closely interlinked.

---

[3] For a detailed analysis of the GHG sources in the EU see European Environment Agency, 'Annual European Union greenhouse gas inventory 1990–2017 and inventory report 2019 – Technical report No 06/2019'; available under https://www.eea.europa.eu//publications/european-union-greenhouse-gas-inventory-2019 accessed 1 December 2019.

[4] See S. Goldhill, 'Greek Drama and Political Theory', in: C. Rowe, M. Schofield (eds), *Cambridge History of Greek and Roman History of Thought* (Cambridge University Press, Cambridge 2000), p.62; R. Balot, 'Greek Political Thought' (Blackwell Publishing, Oxford 2006), pp.2 et seq.

[5] According to the 'principle of conferral' in Article 4(1) of the TEU, the EU may only regulate an area if the competence to do so has been conferred on it by the Member States.

Climate change implies costs for the Member States. As a result of climate change, the Netherlands, for example, has to raise its dykes and Italy, just to name another example, has to deal more often with droughts and extreme weather events. These are tangible 'social costs', also referred to as external costs or negative externalities.[6] The Joint Research Centre of the European Commission calculated that if no further action is taken, the EU welfare loss could amount to 1.9 per cent of GDP (240 billion euros) per year by the end of the century and could be reduced by two-thirds in a 2°C global warming scenario. The highest welfare loss would occur in Southern Europe.[7]

However, there is a free-rider problem to deal with in relation to the costs of climate change. The reason for this is that the reduction of GHG emissions suffers from a 'tragedy of the commons'.[8] Everyone prefers to enjoy the benefits of the emission reductions without having to contribute to the costs of bringing them about.

By implementing climate policies in the EU, GHG emitters are basically forced to take account of those social costs. The polluter pays principle, which is – according to Article 191(2) of the TFEU – one of the fundamental principles of European environmental law, aims to internalize external effects, by attributing (e.g. climate change-related) costs to the polluters. Although not directly applicable, the polluter pays principle does shape EU environmental and climate policies. There are two main approaches to dealing with externalities:

- Direct regulation ('Command-and-Control')

'Command-and-control' deals with external effects by setting and enforcing environmental standards. The regulatory measures usually implement technology-based and/or performance-based standards in order to minimize the environmental impact (i.e. the external effect) of a certain behaviour. Sanctions should make sure that the regulations are effectively complied with.

- Market-based instruments (MBIs)

MBIs do not determine certain technologies or behaviour, but rather give a financial incentive to the polluter to reduce its environmental impact. Examples of MBIs include environmental taxation, tradable emission allowances, liability regimes and subsidies. The most elaborated and prominent example of a MBI in the EU is the emissions trading system (ETS) for GHGs. The recent climate debate centres on the (domestic) taxation of carbon, although an EU-wide CO_2 tax seems unlikely due to the unanimity requirement under Articles 192(2)(a) or 194(3) TFEU.

These regulations limit the emissions of GHGs and entail costs for companies. They will pass on those costs, to the extent possible, to end-consumers, who will consume fewer goods with a high CO_2 footprint (such as energy, steel or cement), which is and should be the ultimate consequence of EU climate mitigation law.

---

[6]   R.H. Coase, 'The Problem of Social Cost', *Journal of Law and Economics* 3 (1960), pp.1–44.

[7]   European Commission, 'Climate Impacts in Europe', Joint Research Centre (JRC, Seville 2018), pp.59 et seq.

[8]   R. Hardin, *Collective Action* (Johns Hopkins University Press, Baltimore 1982).

2.3   INTERNATIONAL CLIMATE POLICY

2.3.1   History of International Climate Policy

In 1988, the United Nations Environment Programme (UNEP) and the World Meteorological Organization (WMO) established the 'Intergovernmental Panel on Climate Change' (IPCC). The IPCC is an independent scientific body assigned to assess the most recent developments in climatology and to provide the public and policymakers with the current state of scientific knowledge about climate change. Thousands of scientists worldwide analyse and review climate-related publications on a voluntary basis. In its first assessment report, published in 1990, the IPCC stated 'with certainty' that the greenhouse effect exists and that human activities contribute to it. Furthermore, the IPCC predicted an increase of global mean temperatures and sea levels.[9] These conclusions urged international decision-makers to engage in global climate action and paved the way for the United Nations Framework Convention on Climate Change (UNFCCC) that was negotiated during the UN Earth Summit held in Rio de Janeiro in 1992 and entered into force in 1994. According to its Article 2, the UNFCCC aims at stabilizing the concentration of GHGs in the atmosphere 'at a level that would prevent dangerous anthropogenic interference with the climate system'; this goal should be achieved 'within a time frame sufficient to allow ecosystems to adapt naturally to climate change, to ensure that food production is not threatened, and to enable economic development to proceed in a sustainable manner'. The overall target to limit the global temperature increases to at most 2 degrees Celsius above pre-industrial levels was adopted 18 years after the conclusion of the UNFCCC.[10] The UNFCCC covers all GHGs (that are not controlled by the Montreal Protocol – cf. Article 4(1) of the UNFCCC[11]), mainly carbon dioxide, methane, nitrous oxide and fluorinated gases.[12]

---

[9]   J.T. Houghton, G.J. Jenkins and J.J. Ephraums (eds), *Climate Change – The IPCC Scientific Assessment* (Cambridge University Press, Cambridge 1990), p.xi; available under: https://www.ipcc.ch/report/ar1/wg1/. The fifth synthesis report, 'Climate Change 2014: Synthesis Report. Contribution of Working Groups I, II and III to the Fifth Assessment Report of the Intergovernmental Panel on Climate Change', Core Writing Team, R.K. Pachauri and L.A. Meyer (eds), was published in 2015. The sixth synthesis report by the IPCC is expected by 2022. All synthesis and assessment reports are available under www.ipcc.ch/reports/ accessed 3 January 2021.

[10]   Outcome of the COP16/CMP6 held in Cancun 2010; see http://unfccc.int/meetings/cancun_ nov_2010/ meeting/6266.php accessed 3 January 2021.

[11]   The substances covered by the Montreal Protocol to the Vienna Convention for the Protection of the Ozone Layer are enumerated in its Annex; the Protocol addresses, inter alia, chlorofluorocarbons (CFCs), halons and hydro-chlorofluorocarbons (HCFCs).

[12]   Cf. the national GHG inventories of Annex I Parties (from 2019) and non-Annex I Parties (from 2005 to most recent), available under https://unfccc.int/process-and-meetings/transparency-and-reporting/reporting-and -review-under-the-convention/greenhouse-gas-inventories-annex-i-parties/national-inventory-submissions-2019 and https://unfccc.int/process/transparency-and-reporting/reporting-and-review-under-the-convention/national -communications-non-annex-i-parties/compilation-and-synthesis-reports-0 accessed 3 January 2021.

The mitigation of GHGs is the primary objective of the treaty, while its secondary objective is adaptation to the changing living conditions caused by climate change (cf. Article 3(2) of the UNFCCC).[13] Article 4 of the UNFCCC assigns various obligations on the signatory states and refers in its first sentence to the principle of *common but differentiated responsibilities* (CBDR). According to this principle, developed countries – which built their wealth on industrial activities and, thus, on the emission of CO_2 – have a greater historical responsibility for climate change and should therefore make more efforts to fight it than developing states. However, this does not mean that developing countries are exempt from any obligation. All signatories – regardless of whether they are considered developed or developing – shall, inter alia, prepare national inventories of the anthropogenic GHG emissions and removals by sinks, set up programmes of measures suitable to mitigate climate change, and develop appropriate plans for climate change adaptation (cf. Article 4(1) of the UNFCCC).[14]

Nevertheless, the UNFCCC itself did not contain any internationally binding emission reduction targets. This situation changed in 1997, when the Conference of the Parties (COP) to the UNFCCC adopted at its third meeting the 'Protocol to the UN Framework Convention on Climate Change'. Since the conference took place in Kyoto, Japan, it is commonly referred to as the Kyoto Protocol. The COP is the supreme body of the UNFCCC. It consists of representatives of all Parties to the Convention and meets regularly for sessions. The COP periodically reviews the implementation of the UNFCCC and takes decisions necessary to promote effective implementation of the treaty (Article 7(2) of the UNFCCC). The yearly meetings take place in different cities and they are consecutively numbered; the first Conference of the (UNFCCC) Parties was held in Berlin (Germany) in 1995 and is referred to as COP1.

2.3.2   The Kyoto Protocol and the Road to Paris

As a result of the 'Berlin Mandate' from COP1 (Decision 1/CP.1), which basically called for a strengthening of the climate commitments of developed countries, COP3 agreed in 1997 on the Kyoto Protocol to the UNFCCC (Decision 1/CP.3). The aim of the Kyoto Protocol is to limit the emissions of certain GHGs in ways 'that reflect underlying national differences in emissions, wealth and capacity'[15] in order to promote sustainable development (cf. Article 2 of the Kyoto Protocol). Due to strong political resistance in some signatory states, it took almost eight years to enter into force. By 2020, 191 states (comprising all UN Member States except Andorra, Canada, South Sudan and the USA) and the EU (as a supranational organization) had committed themselves to the Kyoto Protocol. The Parties met regularly at the 'Conference of the Parties Serving as the Meeting of Parties to the Kyoto Protocol', abbreviated as CMP. The CMP used to take place at the same time and in the place as the COP. The purpose of the

---

[13]   U. Beyerlin and T. Marauhn, *International Environmental Law* (Hart, CH Beck, Nomos, Oxford 2011), p.159.

[14]   The additional duties of developed states and states in transition mentioned in Annex I and of developed states mentioned in Annex II, respectively, are stated in Article 4(3)–(5) of the UNFCCC.

[15]   M. Grubb, 'The Economics of the Kyoto Protocol', *World Economics* 4:3 (2003), p.144; available under https://discovery.ucl.ac.uk/id/eprint/1471215/1/Grubb_BC10.pdf accessed 3 January 2021.

CMP was to assess and review the implementation of the Kyoto Protocol and to take decisions in order to enhance its effectiveness and improve the legal framework.[16]

Following the CBDR principle, the Parties' obligations regarding emission limitation vary. The states listed in Annex I to the UNFCCC have to reduce or limit their GHG emissions by or to a certain amount compared to base year levels as specified in Annex B to the Kyoto Protocol. These countries will therefore be referred to as 'Annex B Parties'.

For most (industrialized) Annex B Parties, 1990 is the base year for the quantified emission limitation or reduction targets regarding the most important GHGs: carbon dioxide (CO_2), methane (CH_4) and nitrous oxide (N_2O). Parties included in Annex B which are undergoing a process of transition to a market economy (for example, Hungary or Bulgaria) use slightly different historical reference years, in accordance with Article 4(6) of the UNFCCC and Article 3(5) of the Kyoto Protocol.[17]

No binding limitation or reduction obligations exist for developing countries (which are Parties that are neither mentioned in Annex I to the UNFCCC nor in Annex B to the Kyoto Protocol). Instead they have to take 'measurable, reportable and verifiable' actions to mitigate climate change.[18]

The binding GHG targets for industrialized nations must be seen in conjunction with the Kyoto commitment periods. For the first commitment period 2008–12, the Netherlands, for example, was obliged to cut emissions by 8 per cent, which means that the total amount of emitted GHGs had to be 8 per cent less than the total amount emitted in the base year 1990. Some other states had to stabilize their emissions (for example, the Russian Federation), whereas some were even allowed to emit more GHGs than in the reference year (for example, Australia). The second commitment period 2013–20 aimed at an 18 per cent GHG emission reduction compared to 1990 levels by 2020.[19] Some large emitters, such as the Russian Federation and Japan, decided to no longer take on binding emission targets. As for the EU Member States, individual national targets are no longer set out in Annex B to the Kyoto Protocol but there is one common goal for the entire EU, namely a reduction of 20 per cent (or 30 per cent under specific conditions[20]) by 2020. With the second commitment period,

---

[16]   Some other bodies have also been established under the UNFCCC to serve the Kyoto Protocol: Secretariat established by Article 8 of the UNFCCC; Subsidiary Body for Scientific and Technological Advice established by Article 9 of the UNFCCC; Subsidiary Body for Implementation established by Article 10 of the UNFCCC.

[17]   The base year for fluorinated gases is 1995 and that for nitrogen trifluoride either 1995 or 2000 (cf. Article 3(7), (8) and (8 bis) of the Kyoto Protocol). For an overview of the various base years see http://unfccc.int/ghg_data/kp_data_unfccc/base_year_data/items/4354.php accessed 3 January 2021.

[18]   Point 1(b)(ii) Bali Action Plan (Decision 1/CP.13) as adopted at COP13/CMP3.

[19]   Cf. Article 1(C) of the Doha amendment to the Kyoto Protocol (amending Article 3(1) of the Kyoto Protocol).

[20]   The EU offered to move to a 30 per cent reduction by the year 2020 under the condition that a global agreement for the period beyond 2012 is concluded in which developing countries commit themselves to 'contribute adequately according to their responsibilities and respective capabilities' and developed states to 'comparable emission reductions'; cf. Article 1(A) of the Doha amendment (amending Annex B) to the Kyoto Protocol. However, no such international agreement has been concluded.

a handful of industrialized nations are finally subject to binding reduction or limitation targets (the EFTA states, Australia, Belarus, Kazakhstan, Monaco and Ukraine). Additionally, more than 70 developing and developed countries have made non-binding pledges to reduce or limit the output of GHGs.

Facing alarming news regarding GHG concentrations in the atmosphere, underlined by the results of the fifth IPCC synthesis report,[21] the international community worked on a universal climate protection agreement with binding GHG targets for all countries, to enter into force in 2020. In 2011, the 'Ad Hoc Working Group on the Durban Platform for Enhanced Action' was founded by COP Decision 1/CP.17 with the goal of developing 'another legal instrument or an agreed outcome with legal force'. A year later, a timetable for such a globally binding instrument was set (at the COP18/CMP11) in Doha, Qatar.

International climate negotiations in the year 2014 were characterized by a will to pave the way for a globally binding agreement. In September 2014, former UN Secretary-General Ban Ki-moon invited relevant stakeholders – from heads of governments to leaders from the private sector and civil society – to the UN Climate Summit in New York in order to mobilize for climate action. The negotiations took place at the second session of the 'Ad Hoc Working Group on the Durban Platform for Enhanced Action' in October 2014 in Bonn, Germany and the following convention conference in Lima, Peru (COP20/CMP10), held in December 2014.[22] The 2015 Paris Agreement was the result.

2.3.3   The Paris Agreement

On 12 December 2015, the Paris Agreement was adopted via UNFCCC Decision 1/CP.21 (the Paris Decision). The agreement itself is contained in an annex to this decision. The negotiation and adoption process of this agreement, together with the analysis of its content, benefited from extensive coverage by academic journals.[23] The international community attitude towards climate change shifted considerably between the failed Copenhagen conference in 2009 (COP15) and the Paris conference (COP21) which resulted in a global agreement. The discourse had changed, focusing on the possibility to limit climate change and to reap economic, social and health co-benefits. In the meantime, investment in renewable energy sources was bigger than ever and the commitment of many non-state actors pushed the negotiators

---

[21]   The fifth IPCC synthesis report was published in November 2014; it brings together the results of the three assessment reports that were issued in 2013.

[22]   A summary of the outcomes of COP20/CMP10 in Lima is available under https://unfccc.int/news/lima-call-for-climate-action-puts-world-on-track-to-paris-2015 accessed 20 April 2021.

[23]   See, for example: C. Streck, P. Keenlyside and M. Unger, 'The Paris Agreement: A New Beginning', *Journal for European Environmental & Planning Law* 13 (2016), pp.3–29; A. Savaresi, 'The Paris Agreement: a new beginning?', *Journal of Energy & Natural Resources Law* 34:1 (2016), pp.16–26; R. Bodle, L. Donat and M. Duwe, 'The Paris Agreement: Analysis, Assessment and Outlook', *Carbon & Climate Law Review*, 10:1 (2016), pp.5–22; M. Doelle, 'The Paris Agreement: Historic Breakthrough or High Stakes Experiment?', *Climate Law* 6 (2016), pp.1–20; M. Wewerinke-Singh and C. Doebbler, 'The Paris Agreement: Some Critical Reflections on Process and Substance', *UNSW Law Journal*, 39:4 (2016), pp.1486–1517.

towards a positive ending.[24] All of this led to the unanimous adoption of the Paris Agreement by the 196 Parties to the UNFCCC.

But once an international agreement is adopted, it must be ratified – a step which can prove lengthy and difficult, as the Kyoto Protocol case had underlined. As written in Article 21.1 of the Paris Agreement, 55 Parties accounting for at least an estimated 55 per cent of global emissions need to ratify it. Fortunately, this threshold was met quickly, on 5 October 2016, and the Agreement entered into force in record time, on 4 November 2016 (only 11 months after its adoption).[25] At the time of this writing, 191 Parties had ratified the Agreement.[26] The USA withdrew from the Paris Agreement on 4 November 2020 after a decision taken on 1 June 2017. This move cast doubt on the implementation of some provisions of the Paris Agreement.[27] However, the country rejoined and became a Party again on 19 February 2021.[28]

In terms of the legal nature of the Paris Agreement, it is a treaty under international law.[29] Additionally, it 'does not replace but instead complements the UNFCCC. It refers to and incorporates existing elements of the climate regime through different legal techniques'.[30] Hence, its legal nature implies that the Paris Agreement is binding for its Parties.[31] However, the situation is not so straightforward: 'The agreements adopted in Paris mark the completion of a decade-long transition from a top-down binding regime focused on developed country mitigation to a bottom-up and substantively non-binding approach to global cooperation on climate change with binding process elements.'[32]

Indeed, the prescriptive provisions in the Paris Agreement articles (characterized by the use of 'shall' and precise writing) are limited to 'primarily procedural' [obligations] and focused on 'nationally determined contributions' (on mitigation) and a core transparency framework,

---

[24]   A total of six factors have been identified by Ivanova. See M. Ivanova, 'Politics, Economics, and Society', in: D. Klein et al. (eds), *The Paris Agreement on Climate Change: Analysis and Commentary* (Oxford University Press, Oxford 2017), pp.22–26.

[25]   See https://unfccc.int/process/the-paris-agreement/status-of-ratification accessed 18 April 2021.

[26]   Ibid.

[27]   See J. Urpelainen and T. Van de Graaf, 'United States Non-cooperation and the Paris Agreement', *Climate Policy* 18:7 (2018), pp.839–51.

[28]   See www.state.gov/the-united-states-officially-rejoins-the-paris-agreement/ accessed 18 April 2021.

[29]   See R. Bodle and S. Oberthür, 'Legal Form of the Paris Agreement and Nature of Its Obligations', in: D. Klein et al. (eds), *The Paris Agreement on Climate Change: Analysis and Commentary* (Oxford University Press, Oxford 2017), p.92.

[30]   Ibid, p.94.

[31]   On international treaties' adoption, ratification and bindingness (for climate change law), see D. Bodansky, J. Brunnée and L. Rajamani, *International Climate Change Law* (Oxford University Press, Oxford 2017), pp.72–95.

[32]   M. Doelle, 'The Paris Agreement: Historic Breakthrough or High Stakes Experiment?', *Climate Law* 6 (2016), p.2.

plus collective obligations regarding finance'.[33] As a result, most of the Paris Agreement provisions constitute an obligation of conduct rather than an obligation of result.

Due to the limited length of this chapter, we cannot conduct an in-depth analysis of the Paris Agreement's content,[34] but some parts, such as the preamble and articles concerning objectives, mitigation, adaptation, or loss and damage, deserve a few words.

The Paris Agreement uses strong wording and important concepts in the preamble, especially when it refers to 'the urgent threat of climate change', 'equitable access to sustainable development and eradication of poverty', 'the fundamental priority of safeguarding food security and ending hunger', 'the imperatives of a just transition of the workforce', as well as aspects of human rights, Mother Earth and climate justice. Despite its lack of legal strength, an ambitious preamble such as this one can only improve the interpretation of the agreement's provisions.

Coming to the objectives, according to Article 2, the Paris Agreement aims to hold 'the increase in the global average temperature to well below 2°C above pre-industrial levels and [pursue] efforts to limit the temperature increase to 1.5°C above pre-industrial levels'. This 'below 2 degrees Celsius and as close as possible to 1.5 degrees Celsius' goal marks significant progress in comparison with the former climate discussions, by officially aiming below the 2 degrees Celsius threshold which was recognized as a threat to vulnerable countries.[35] Yet, it must be noted that with the current national commitments made in the framework of the Paris Agreement, the global average temperature will rise well over 2 degrees Celsius.[36] The objectives also include adaptation to climate change and a switch in financial flows towards low GHG emissions and 'climate-resilient development'.

Regarding mitigation, Article 4 is dedicated to the Nationally Determined Contributions (NDCs). This is the most obvious translation of what 'bottom-up' means for this agreement. Instead of a global objective set from above and split between the Parties – the model of the Kyoto Protocol – the Paris Agreement has set an overall goal for the Parties to reach, each doing their best, incentivized by 'naming-and-shaming'. For their NDCs, which have to be updated (only upward) every five years, the Parties have to commit to their 'highest possible ambition', taking into account their 'common but differentiated responsibilities and respective capabilities' (CBDRRC). In addition, Article 5 requires the Parties to conserve and enhance

---

[33]   See R. Bodle and S. Oberthür, 'Legal Form of the Paris Agreement and Nature of Its Obligations', in: D. Klein et al. (eds), 'The Paris Agreement on Climate Change: Analysis and Commentary' (Oxford University Press, Oxford 2017), pp.97–103.

[34]   For this purpose, the reader can find a wealth of detailed information in: D. Klein et al. (eds), *The Paris Agreement on Climate Change: Analysis and Commentary* (Oxford University Press, Oxford 2017).

[35]   H. Thorgeirsson, 'Objective (Article 2.1)', in: D. Klein et al. (eds), *The Paris Agreement on Climate Change: Analysis and Commentary* (Oxford University Press, Oxford 2017), pp.124–7.

[36]   As of April 2021, the pledges would lead to at least 2.6 degrees Celsius of warming above pre-industrial levels by the end of the century. See 'Addressing Global Warming', Climate Action Tracker, https://climateactiontracker.org/global/temperatures/ accessed 18 April 2021.

GHG sinks and reservoirs (notably forests), while Article 6 recognizes the possibility to transfer 'mitigation outcomes' between Parties.

On adaptation, Article 7 is much weaker, even as it represents a culmination of adaptation to climate change in international agreements. In a nutshell, adaptation is now seen as measurable. Each Party has to adopt an adaptation plan and a special focus is placed on 'its potential implications from the perspective of social development' (such as population displacement).[37]

Both the transparency rules and the mechanism used to compile (and control) progress towards the Paris Agreement's objectives – the 'global stocktake' included in Articles 13 and 14 respectively – deal with mitigation and adaptation. These two aspects of the fight against climate change are placed in parallel by the Paris Agreement.

Article 8 deals with loss and damages, which is sometimes confused with adaptation. As Viñuales explains, 'adaptation is (still) about prevention whereas loss and damage is about response (and potentially reparation)'.[38] Importantly, this provision deals with the prevention of loss and damages due to climate change (like extreme weather events and sea level rise) and with cooperation between countries to address them, but not with the central question of liability and compensation.[39]

Regarding the institutions of the Paris Agreement, the main one is the Conference of the Parties serving as the meeting of the Parties to this Agreement (CMA). It is the equivalent of the CMP for the Kyoto Protocol and it is the body which takes the decisions for a proper implementation of the Paris Agreement. Almost all of the other institutions which will enforce the Agreement's provisions are pre-existing ones.[40] However, we can mention here the new Paris Committee on Capacity Building (PCCB) 'to address current and emerging gaps and needs in implementing capacity-building in developing country parties and to enhance capacity-building efforts further'.[41]

Finally, COP22 to COP25 took place from 2016 to 2019 (at the same time as the CMP and the CMA), mainly in order to elaborate and find an agreement on the 'Paris rulebook', which is 'the operating manual needed for when the global deal enters into force in 2020', encompassing issues such as NDC accounting, transparency, loss and damage and financing.[42] COP26,

---

[37]   J. Viñuales, 'The Paris Climate Agreement: An Initial Examination', in C-EENRG Working Papers, no. 6, 15 December 2015, pp.6–7.

[38]   Ibid, p.7.

[39]   Ibid, pp.7–8; and L. Siegele, 'Loss and Damage (Article 8)', in: D. Klein et al. (eds), *The Paris Agreement on Climate Change: Analysis and Commentary* (Oxford University Press, Oxford 2017), pp.232–3.

[40]   See Figure 2.1, 'The institutional structure of the climate change regime at the time of writing', in J. Depledge, 'The Legal and Policy Framework of the United Nations Climate Change Regime', in: D. Klein et al. (eds), *The Paris Agreement on Climate Change: Analysis and Commentary* (Oxford University Press, Oxford 2017), p.38.

[41]   C. d'Auvergne and M. Nummelin, 'Capacity-building (Article 11)', in: D. Klein et al. (eds), *The Paris Agreement on Climate Change: Analysis and Commentary* (Oxford University Press, Oxford 2017), p.285.

[42]   For a summary of each conference's outcome, see the reports of the organization Carbon Brief, available under www.carbonbrief.org/cop22-key-outcomes-agreed-at-un-climate-talks-in-marrakech, www.carbonbrief.org/cop23 -key-outcomes-agreed-un-climate-talks-bonn, www.carbonbrief.org/cop24-key-outcomes-agreed-at-the-un-climate

which was postponed from 2020 to 2021 due to the COVID-19 pandemic, will constitute a strong landmark, as the Parties to the Agreement are expected to submit their climate pledges for 2030 at this point.[43]

2.3.4 Instruments to Combat Climate Change

Reduction of GHG emissions can be achieved in several ways. The first distinction is between national instruments (such as a domestic carbon tax) and international instruments (such as the Green Climate Fund, which operates globally to mobilize climate finance for developing countries[44]). The second distinction is between direct regulation (such as emission standards for cars) and market-based instruments (such as emissions trading for industries). Market-based instruments are also provided by the Kyoto Protocol and the Paris Agreement.[45] These instruments aim at cooperation between Parties in order to jointly achieve GHG emission reductions at the lowest cost possible. The main national instruments are standards and subsidies for saving energy and using renewable energy sources as an alternative to hydrocarbons as the primary fuel. Influenced by the energy crisis in the 1970s, the EU had already put in place legislation aiming at promoting energy efficiency and energy savings. Legislation at EU level promoting the use of renewable energy sources is more recent.

Although the Kyoto Protocol did not specify what kind of instruments the Parties have to use, it did introduce a new set of instruments generally referred to as 'flexibility' mechanisms. The main idea behind them is to stimulate cost-effectiveness by enabling Annex B countries to use achievements in GHG mitigation made in other countries for the fulfilment of their own reduction or limitation obligation. In case of emission reduction projects, approval of the host Party is required. Moreover, (public or private) participants have to be authorized to participate by a Party involved in the project.

Three different flexibility mechanisms are established under the Kyoto Protocol:[46]

- Joint Implementation (JI) based on Article 6: An Annex B Party can earn so-called Emission Reduction Units (ERUs) by financing a project in another Annex B country that will result in a reduction or removal of GHGs in the atmosphere (for example, replacing a coal-fired power plant with a gas-fired power plant). The achieved ERUs can be counted towards meeting the country's emission target as set out in the Kyoto Protocol.

-talks-in-katowice and www.carbonbrief.org/cop25-key-outcomes-agreed-at-the-un-climate-talks-in-madrid all accessed 21 December 2019. For the rulebook itself, see the decisions gathered in documents FCCC/PA/CMA/2018/3/Add.1 and FCCC/PA/CMA/2018/3/Add.2, available under https://unfccc.int/katowice accessed 13 June 2019.

[43]   www.carbonbrief.org/cop24-key-outcomes-agreed-at-the-un-climate-talks-in-katowice accessed 29 January 2019.

[44]   The progression of the pledges is regularly tracked by the dedicated website, available under www.greenclimate .fund/ accessed 11 March 2020.

[45]   Not all instruments in the Paris Agreement are market-based, however.

[46]   See http://unfccc.int/kyoto_protocol/items/2830.php accessed 20 April 2021.

- Clean Development Mechanism (CDM) based on Article 12: for every tonne of GHGs reduced or removed by a financed project, the funding country receives Certified Emission Reduction credits (CERs) which can be used to fulfil the emission obligations under the Kyoto Protocol. This is similar to the JI mechanism, but JI projects are realized in other Annex B states, whereas CDM projects are implemented in developing countries.
- International Emissions Trading (IET) based on Article 17: countries emitting less GHGs than allowed under Annex B can sell the remaining amount (in the form of 'assigned amount units', AAUs) to other Annex B states that face the risk of exceeding their individual cap. Besides unused AAUs, Parties may also trade ERUs, CERs and removal units from land-use, land-use change and forestry activities.

The flexibility mechanisms of JI, CDM and IET promote the cost-effectiveness of climate action since they enable Annex B Parties to fulfil their reduction or limitation obligations where abatement costs are lower.[47] Another feature of the Kyoto Protocol, giving Parties some degree of flexibility in achieving their emission goals, is the inclusion of Land-Use, Land-Use Change and Forestry (LULUCF). By promoting natural carbon sinks such as forests, these countries may offset the CO_2 removed and sequestered by soil and fauna ('removal units'; RMUs) with the assigned GHG emissions amounts.

With the adoption of the Paris Agreement, a new but somewhat comparable set of international instruments to facilitate GHG emission reductions is provided in its Article 6:

- Cooperative approaches: provisions 6.2 and 6.3 of the Agreement aim to open the door for voluntary initiatives from subnational level to groups of countries in order to create platforms for the exchange of internationally transferred mitigation outcomes (ITMOs), hence mostly through market-based instruments (MBIs). To implement these, the Parties have to follow some rules specified in Article 6.2, so as to 'promote sustainable development and ensure environmental integrity and transparency' and also to avoid double counting.[48]
- UNFCCC-governed crediting mechanism: provisions 6.4 to 6.7 create the so-called Sustainable Development Mechanism (SDM), comparable to JI and the CDM.[49] An activity reducing GHG emissions is financed by a public or private entity from a country and installed in another country, resulting in carbon credits being transferred to the financer. However, some important differences exist: both developed and developing countries can be the home of the financing entity or host a project; the final goal of

---

[47]   Cf. R. Pereira and C. Jourdain, 'International and EU Climate Change Law', in: K.E. Makuch and R. Pereira, *Environmental and Energy Law* (Wiley-Blackwell, Oxford 2012), p.150.

[48]   See more in A. Howard, 'Voluntary Cooperation (Article 6)', in: D. Klein et al. (eds), *The Paris Agreement on Climate Change: Analysis and Commentary* (Oxford University Press, Oxford 2017), pp.184–7.

[49]   Carbon Market Watch, 'Good-bye Kyoto: Transitioning Away from Offsetting after 2020' (2017), p.4. https://carbonmarketwatch.org/wp-content/uploads/2017/04/Good-bye-Kyoto_Transitioning-away-from-offsetting-after-2020_WEB_1final.pdf accessed 29 January 2019.

these projects has to be the mitigation of global emissions (and not merely offsetting GHG emissions by the financers); and a stronger emphasis is placed on sustainable development (hence not only GHG emissions reduction).[50] This will hopefully generate a substantial change in the design of those GHG emission reduction projects, which in numerous instances in the past were plagued by controversies related to environmental integrity and human rights abuses.[51]

– Non-market approaches: provisions 6.8 and 6.9 recognise the importance of non-market mechanisms for fighting climate change. These provisions are an open door to future policy instruments focusing on public participation and climate coordination between organisations and States.[52]

As a whole, the new 'voluntary cooperation' instruments that countries can use to reach their NDCs with the help of their partners are another proof of the bottom-up rationale behind the Paris Agreement: they offer a diversity of opportunities, considering both market and non-market approaches, so that every Party can find the tool which it deems to be the most effective and efficient for its situation.

However, a lot still remains to be decided before the implementation of these mechanisms in 2021, and their development will have to be monitored and controlled. One of the points still being debated is the switching period from CDM and JI to SDM, with questions about the continuity of the first tools after 2020 and concerns about double counting.[53]

2.4 HISTORY AND ORIGINS OF EU CLIMATE POLICY

2.4.1 EU Climate Policy before Kyoto

The roots of European climate action go back to the 1980s. After a first 'wake-up call' by the European Parliament in 1986, the Commission acknowledged in a Communication to the Council that the greenhouse effect as a global problem needed to be tackled internationally.[54]

---

[50] See more in A. Howard, 'Voluntary Cooperation (Article 6)', in: D. Klein et al. (eds), *The Paris Agreement on Climate Change: Analysis and Commentary* (Oxford University Press, Oxford 2017), pp.187–9.

[51] See P. Villavicencio Calzadilla, 'Human Rights and the New Sustainable Mechanism of the Paris Agreement: A New Opportunity to Promote Climate Justice', *Potchefstroom Electronic Law Journal* 21 (2018), pp.1–39.

[52] See more in A. Howard, 'Voluntary Cooperation (Article 6)', in: D. Klein et al. (eds), *The Paris Agreement on Climate Change: Analysis and Commentary* (Oxford University Press, Oxford 2017), pp.189–90.

[53] On this topic, see Carbon Market Watch, 'Good-bye Kyoto: Transitioning Away from Offsetting after 2020' (2017) https://carbonmarketwatch.org/wp-content/uploads/2017/04/Good-bye-Kyoto_Transitioning-away-from -offsetting-after-2020_WEB_1final.pdf accessed 29 January 2019; and UNFCCC, 'Countries Urge Continued Use of Clean Development Mechanism' (2018) https://unfccc.int/news/countries-urge-continued-use-of-clean-development -mechanism accessed 29 January 2019.

[54] Resolution of the EP of 12.9.1986 on measures to be taken in research and energy policy to combat the increasing concentration of CO_2 in the atmosphere, A2-68/86, OJ 12.9.1986, C 255/225; Commission, Communication to the Council 'Greenhouse Effect and the Community', COM(88) 656 final.

The Commission stressed the necessity of European climate action, including the promotion of alternatives to fossil fuels, research programmes, adaptation measures and cooperation with developing countries. Correspondingly, the Council formally invited the Commission to reconsider existing Community (now: Union) policies which may impair the climate. The Council also wanted Member States to take 'urgent action to increase energy savings; to improve energy efficiency; to promote the development and use of energy sources, such as non-fossil fuels, which will not contribute to the greenhouse effect'.[55] Before the second World Climate Conference in 1990, the European Council called for 'concrete action and, in particular, measures relating to CO_2 emissions'[56] in order to have a strong Community position in the international negotiations for a climate convention.

Despite these declarations of intent, the first attempt to implement comprehensive legislation to combat the greenhouse effect failed. The initiative for a tax on CO_2 and energy proposed by the Commission in 1992[57] did not reach unanimity in the Council and was dismissed.[58] More successful were directives aiming at the improvement of energy efficiency. Based on the internal market competence according to what is now Article 114 of the TFEU, product standards were established for certain consumer goods, such as hot water boilers (Council Directive 92/42/EEC) or refrigerators (Directive 96/57/EC). The so-called 'SAVE' Directive 93/76/EEC required Member States in 1993 to implement energy saving programmes with the overarching goal of limiting CO_2 emissions. In the same year the Council enacted Decisions 93/500/EEC (the so-called Altener programme) and 93/389/ EEC. Whereas the former concerned the (financial) promotion of RES, the latter obliged the Member States to establish a mechanism for monitoring all anthropogenic GHG emissions not controlled by the Montreal Protocol[59] and to implement national programmes containing GHG reduction and limitation measures in order to stabilize CO_2 emissions in 2000 at 1990 levels in the Community as a whole.[60] These

---

[55]   Resolution of the Council of 21.6.1989 on the greenhouse effect and the Community, 89/C 183/03, OJ 20.7.1989, C 183/4.

[56]   European Council in Dublin 25–26 June 1990, Annex II to the Conclusions of the Presidency – Declaration by the European Council on the environmental imperative; available under www.consilium.europa.eu/media/20562/1990_june_-_dublin__eng_.pdf accessed 3 January 2021.

[57]   Commission, 'Proposal for a Council directive introducing a tax on carbon dioxide emissions and energy', COM(92) 226 final.

[58]   Cf. S. Oberthür and M. Pallemaerts, 'The EU's Internal and External Climate Policies: An Historical Overview', in: S. Oberthür and M. Pallemaerts (eds), *The New Climate Policies of the European Union* (VUBPRESS, Brussels 2010), p.31.

[59]   The Montreal Protocol to the Vienna Convention for the Protection of the Ozone Layer was concluded in 1986 with the goal of reducing production and consumption of ozone-depleting substances. Since these substances are also very potent GHGs, the legislation contributes to the fight against global warming. There are several EU acts implementing the provisions of the Montreal Protocol in Europe, see https://ec.europa.eu/clima/policies/ozone/regulation_en accessed 3 January 2021.

[60]   Council Decision of 24.6.1993 for a monitoring mechanism of Community CO_2 and other greenhouse gas emissions, 93/389/EEC, OJ 9.7.1993, L 167/31.

first steps towards a harmonized European climate policy obtained only limited success: GHG emissions even increased in the following years.[61] With the adoption of the Kyoto Protocol in 1997, the EU legislature intensified its efforts to protect the climate.

2.4.2   EU Climate Policy Post-Kyoto – the Road to 2020

Early instruments to reduce CO_2 emissions

The adoption of the Kyoto Protocol gave a boost to climate action at the European level.[62] The EU committed itself to a common emission reduction goal of 8 per cent compared to 1990 levels in the first commitment period 2008–12 and a reduction of 20 per cent (30 per cent under certain conditions) in the second period 2013–20. The EU legislature decided upon national contributions to ensure the achievement of the overall EU commitments under the first and second Kyoto periods. A Member State not fulfilling its specific national commitment is thus infringing EU legislation.

Under the European Climate Change Programme (ECCP), the EU legislature was active in all of the three main fields of climate protection: energy efficiency, renewable energy sources and GHG mitigation. Regarding the latter, ETS Directive 2003/87/EC introduced a market-based concept of regulation. As a result of the ETS, CO_2 is priced, which provides a financial incentive to reduce GHG emissions, while emitters operate under an emissions cap.

Picking up pace: The EU's forerunner role towards 2020

Driven by the intention to retain its leadership role in the fight against global warming, the EU came up with unilateral climate targets in 2007.[63] In order to achieve the long-term goal of a maximum temperature increase of 2 degrees Celsius above pre-industrial levels, the European Council agreed on the so-called 20-20-20 targets for 2020:

- Reducing GHG emissions by 20 per cent;
- Increasing the share of RES in the energy mix to 20 per cent;
- Improving energy efficiency by 20 per cent.[64]

---

[61]   EEA, Trends and projections in Europe 2014 – Tracking progress towards Europe's climate and energy targets for 2020 (2014) p.44.

[62]   S. Oberthür and M. Pallemaerts (n. 58, p.53) argue that the inactivity of the EU regarding climate policies in the 1990s created a 'credibility gap […] that has gradually been closed in the 2000s'. The EU ratified the Kyoto Protocol on 31 May 2002.

[63]   This unilateral step forward strengthened the EU's position in the COP13/CMP3 in Bali 2007 and enabled it to 'play a pivotal role in securing agreement on the roadmap towards a new comprehensive agreement': Commission, Communication to the European Parliament, the Council, the European Economic and Social Committee and the Committee of the Regions '20 20 by 2020 – Europe's climate change opportunity', COM(2008) 30 final, p.3.

[64]   European Council, Presidency Conclusions of 8th and 9th March 2007, 7224/1/07; available under www.consilium.europa.eu/ueDocs/cms_Data/docs/pressData/en/ec/93135.pdf accessed 20 April 2021.

The 20-20-20 targets go beyond mere reduction of GHG emissions. The inclusion of goals relating to energy efficiency and RES shows the EU's drive to shift to a low-carbon economy and to enhance security of energy supply.

The strengthened ambitions of the EU on climate and energy action were also reflected in the Lisbon Treaty, which entered into force in 2009. The newly incorporated provisions in the field of energy policy (Article 194 of the TFEU) specify that the EU shall aim, inter alia, to promote energy efficiency, energy saving and the development of renewable energy. Moreover, Article 191(1) of the TFEU recognized for the first time the international fight against climate change as one the priorities of EU environmental policy. Correspondingly, Article 191(4) of the TFEU obliges the Union and its Member States to cooperate '[w]ithin their respective spheres of competence' with third countries and international organizations for the purpose of international environmental protection.[65]

The EU's active role in the negotiations for a new global agreement combating climate change proved the Union's strong commitment to worldwide climate action. The conclusion of the Paris Agreement can be considered as one of the biggest achievements in EU foreign environmental policy. In order to effectuate and sustain its forerunner role, the European Council had already agreed at an early stage – before the Paris Agreement was signed – on a long-term strategy for the GHG reductions towards 2030.[66] Similar to the 20-20-20 goals for 2020, the (strengthened) 2030 targets are again divided into GHG mitigation, renewable energy sources and energy efficiency, and aim to ensure that the EU meets its international climate obligations.

2.5   EU CLIMATE POLICY TOWARDS 2030 AND BEYOND

2.5.1   Climate Policy Objectives

Primary objectives
EU climate policy, as set in the 2030 Climate and Energy Framework, aims to achieve two primary objectives:

- reducing GHG emissions (for a 'low-carbon society') and
- securing energy supply.

First, the switch to 'green' energy sources and improvements regarding energy efficiency in connection with binding emission reduction targets for all sectors should mitigate the total amount of GHG emissions, to fight climate change. The second main objective is to secure energy supply by increasing the share of RES in the energy mix and improving the energy

---

[65]   The competence of the Union to conclude international agreements in the field of international environmental protection is based on Article 192 of the TFEU (external competence by virtue of implied powers; cf. the CJEU's AETR case law). The procedure for concluding agreements is stipulated in Article 218 of the TFEU.

[66]   European Council, Conclusions of 23rd and 24th of October 2014, EUCO 169/14.

consumption of products, installations and buildings.[67] In 2018, 58 per cent of the EU's gross domestic energy consumption was met by imports.[68] The promotion of low-carbon energy projects in conjunction with higher energy efficiency diminishes the reliance on fossil fuels as well as on foreign energy suppliers. Consistently, decarbonising the economy and moderating energy demand are two of the five dimensions of the Energy Union Strategy.[69]

Secondary objectives

Besides the two main purposes of combating climate change and securing energy supply, the EU's low-carbon policy as set out in the 2020 package and the 2030 framework has several secondary objectives:

- developing and exporting low-carbon technologies ('green growth');
- safeguarding solidarity between EU Member States;
- ensuring cost-effectiveness;
- ensuring competitiveness.

'Secondary objectives' are aspects that the legislature takes into account when designing a policy which are of minor importance compared to the primary objectives, but which still shape the legislative action to a certain extent. In other words, their achievement is not indispensable but is desirable, and they are therefore reflected in the preambles and content of the legislation. The most important secondary objectives of EU climate policy towards 2030 are described hereafter.

Developing and exporting low-carbon technologies ('green growth')

First and foremost, encouraging industries and the energy sector to use and develop low-carbon technologies is deemed to have positive impacts on the European economy. Under the headline of 'green growth', the Commission argues that the implementation of the climate targets would create a win–win situation for companies and the environment since a new and promising market is created in which European players could take a leading role and, consequently, export green technology to non-European countries.[70]

The economic advantages resulting from the emergence of new ecological business branches, the 'greening' of the traditional industries and the creation of 'green jobs' could offset the costs incurred by the transition towards a low-carbon society and justify respective investments – especially when considering the costs that would arise in case of a failure to act.[71]

---

[67]   Cf. Commission, European Energy Security Strategy, COM(2014) 330 final, especially pillar 3 ('Moderating energy demand') and pillar 5 ('Increasing energy production in the European Union').

[68]   Eurostat, 'Shedding Light on Energy in the EU – A Guided Tour of Energy Statistics', 2020 edition, Section 2.3; available under https://ec.europa.eu/eurostat/cache/infographs/energy/bloc-2c.html accessed on 3 January 2021.

[69]   Commission, 'A Framework Strategy for a Resilient Energy Union with a Forward-Looking Climate Change Policy', COM(2015) 80 final.

[70]   Commission (n. 63), pp.3 et seq.

[71]   Commission (n. 63), p.10.

Safeguarding solidarity between EU Member States

Another consideration shaping the EU's legal and regulatory climate framework is solidarity between the Member States.[72] The contributions to the EU's climate targets are distributed between the Member States taking into account their relative wealth, measured by GDP per capita. This approach is reflected in several provisions. The ETS Directive, for instance, contains a 'solidarity mechanism' introducing an allocation formula for the auctioning of tradable emission rights. According to Article 10(2) of the ETS Directive, only 90 per cent of the total quantity of allowances is distributed among the Member States in shares identical to the share of verified emissions, while 10 per cent is distributed among lower-income Member States (mentioned in Annex IIa) for the purpose of solidarity, growth and improvement of electricity network interconnections. A disproportional higher share of emission allowances gives the less affluent Member States a larger emission space and some additional revenues from auctioning allowances to bear the costs of making a transition to a low-carbon economy. Additionally, a so-called Modernisation Fund is established to finance energy system and energy efficiency projects in lower-income Member States (cf. Article 10d and Annex IIb). Also, the Effort Sharing Regulation (ESR), regulating GHG emissions from non-ETS sectors, is based on the idea of inter-European solidarity: the national emission targets are set on the basis of the EU countries' GDP per capita, so that less wealthy Member States have lower reduction targets.[73]

Ensuring cost-effectiveness

Another policy aspect that is taken into account when formulating EU climate legislation is cost-effectiveness. A climate policy measure is cost-effective if the desired goal (for example, reducing GHG emissions by 20 per cent) is achieved at minimum costs.[74] EU climate law establishes several flexibility mechanisms and economic instruments to ensure that GHG mitigation and renewable energy targets are met cost-effectively. In particular, the tradability of CO_2 emission allowances facilitates a cost-effective achievement of the GHG emission reduction targets in the ETS sectors. As regards the non-ETS sectors, the ESR allows eligible Member States to achieve their national reduction targets by covering a certain amount of emissions with ETS allowances or credits from the land-use sector. Moreover, under the RES Directive, three cooperation mechanisms (statistical transfers, joint projects and joint support schemes) ensure flexibility to enhance cost-effectiveness.

Ensuring competitiveness

Despite the introduction of market-based instruments and flexibility mechanisms, there are sectors and sub-sectors that may incur competitive disadvantages from the implementation of

---

[72]　Commission (n. 63), p.5; European Council, Conclusions of 23rd and 24th of October 2014, EUCO 169/14, margin numbers 2.2. and 2.8.

[73]　Article 4 in conjunction with Annex I of the Effort Sharing Regulation.

[74]　Imagine there are two options to abate 10 Mt CO_2. The realization of option A would result in total costs of €900 million; the realization of option B would cost only €850 million. The latter alternative would be considered more cost-effective since the same reduction amount is achieved at lower costs.

climate policy. In particular, the pricing of GHG emissions pursuant to the ETS Directive may have detrimental impacts on heavy industries and other large emitters of CO_2 in Europe, which creates a risk of 'carbon leakage'. This term refers to a situation in which companies transfer their manufacturing location to a non-EU country with a more lenient climate policy to save production costs. To avoid this, EU policymakers agreed on certain exceptions for sectors and sub-sectors exposed to a significant risk of carbon leakage, including free (instead of auctioned) allowances under the ETS. High energy prices could also negatively affect the competitiveness of European manufacturers, since competitors in the US pay much less for energy as a result of the shale gas boom in recent years. In order to provide for affordable energy supply in a low-carbon economy, the EU takes steps to complete the internal energy market for electricity and promotes more market-oriented approaches in the subsidisation of RES.[75] This underscores that climate policy cannot be dissociated from other policy objectives, including economic policy and the preservation of the competitiveness of European industries.

A new instrument to protect the EU's competitiveness and its efforts to achieve climate-neutrality by 2050 is discussed in the context of the European Green Deal under what is called a Carbon Border Adjustment Mechanism (CBAM). The idea is to price the imports from less ambitious countries in order to reflect more accurately the imports' carbon content.[76] The CBAM would tackle the problem of carbon leakage while encouraging the EU's trading partners to implement adequate CO_2 regulations or pricing schemes.

2.5.2 Climate Targets for 2030

In October 2014, the European Council agreed on a new climate and energy framework including EU-wide targets for the period between 2020 and 2030.[77] The targets for renewable energy production and energy efficiency were strengthened in the course of the legislative procedure.[78] For 2030, the EU aims to:

- reduce GHG emissions by at least 40 per cent compared to 1990 levels;
- increase the share of RES in the energy mix to at least 32 per cent;
- improve energy efficiency by at least 32.5 per cent.

The Commission under Ursula von der Leyen intends to raise the climate ambition for 2030. Under the European Green Deal, the EU should not only strive for a carbon-neutral economy by 2050, but also set new climate targets for 2030 (see in more detail 2.5.3). In its proposal for a European Climate Law, the Commission recommends a net-emission target of at least 55 per

---

[75] According to the 2018 RES Directive, state support for renewable electricity shall be granted in the form of a market premiums and the level of support shall be determined by market-based systems, e.g. tendering procedures (cf. Article 4 and recitals 16 and 19).

[76] Communication from the European Commission, The European Green Deal, COM(2019) 640 final, p.5.

[77] European Council, Conclusions of 23rd and 24th of October 2014, EUCO 169/14; available under www .consilium.europa.eu/uedocs/cms_data/docs/pressdata/en/ec/145397.pdf accessed 3 January 2021.

[78] Cf. recital 6 of the Energy Efficiency Directive 2018/2002.

cent by 2030.[79] Strengthening the greenhouse gas emission reduction target will also require a revision of the RES and energy efficiency goals for 2030.[80]

Another important target set out by the 2030 framework[81] concerns the achievement of a fully functioning and connected internal energy market.[82] The promotion of interconnected electricity networks facilitates the transportation of variable renewable energy and could therefore have a positive effect on investments in RES-projects. Moreover, the integration of the energy markets is expected to lead to lower energy prices, which also strengthens the competitiveness of EU industries. The 2030 climate and energy framework is based on regulatory strategies already known from the 2020 package. The concepts have been refined and a new governance structure has been implemented in order to unify and simplify the administration and monitoring of the Member States' efforts to reach the 2030 targets. In the next section, we will explain the most important amendments of EU climate policy.

2.5.3   Key Instruments to Achieve the 2030 Climate Targets

GHG reduction legislation

In 2018, the legal framework for 2021–30 was established with a view to achieving the overall objective of reducing the EU's CO_2 emissions by at least 40 per cent by 2030. By the end of 2020, however, the European Council had agreed with the Commission's Green Deal proposal to strengthen the EU's reduction target for 2030 to 55 per cent. All sectors having a greenhouse gas footprint, including land use and forestry, must contribute to achieving this objective. The EU legislature has thus enhanced and expanded the legal framework for cutting CO_2 emissions, as compared with the legal framework applicable until 2020.

European law on the reduction of greenhouse gases for the period 2021–30 consists of three pillars: the ETS Directive applicable to all sectors subject to emissions trading; the LULUCF Regulation concerning land use, land use change and forestry sectors; and the ESR for all other sectors subject to neither the LULUCF Regulation nor the ETS Directive. Rules on emissions from international aviation fall outside of the EU acquis; such rules are included in regulations

---

[79]   Proposal for a Regulation of the EP and of the Council establishing the framework for achieving climate neutrality and amending Regulation (EU) 2018/1999 (European Climate Law), COM(2020) 563 final, amending Commission proposal COM(2020) 80 final.

[80]   See Commission, A Union of Vitality in a World of Fragility (Work Programme 2021), COM(2020) 690 final.

[81]   In order to achieve a fully integrated EU energy market, the 'Clean Energy for All Europeans' package contains four pieces of legislation adapting EU market rules to the new market realities that are characterized by RES: the Directive on common rules for the internal market for electricity (EU) 2019/944 which replaces Electricity Directive (2009/72/EC), the new Regulation on the internal market for electricity (EU) 2019/943 which replaces the Electricity Regulation (EC/714/2009), the Regulation on risk preparedness in the electricity sector (EU) 2019/1941 and Regulation (EU) 2019/942 establishing an EU Agency for the cooperation of energy regulators, recasting the Regulation 713/2009.

[82]   Cf. the five dimensions of the Energy Union as described by the European Commission in its Communication on the Energy Union Package, COM(2015) 80 final, p.7 et seq.

of the International Civil Aviation Organization (ICAO). These aspects are examined in detail below.

Emissions Trading System

The ETS is said to be 'a cornerstone of the European Union's policy to combat climate change and its key tool for reducing industrial greenhouse gas emissions cost-effectively'.[83] The ETS is based on a cap-and-trade system, which means that the total amount of GHG emission allowances (the 'cap') is determined *a priori*. In principle, the sectors covered by the ETS Directive are not allowed to release more CO_2 equivalent than scheduled. The right to emit one tonne of GHGs is transferable so that these emission allowances can be freely traded between the regulated companies. Each year the companies falling under the ETS Directive have to surrender allowances equivalent to the amount of their GHG emissions, otherwise they risk severe fines.

The ETS also underwent an 'update' with regard to the fourth trading period 2021–30 with the intention to remedy some inadequacies. Accordingly, the purpose of Directive 2018/410,[84] amending the ETS Directive (Directive 2003/87/EC[85]), is to increase the cost-effectiveness of emission reduction measures and to promote investments in carbon-efficient technologies. The principal functioning of the system remains largely unchanged, but the requirements are more stringent. The cap on emissions will be reduced by 2.2 per cent annually, thereby increasing the pressure to reduce emissions. Although a proportion of the allowances will continue to be allocated free of charge, the benchmark criteria will nevertheless be stricter and from 2026 onwards free allowances will gradually be phased out, unless such companies are affected by carbon leakage. While free allocation of emission allowances to industries at risk of carbon leakage will continue to be possible up to 100 per cent, the eligibility criteria will be tightened, meaning that in the future only 44 carbon leakage sectors (previously: 175) will benefit from free allocations.[86] Furthermore, transfers of the allowance surplus to the Market Stability Reserve (MSR), which started in 2019, will double, rising from 12 per cent of surplus allowances to 24 per cent until 2023.

---

[83]   Directive 2009/31/EC of the European Parliament and the Council of 23.4.2009 on the geological storage of carbon dioxide and amending Council Directive 85/337/EEC, European Parliament and Council Directives 2000/60/EC, 2001/80/EC, 2004/35/EC, 2006/12/EC, 2008/1/EC and Regulation (EC) No 1013/2006, OJ 5.6.2009, L 140/114.

[84]   OJ 19.3.2018, L 76/3.

[85]   Directive 2003/87/EC of the European Parliament and of the Council of 13.10.2003 establishing a scheme for greenhouse gas emission allowance trading within the Community and amending Council Directive 96/61/EC, OJ 25.10.2003, L 275/32, amended by Directive 2009/29/EC of the European Parliament and of the Council of 23.4.2009 amending Directive 2003/87/EC so as to improve and extend the greenhouse gas emission allowance trading scheme of the Community, OJ 5.6.2009, L 140/63.

[86]   Cf. Commission Notice, Preliminary Carbon Leakage List, OJ 8.5.2018, C 162/1.

Covering non-ETS sectors via the Effort Sharing Regulation

The ETS Directive covers around 40 per cent of the total GHGs released by the EU Member States.[87] For sectors not addressed by the ETS (especially waste, transport, agriculture and housing), Regulation 2018/842 lays down binding greenhouse gas reduction targets for the period 2021–30. This so-called Effort Sharing Regulation supersedes the Effort Sharing Decision (Decision No 406/2009/EC). It sets the target to reduce EU-wide emissions by 30 per cent below 2005 levels in the sectors covered by 2030 (Article 1). The Regulation requires Member States to take appropriate measures to achieve the respective reduction targets at the national level, taking into account the economic capacity of the Member States (Article 4 in conjunction with Annex I). A linear reduction factor should ensure that Member States intensify their efforts to achieve their emission targets year after year. In addition, this mechanism will ensure that the risk of not achieving the national target is recognized at an early stage and that corresponding national compliance measures can be taken.

The ESR also provides flexibility mechanisms to maximize cost-effectiveness. First, Article 5 provides for the possibility of 'banking' (carrying over emission allocations to subsequent years) and 'borrowing' (borrowing emission allocations from a subsequent period). Member States can also reach their national targets by purchasing emissions from other Member States that have excess emission allocations. Furthermore, some Member States (listed in Annex II) may cancel a limited number of ETS allowances as a way to comply with the ESR (Article 6).

The Commission monitors compliance with the requirements of the ESR during its annual evaluation under Article 21 of the Emissions Reporting Regulation (Regulation 525/2013). If it reaches the conclusion that a Member State is making insufficient progress, this Member State will have to submit to the Commission a suitable corrective action plan (including a specific timetable) within three months. If the emissions of the Member State exceed the allocated amount, the excess emissions multiplied by a factor of 1.08 are incurred as a penalty in the following year (Article 9).

Land use, land-use change and forestry (LULUCF)

Whereas the ETS Directive and the ESR have been part of the EU acquis for some time and are widely known, the legal requirements concerning LULUCF have thus far led a rather shadowy existence. Although accounting rules on LULUCF emissions had already been established by Decision 529/2013/EU, it was not until the entry into force of Regulation 2018/841 on the inclusion of LULUCF emissions and removals (the 'LULUCF Regulation') that a detailed legal basis was created. The aim of the LULUCF Regulation is to preserve forests and plant stocks that serve as CO_2 sinks. The 'no-debit rule' is its central commitment: the total amount of greenhouse gas emissions from land use (including agricultural land), land-use change and forestry sub-sectors must not exceed the total amount of removals by soil, plants and trees in both periods 2021–5 and 2026–30 (Article 4). To express it with a simple example: any deforestation (e.g. in order to create space for an infrastructure project) needs to be compen-

---

[87]   Commission, EU Emissions Trading System; available under https://ec.europa.eu/clima/policies/ets_en accessed 3 January 2021.

sated by an equivalent afforestation effort. Enhancing the share of natural carbon sinks even leads to negative emissions – the LULUCF sector removes more CO_2 from the atmosphere than it emits.

The Regulation lays down the relevant accounting rules that build on the requirements set out in Decision 529/2013/EU relating to land use, land-use change and forestry. To ensure compliance with the requirements, the Regulation imposes reporting obligations on the Member States vis-à-vis the Commission (Article 14).

Under the 'no-debit rule', the LULUCF Regulation provides EU Member States with certain flexibilities. Flexibility mechanisms should help them achieve a reduction in emissions in the most cost-effective manner. If the amount of greenhouse gases removed from the atmosphere is greater than the amount emitted in the LULUCF sector (for instance due to afforestation) in the first period from 2021 to 2025, the Member State can transfer the net reduction in greenhouse gases to the following period from 2026 to 2030 ('banking'). Deficits in one period can be repaired by purchasing the excess part of emission allocations from other Member States or by transferring surpluses from the system established by the ESR.

Attaching legal weight to natural developments will pose significant challenges to the Member States. It remains to be seen whether the positive effects of these flexible mechanisms will be proportionate to their administrative burden.

Carbon capture and storage

Alongside the promotion of natural sinks (LULUCF Regulation), the EU facilitates the implementation of technologies that store CO_2 in artificial carbon sinks to prevent its emission into the atmosphere. One of these techniques is CCS, in which CO_2 gets captured at its source (for example, a coal-fired power plant) and is transported via a pipeline or vehicle to a suitable storage site where it is injected in deep subsoil layers. It is assumed that the gas will stay there for a very long period – hundreds or even thousands of years.[88] The CCS Directive 2009/31/EC[89] offers a comprehensive legal framework especially for the last step of CCS, sequestration underground, in order to ensure that this GHG is kept away from the atmosphere in an environmentally safe and permanent manner (cf. Article 1 of the CCS Directive). Apart from a few subsidized pilot projects, CCS is not yet deployed at a commercial scale.[90]

Transport sector

Almost a quarter of the EU's GHG emissions stem from the transport sector. Obviously, clean mobility is a cornerstone of the efforts to achieve the 2030 GHG reduction target. While the EU strives for zero-emission vehicles in the long term, the first steps towards low-emission mobility are the introduction of emission standards for new cars (37.5 per cent by 2030 com-

---

[88]   B. Metz et al (eds), 'IPCC – Report Carbon Capture and Storage' (2005), p.246.

[89]   Directive 2009/31/EC of the European Parliament and the Council of 23.4.2009 on the geological storage of carbon dioxide and amending Council Directive 85/337/EEC, European Parliament and Council Directives 2000/60/EC, 2001/80/EC, 2004/35/EC, 2006/12/EC, 2008/1/EC and Regulation (EC) No 1013/2006, OJ 5.6.2009, L 140/114.

[90]   See Commission, Implementation of Directive 2009/31/EC on the Geological Storage of Carbon Dioxide (Report to the EP and the Council), COM(2019) 566 final.

pared to 2021), vans (31 per cent by 2030 compared to 2021) and trucks (30 per cent by 2030 compared to 2019).

International aviation

Although the future impact of the COVID-19 pandemic on people's travel behaviour is unclear, the aviation sector's continuous growth in past years has resulted in a sharp rise in aviation-related greenhouse gas emissions. To curb this trend, the ICAO has been developing a market-based mechanism to reduce CO_2. The envisaged Carbon Offsetting and Reduction Scheme for International Aviation (CORSIA) is meant to make CO_2-neutral growth of the aviation industry possible by employing an offsetting system. The EU has extended the existing moratorium on the inclusion of international aviation in the EU ETS in order to allow some time for the implementation of CORSIA. European aviation continues to be subject to the EU ETS.

Renewable energy

In 2016 the Commission presented its 'Clean energy for all Europeans' package, consisting of eight legislative proposals intended to facilitate the transition to a clean energy economy and to make the Energy Union work.[91] The package follows a holistic approach: it not only amends climate-related energy legislation, but also provides a revision of energy market design and introduces a new governance structure for climate and energy policies in the Member States.

The recast of the RES Directive 2009/28/EC, Directive 2018/2001,[92] revises the policy framework for the production and promotion of RES in the EU. This 2018 RES Directive establishes an overall target of a 32 per cent share of RES in the gross final energy consumption by 2030 (Article 3). In contrast to its predecessor, the 2018 RES Directive does not determine binding targets for each Member State. Rather it allows the Member States to define their contributions on a national level, while they shall collectively ensure that the sum of their contributions adds up to the overall RES target. The Commission monitors the national trajectories under the monitoring system established by the Energy Governance Regulation (see below). Besides the overall 32 per cent target, the 2018 RES Directive also introduces a sub-target for biofuels. Fuel suppliers are required to supply a minimum of 14 per cent of the energy consumed in road and rail transport from RES.

Energy efficiency

The revised Energy Efficiency Directive 2018/2002[93] establishes a new energy efficiency target of 32.5 per cent by 2030, to be achieved by means of indicative national energy efficiency contributions. Translated in million tonnes of oil equivalent, this means that the (initially)

---

[91]   See https://ec.europa.eu/energy/en/topics/energy-strategy/clean-energy-all-europeans accessed 3 January 2021.

[92]   Directive (EU) 2018/2001 of the European Parliament and of the Council of 11 December 2018 on the promotion of the use of energy from renewable sources, OJ 21.12.2018, L 328/82.

[93]   Directive (EU) 2018/2002 of the European Parliament and of the Council of 11 December 2018 amending Directive 2012/27/EU on energy efficiency, OJ 21.12.2018, L 328/210.

28 Member States may not consume more than 1,273 Mtoe of primary energy and 956 Mtoe of final energy in 2030. These numbers were adapted for the EU 27 to, respectively, 1,128 Mtoe and 846 Mtoe. Article 1 of the revised Energy Efficiency Directive refers to the 'energy efficiency first' principle, though the content of this general policy principle remains vague. Recital 2 of the Energy Efficiency Directive gives some general indications of what 'energy efficiency first' means, including that 'energy efficiency improvements need to be made whenever they are more cost-effective than equivalent supply-side solutions'. The structure and content of the Directive itself remains largely unchanged. In addition, the Energy Performance of Buildings Directive 2018/844[94] has been revised. Member States have to develop long-term renovation strategies with a time horizon of 2050, including a roadmap with measures and domestically established measurable progress indicators.

New energy and climate governance system

The Energy Governance Regulation 2018/1999[95] is the parenthesis of the 'Clean energy for all Europeans' package and the GHG emission reduction legislation (with the exception of the ETS Directive) and brings together the scattered planning and reporting obligations. In order to ensure that the EU meets its energy and climate targets for 2030, the Regulation implements a system in which the Member States and the Commission work closely together. The Member States describe their contributions and elaborate the measures to implement them in their National Energy and Climate Plans (NECP); the draft of the first NECP was due by 31 December 2019. The Commission monitors the NECP and evaluates whether the Member States are on track to meet the 2030 climate goals. If the Commission identifies shortcomings it may give recommendations to the Member States, take measures at EU level or even trigger the so-called gap-filler mechanism – a mechanism obliging those Member States that fall behind to implement additional measures on a national level.

2.5.4 Roadmap to Carbon Neutrality by 2050

The Juncker Commission developed a long-term vision for a decarbonized EU. In its 2018 communication 'A Clean Planet for All'[96] the Commission provides the cornerstones of a strategy that envisages a carbon-neutral but nonetheless prosperous, modern and competitive economy by the year 2050. The concept of zero net emissions can be traced back to the Paris Agreement, which states in Article 4(1) that the Parties should 'achieve a balance between anthropogenic emissions by sources and removals by sinks of greenhouse gases in the second half of this century' to hold the increase in global average temperature to well below 2°C above

---

[94]   Directive (EU) 2018/844 of the European Parliament and the Council of 30 May 2018 amending Directive 2010/31/EU on the energy performance of buildings and Directive 2012/27/EU on energy efficiency, OJ 19.6.2018, L 156/75.

[95]   OJ 21.12.2018, L 328/1.

[96]   Communication from the European Commission, Clean Planet for All: A European Strategic Long-term Vision for a Prosperous, Modern, Competitive and Climate Neutral Economy, COM(2018) 773 final.

pre-industrial levels. The Commission identifies seven strategic building blocks that should pave the way to a decarbonized EU economy by 2050:

- maximize the benefits from energy efficiency, including zero emission buildings;
- maximize the deployment of RES and the use of electricity to fully decarbonize energy supply;
- embrace clean, safe and connected mobility;
- conceive a competitive EU industry and the circular economy as a key enabler to reduce GHG emissions;
- develop an adequate smart network infrastructure and interconnections;
- reap the full benefits of the bio-economy and create essential carbon sinks;
- tackle remaining CO_2 emissions with carbon capture and storage.

The Commission is aware of the fact that the development and improvement of carbon-neutral technology is a costly issue. Substantial investment will be needed, but this is also considered to increase the productivity of key industrial sectors in the EU and, thus, to give the European economy and employment a boost. Moreover, reduced dependency on foreign imports of primary energy sources and improved air quality could be positive side effects of the EU's climate targets for 2050.

In late 2019 the new Commission, presided over by Ursula von der Leyen, took office. Under what is called the 'European Green Deal', EU climate action seems to be a pivotal point for internal and external EU politics in general.[97] The European Green Deal is seen as a 'new growth strategy that aims to transform the Union into a fair and prosperous society, with a modern, resource-efficient and competitive economy, where there are no net emissions of greenhouse gases in 2050 and where economic growth is decoupled from resource use'.[98] In this context the Commission proposed to increase the EU's 2030 climate target in order to become genuinely carbon-neutral by 2050, as specified by the Commission in a proposal for a European Climate Law in March 2020.[99] By the end of 2020, the European Council had agreed with the Commission's plan to strengthen the EU's greenhouse gas emissions reduction target for 2030 to net 55 per cent (considering removals by sinks).[100]

---

[97]　Cf. the Political Guidelines for the European Commission 2019–2024, A Union that Strives for More; available under https://ec.europa.eu/commission/sites/beta-political/files/political-guidelines-next-commission_en.pdf accessed 3 January 2021.

[98]　Cf. Communication from the European Commission, The European Green Deal, COM(2019) 640 final, p.2.

[99]　Proposal from the European Commission for a Regulation of the European Parliament and of the Council establishing the framework for achieving climate neutrality and amending Regulation (EU) 2018/1999 (European Climate Law), COM(2020) 80 final. The Commission's final proposal for the stepped-up 2030 target (net 55 per cent) was established on 17 September 2020, COM(2020) 563 final, after conducting an impact assessment: see the Communication from the European Commission, Stepping Up Europe's 2030 Climate Ambition, COM(2020) 562 final.

[100]　European Council meeting (10 and 11 December 2020) – Conclusions, EUCO 22/20, Brussels, 11 December 2020.

The negotiations for the European Climate Law, the Multiannual Financial Framework 2021–2027 (the long-term EU budget) and the coronavirus recovery fund ('Next Generation EU') were closely related. In order to support Member States whose economies are particularly dependent on fossil fuels, a Just Transition Mechanism (JTM) will be implemented. The JTM is a new instrument expected to help mobilize at least 150 billion euro over the period 2021–27 through grants, guarantees and loans and is composed of three pillars: the Just Transition Fund (JTF), the Just Transition Scheme (JTS) and the public sector loan facility.[101] The JTM's aim is to financially alleviate the transition towards a climate-neutral economy through public investments and reskilling or upskilling of impacted workers.

At the time of writing this chapter, the inter-institutional negotiations between Commission, Council and EP on the final version of the European Climate Law were under way. The EP asked for a greater tightening of the 2030 target (60 per cent emission reduction regardless of CO_2 removals) and proposed several other amendments to the Commission's draft in order to strengthen the regulatory backbone of climate law governance.[102]

2.6   THE ROLE OF CITIES IN CLIMATE ACTION

Alongside states, various other actors play a role in climate change mitigation and adaptation, including regions, cities, businesses and NGOs. In the literature, all these actors are often conflated under the terms 'non-state' and/or 'subnational' actors.[103] However, as Smeds and Acuto highlight, 'equating a city with a nonstate actor such as a multinational corporation is incomprehensible and analytically useless'.[104] They argue for 'clear distinctions between local governments and other actors'. Cities have an indisputable role to play in climate change action due to their prominent share of global energy consumption and GHG emissions.[105] As a consequence, they cannot be left out of the debate on solutions to the climate crisis.

---

[101]   See R.C. Fleming and R. Mauger, 'Green and Just? An Update on the "European Green Deal"', *Journal for European Environmental & Planning Law* 18:1–2 (2021), pp.1702.

[102]   The EP proposes the implementation of a climate advisory board ('European Climate Change Council') and the establishment of an EU greenhouse gas budget which represents the Union's fair share of the remaining global emissions: see www.europarl.europa.eu/doceo/document/TA-9-2020-0253_EN.pdf accessed 3 January 2021.

[103]   See e.g. A. Hsu, A. J. Weinfurter and K. Xu, 'Aligning Subnational Climate Actions for the New Post-Paris Climate Regime', *Climatic Change* 142 (2017), pp.419–32; L. Hermwille, 'Making Initiatives Resonate: How Can Non-state Initiatives Advance National Contributions under the UNFCCC?' *International Environmental Agreements* 18 (2018), pp.447–66; A. Hsu et al., 'A Research Roadmap for Quantifying Non-State and Subnational Climate Mitigation Action', *Nature Climate Change* 9 (2019), pp.11–17.

[104]   E. Smeds and M. Acuto, 'Networking Cities after Paris: Weighing the Ambition of Urban Climate Change Experimentation', *Global Policy* 9:4 (2018), p.557.

[105]   M. Ivanova, 'Good COP, Bad COP: Climate Reality after Paris', *Global Policy* 7:3 (2016), p.417; P. Bertoldi et al., 'Towards a Global Comprehensive and Transparent Framework for Cities And Local Governments Enabling an Effective Contribution to the Paris Climate Agreement', *Current Opinion in Environmental Sustainability* 30 (2018), p.67.

2.6.1   The Growing Involvement of Cities in Climate Action

Cities' involvement in environmental and climate action dates back decades. Fuhr et al. refer to the first 'engagement in climate policy-making' by municipalities occurring in the late 1980s.[106] The role of cities was acknowledged in various landmark international environmental protection texts such as the Brundtland Report (1987) and the Rio Earth Summit's Agenda 21 (1992). More recently, cities constituted the specific focus of diverse UN-supported programmes, including the UN Habitat New Urban Agenda (2016) and the 2030 Development Agenda with its SDG11 on sustainable cities (2015).[107]

Cities can provide valuable help to nation states in more than one way: they can reduce their own GHG emissions, but can also serve as 'originators or contributors of transformational change'.[108] Novel technologies and policies can be tested at a local level and best practices can then be shared with other cities. Networks of cities have been created to facilitate exchange of information, capacity-building and rule setting.[109] The C-40 network or the International Council for Local Environmental Initiatives (ICLEI) are examples of these networks that allow their members to work together but also to establish 'a two-way dialogue between cities and the UNFCCC'.[110] The intrinsic value of cities and their networks for climate action was reinforced when 'conventional strategies of addressing climate change through universal, interstate negotiations'[111] failed at COP15 in Copenhagen in 2009.[112]

As a consequence, the road to Paris had been paved with bottom-up initiatives including a stronger focus on cities, to the extent that Chan et al. state that 'the 2014 UN Climate Summit was the first UN summit dedicated to state and non-state climate action'.[113] During COP20 in Lima in 2014 the Non-State Actor Zone for Climate Action, a web-based platform abbreviated as NAZCA, was launched, 'showcasing climate commitments by companies, cities and investors'.[114] A year later in Paris, 11,000 commitments were registered, including 2,250 cities.[115]

---

[106]   H. Fuhr, T. Hickmann and K. Kern, 'The Role of Cities in Multi-level Climate Governance: Local Climate Policies and the 1.5°C Target', *Current Opinion in Environmental Sustainability* 30 (2018), p.1.

[107]   Ibid.

[108]   L. Hermwille (n. 103), p.454.

[109]   A. Hsu, A. J. Weinfurter and K. Xu (n. 103), p.422.

[110]   J. Kuyper, K. Bäckstrand and H. Schroeder, 'Institutional Accountability of Nonstate Actors in the UNFCCC: Exit, Voice, and Loyalty', *Review of Policy Research* 34:1 (2017), p.100.

[111]   M.J. Hoffmann, 'Climate Governance at the Crossroads: Experimenting with a Global Response after Kyoto' (2011) Oxford, abstract, available under www.oxfordscholarship.com/view/10.1093/acprof:oso/9780195390087.001.0001/acprof-9780195390087 accessed 11 March 2020.

[112]   E. Smeds and M. Acuto (n. 104), p.550.

[113]   S. Chan et al., 'Effective and Geographically Balanced? An Output-based Assessment of Non-state Climate Actions' Climate Policy 18:1 (2018) p.26.

[114]   M. Ivanova (n. 105), p.413.

[115]   Ibid.

Even though cities, alone or through networks, were already increasingly involved in climate action before Paris, they 'cannot "save the planet" alone'.[116] Fortunately, the 2015 Paris Agreement brought them recognition.

2.6.2   The Paris Agreement's Recognition of the Role of Cities

The Paris Agreement's bottom-up logic, with its 'shift in strategy from "name and shame" to "name and acclaim"',[117] incentivized cities to take part to global climate action to the point that cities 'are increasingly becoming the engine of both mitigation and adaptation action'.[118] In this regard, the Paris Agreement marks a move from a 'regulatory' and centralized approach to a polycentric model where cities are a necessary part of the solution.[119]

Although the Paris Agreement itself does not refer explicitly to cities, its preamble recognizes 'the importance of the engagements of all levels of government'. Cities are, in fact, mentioned in the preamble of the accompanying decision 1/CP.21 and grouped with other actors under the term 'non-Party stakeholders'. Non-Party stakeholders benefit from various paragraphs of their own (§§ 116–122 and 133–136) wherein their efforts are 'welcomed' (§§ 117 and 133) and where Parties are encouraged to 'work closely with [them] to catalyse efforts to strengthen mitigation and adaptation action' (§ 118). Additionally, non-Party stakeholders are invited to 'scale up their efforts and support actions to reduce emissions and/or to build resilience and decrease vulnerability to the adverse effects of climate change and demonstrate these efforts via the [NAZCA] platform' (§ 134). Cities are not just anecdotal actors but are expected to take action and to report it to the NAZCA web-based platform.

As of April 2021, there were 10,748 cities registered on the platform for a total of 11,970 actions.[120] An event that certainly increased the number of cities joining was the US withdrawal from the Paris Agreement, announced by the Trump administration in 2017. According to Ahmad et al., by 2017 there were 'over 100 cities that have pledged to a goal of 100 per cent renewable energy by 2035 and over 300 cities that have publicly supported the Paris Agreement goals'.[121] It was argued that the size of this pro-Paris movement of US states, cities and businesses is such that the withdrawal would only 'have a relatively mild effect on US emissions'.[122]

---

[116]   E. Smeds and M. Acuto (n 104), p.556.

[117]   M. Ivanova (n. 105), p.413.

[118]   R. Kinley, 'Climate Change after Paris: From Turning Point to Transformation', *Climate Policy* 17:1 (2017), p.12.

[119]   A. Hsu, A.J. Weinfurter and K. Xu (n. 103), p.419.

[120]   Global Climate Action – NAZCA, https://climateaction.unfccc.int/views/stakeholders.html?type=cities accessed 18 April 2021.

[121]   F.M. Ahmad, J. Huang and B. Perciasepe, 'The Paris Agreement Presents a Flexible Approach for US Climate Policy', *Carbon & Climate Law Review* 11:4 (2017), p.291.

[122]   L. Hermwille (n. 103), p.458.

Although the literature underlines the key role that cities have to play in reaching the 2 degrees Celsius or even the 1.5 degrees Celsius goal,[123] a fundamental question remains: how can the outcome of cities' climate actions be assessed, and how can an overlap with states' pledges be avoided?

2.6.3   The Difficulty of Assessing Climate Action by Cities

As non-Party stakeholders to the Paris Agreement, cities are not bound by the measurement, reporting, verification and transparency provisions that apply to the Parties.[124] Although the literature is advocating for a 'comprehensive and transparent reporting framework'[125] dedicated to cities, which would also avoid overlap with other actors' action,[126] Hale argues that, 'to the extent they overlap, nonstate actions reinforce, implement, and give credibility to the national pledges; to the extent they do not overlap, they help close the emissions gap'.[127]

City networks and the NAZCA platform were mentioned above as options to facilitate this. However, there certainly is a necessity for the UNFCCC to orchestrate[128] this newly recognized layer of actors.[129] Nevertheless, a couple of studies on cities' climate pledges and their first results tend to show ambitious targets and even over-performance.[130]

According to Fuhr et al., these pro-active cities combine five key drivers: high capacities combined with high problem pressure, local democracy, enabling policy framework, socio-economic environment and local leadership.[131] As one can see, all these drivers are not in the hands of the cities themselves. Therefore, in addition to pushing cities to take climate actions and to organize a reporting system, it appears that a successful implementation of the Paris Agreement will also depend on states leaving enough room for their subnational governments to get on board.

---

[123] N. Höhne et al., 'The Paris Agreement: Resolving the Inconsistency between Global Goals and National Contributions' *Climate Policy* 17:1 (2017), p.25; A. Hsu, A.J. Weinfurter and K. Xu (n. 103), p.420.

[124] For the literature about this lack of compelled feedback for cities, see N. Höhne et al. (ibid.), p.26.

[125] P. Bertoldi et al. (n. 105), p.67.

[126] A. Hsu et al. (n. 103), p.13.

[127] T. Hale, '"All Hands on Deck": The Paris Agreement and Nonstate Climate Action', *Global Environmental Politics* 16(3) (2016), p.20.

[128] On orchestration, see S. Chan, P. Ellinger and O. Widerberg, 'Exploring National and Regional Orchestration of Non-state Action for a <1.5°C World', *International Environmental Agreements: Politics, Law and Economics*, 18:1 (2018), p.135; L. Hermwille (n. 103), p.447; and S. Chan et al. (n. 113), p.27.

[129] On scaling up/out and vertical/horizontal alignment, see A. Hsu, A.J. Weinfurter and K. Xu (n. 103), p.419; and H. Fuhr, T. Hickmann and K. Kern (n. 106), p.1.

[130] S. Chan et al. (n. 113), p.24; and A. Kona et al., 'Covenant of Mayors Signatories Leading the Way Towards 1.5 Degree Global Warming Pathway', *Sustainable Cities and Society*, 41 (2018), pp.568–75.

[131] H. Fuhr, T. Hickmann and K. Kern (n. 106), pp.1–2.

2.7 CONCLUSION

Climate policy is an aggregate of binding and non-binding measures to combat global warming and its consequences. In the EU, we see a clear trend towards harmonizing legislation and integrating climate and energy policy. This development was heavily influenced by international agreements, in particular the UNFCCC, the Kyoto Protocol and the Paris Agreement.

From the very beginning, the EU has considered itself as a leader in international climate action. This ambition is reflected in the 20-20-20 goals for 2020, the 2030 Climate and Energy Framework and the roadmap for 2050 striving for carbon neutrality. Their implementation must ensure the achievement of the climate targets and, moreover, enhance energy security, while taking into account cost-effectiveness of the measures and inter-European solidarity. Current and future EU climate policy should lead to 'green growth' and prosperity in a decarbonized society.

At an international level, the temporary US withdrawal from the Paris Agreement threatened the achievement of the goal of an increase above pre-industrial temperature levels of 2 or preferably 1.5 degrees Celsius, because it is one of the largest GHG emitters. However, the US government re-entered the Paris Agreement in February 2021, which has sent an important signal to other major actors, such as China, India and Russia. As a rule of thumb, all Parties to the Paris Agreement have to abide by their commitments and do even more, as the sum of their pledges is far from guaranteeing the 2 or 1.5 degrees Celsius target by the end of the century. Implementation promises to be more demanding than the negotiation of the Paris Agreement itself.

Also in the future, EU climate policy will be subject to changes. All implemented instruments need to be reviewed regularly, with regard to, among other things, their effectiveness, fairness and efficiency. This will undoubtedly lead to amendments and reinforcements of EU climate law in order to reach a decarbonized society by 2050.

CLASSROOM QUESTIONS

1. What are the two primary objectives of European climate policy and which key instruments in EU climate law shall ensure that each respective objective is achieved?
2. What is the level of bindingness and prescriptiveness of the Paris Agreement?
3. To what extent can cities act against climate change?

SUGGESTED READING

Books

Delbeke J and Vis P, *Towards a Climate-Neutral Europe: Curbing the Trend (Routledge 2019).*
Farber DA and Peeters M (eds), *Climate Change Law (Edward Elgar Publishing 2016).*
Klein D, Carazo MP, Doelle M, Bulmer J and Higham A (eds), *The Paris Agreement on Climate Change: Analysis and Commentary (Oxford University Press 2017).*

Articles and chapters

Bertoldi P, Kona A, Rivas S, Dallemand JF, 'Towards a Global Comprehensive and Transparent Framework for Cities and Local Governments Enabling an Effective Contribution to the Paris Climate Agreement' (2018) 30 *Current Opinion in Environmental Sustainability*, 67–74.

Bogojevic S, 'Climate Change Law and Policy in the European Union' in C Carlarne, K Gray and R Tarasofsky (eds), *The Oxford Handbook of International Climate Change Law* (Oxford University Press 2016).

Peeters M, 'EU Climate Law: Largely Uncharted Legal Territory' (2019) 9(1) *Climate Law*, 137–47.

Policy documents

European Commission website on EU climate policy https://ec.europa.eu/clima/policies/eu-climate -action_en.

IPCC website on scientific knowledge about climate change www.ipcc.ch.

UNFCCC website https://unfccc.int/.

PART II
ESSENTIAL EU CLIMATE LAW

3

EU emissions trading system

ABSTRACT

- The centrepiece of the EU Emissions Trading System (EU ETS) is Directive 2003/87/EC, which was amended in 2009, among others, and revised in 2018;
- The EU ETS Directive is based on 'cap-and-trade': it (a) caps the greenhouse gas emissions of large emitters by allocating emission allowances and (b) makes the trading of those allowances possible to achieve low-cost emission reductions;
- From a cost-effectiveness point of view, the EU ETS succeeds in meeting the emissions caps at a low cost. However, the economic crisis of 2008 led to an allowance surplus and triggered a low allowance price, weakening investment in low-carbon technology. A Market Stability Reserve (MSR) takes in some of those surplus allowances, which raises the allowance price;
- From a solidarity point of view, part of the allowances to be auctioned is redistributed from Western to Eastern European Member States, which does not affect the cost-effectiveness of the EU ETS;
- The EU ETS faces three fundamental regulatory trade-offs:
  - Climate ambition versus international competitiveness: a more stringent EU emissions cap could weaken the competitiveness of internationally operating industries;
  - Structural reform versus cost-effectiveness: preventing allowance oversupply stimulates low-carbon technologies but could also undermine the cost-effectiveness of the EU ETS;
  - Keeping it simple versus meeting multiple goals: actors prefer a simple EU ETS but also want it to stimulate low-carbon technology, protect competitiveness, prevent carbon leakage and improve solidarity, which increases complexity.

3.1 INTRODUCTION

This chapter explores the 'EU ETS': the European Union (EU) Emissions Trading System (ETS) for greenhouse gases. This scheme mainly targets emissions of carbon dioxide (CO_2), nitrous oxide (N_2O) and perfluorocarbons (PFCs). The EU ETS is a 'cornerstone' and 'key tool' of EU climate policy, according to the European Commission.[1]

---

[1] https://ec.europa.eu/clima/policies/ets_en accessed 11 December 2020.

Initially, the 2030 goal of the EU was to reduce greenhouse gas emissions by at least 40 per cent below 1990 levels. To meet this target, ETS sectors would have to reduce their emissions by 43 per cent compared to 2005 (while non-ETS sectors have a 30 per cent reduction target). At the end of 2020, however, the European Council agreed with the European Commission's plan to strengthen the EU's reduction target to 55 per cent.[2] Emissions trading is intended to lower the costs of meeting this emission reduction target. For some companies that emit greenhouse gases, called 'emitters', it is relatively cheap to reduce emissions, whereas for others it is relatively expensive. If the latter can pay the former to further reduce emissions on their behalf, the same emission reductions are achieved at lower costs. If a legal framework is established to facilitate such transfers, emissions trading is born.

Emissions trading started in the EU in 2005 with Directive 2003/87/EC. This 'EU ETS Directive' was amended a few times, mainly in 2009 by Directive 2009/29/EC and in 2018 by Directive 2018/410. Emission rights, called 'allowances', can be traded between large emitters that are covered under the scheme, such as electricity producers, steelmakers and airlines.

In this chapter we examine the EU ETS in the light of the overarching EU climate policy principles of cost-effectiveness and solidarity. Section 3.1 explains the basics of greenhouse gas emissions trading and section 3.2 discusses its possible design variants. Section 3.3 examines the EU ETS Directive. Section 3.4 looks into the possibility of linking the EU ETS to non-EU emissions trading systems around the globe. Section 3.5 analyses the main implementation problems of the EU ETS and their solutions. Finally, section 3.6 answers the question of the extent to which the overarching principles of EU climate policy are honoured in the EU ETS Directive, and what balance has been struck between them.

3.2   BASICS OF GREENHOUSE GAS EMISSIONS TRADING

The concept of emissions trading was developed by John Dales in the 1960s.[3] Each emissions trading system starts with an emissions target for certain companies. This target puts a collective limit on their emissions. Each emitter is allocated a small piece of this collective limit, in the form of emission rights called 'allowances'. In the case of climate change, each allowance gives the holder the right, during a specified period, to emit one tonne of CO_2 (or, in the case of other greenhouse gases, one tonne of CO_2-equivalent).[4] The limited number of allowances defines the emissions cap of the emitter. The number of allowances decreases every year to

---

[2]   European Council meeting (10 and 11 December 2020) – Conclusions, EUCO 22/20, Brussels, 11 December 2020.

[3]   JH Dales, *Pollution, Property and Prices: An Essay in Policy-Making and Economics* (Toronto University Press 1968).

[4]   Carbon dioxide equivalency describes, for a certain greenhouse gas, the amount of CO_2 that would have the same global warming potential (GWP) over 100 years. For example, the GWP for nitrous oxide (N_2O) over 100 years is 298. This means that the emission of 1 million metric tonnes of N_2O is equivalent to the emission of 298 million metric tonnes of CO_2.

reduce emissions. Emitters either get the allowances free from the government or have to buy them at auction.

Why would a company choose not to reduce emissions itself but instead to buy allowances from another company? The reason is cost savings. If the government forces emitters to reduce their emissions by imposing emissions caps, there will be some emitters for whom it is relatively cheap to reduce emissions and others for whom it is relatively expensive. If the latter can pay the former to further reduce emissions on their behalf, and thus buy allowances, the same emission reductions are achieved at lower costs.[5]

A numerical example, visualized in Figure 3.1, helps to illustrate how emissions trading works. First there is the potential seller: firm A. If this firm reduces emissions below its emissions cap, it frees up allowances that can be sold. It will sell those allowances if the price it gets for them (say, €10 per allowance) is higher than what it costs to reduce the emissions and free up those allowances (say, €5 per tonne of CO_2). Then there is the potential buyer: firm B. An emitter that emits more than its emissions cap has a number of options to cover the emissions deficit, including the possibility to purchase allowances. Firm B will actually buy allowances if it has to pay less (say, the aforementioned €10 per allowance) than it would cost to reduce the emissions itself (say, €20 per tonne of CO_2). Both firms will gain from emissions trading (by moving, in Figure 3.1, from situation 1 to situation 2). The seller of the allowances earns (10 minus 5 equals) €5 per tonne of CO_2 and the buyer of the allowances saves (20 minus 10 equals) €10 per tonne of CO_2. Hence, the transaction will take place. This example clarifies that the instrument of emissions trading is literally cost-effective: 'effective' because the emission reductions are achieved (provided that monitoring and compliance are in order), while 'costs' are saved by allowing emissions to be traded.

When emissions trading is enabled, some firms are incentivized to become more active in reducing emissions than others. This is also illustrated in Figure 3.1. Firm A reduces twice as much emissions in situation 2 compared to a scenario without emissions trading (in situation 1). Where firm B would have been reducing emissions (in situation 1), it now decides not to reduce emissions at all but to buy allowances instead (in situation 2). Importantly, the end result is that the total of the envisaged emission reductions is still achieved. In situation 1, two 'blocks' of emission reductions need to be realized, one by each firm. In situation 2, however, firm A prefers to realize the two 'blocks' of emission reductions itself.

It pays for firm A to reduce emissions at €5 and sell emissions at €10 per tonne of CO_2. At the same time, firm B saves costs by paying €10 for the required emission reductions instead of €20 per tonne of CO_2. Both firms gain, and society at large benefits as well. Since allowances are traded at €10 per tonne of CO_2, the unnecessary expenditure of €20 per tonne of CO_2 abatement is avoided. This is desirable from a social welfare point of view.

---

[5]　Emissions trading is sometimes compared with a waterbed. Suppose that you would like to raise the water level at the head of the bed: then you would have to push down the water level at its foot. This is also how emissions trading works. If an emitter would like to increase his emission level, another emitter will have to decrease his emissions. After all, the total level of allowed emissions is fixed.

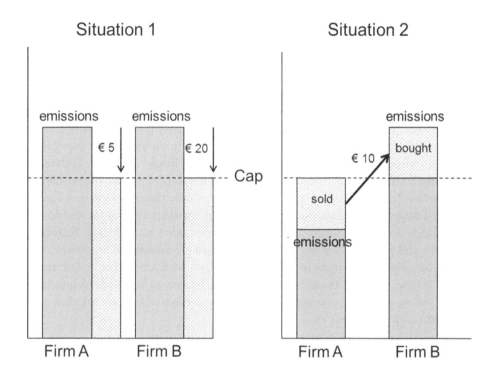

Source: Own. Euros (€) mentioned are euros per tonne of CO_2.

Figure 3.1 How emissions trading works

Emitters that surpass their emissions cap have a menu of options at their disposal, including:

- buying allowances to cover the emissions deficit;
- reducing emissions within the company itself, for instance by saving energy or by switching to renewable energy;
- storing emissions underground by making use of carbon capture and storage (CCS);
- paying a fine for non-compliance.

The first and fourth options are treated in this chapter, whereas the second option is discussed in Chapters 5 and 6 and the third option in Chapter 7 of this book. An emitter will choose the cheapest option to comply with his emissions cap. At the time of writing this chapter, we observe that buying allowances is cheaper than some but not all other options: the allowance price is about €30, whereas the penalty for non-compliance under the EU ETS is €100 per tonne of CO_2-equivalent. Depending on the technology used and the scale of its application, the costs of reducing greenhouse gas emissions by saving energy within the firm are roughly €10–30 per tonne of CO_2-equivalent (although this ranges widely from negative costs to several hundreds of euros per tonne), whereas the costs of CCS are typically around €50–60 per tonne of CO_2 (although they vary from €20 to more than €150 per tonne).[6]

In the future, however, these cost figures are likely to change as a result of (uncertain) technological developments and (uncertain) regulatory interventions. In principle, the allowance price could rise in the long run, since the overall emissions cap of the EU ETS decreases by 2.2 per cent every year from 2021 onwards (and even more to reflect the 55 per cent emission reduction target for 2030). However, various companies try to develop technologies that decrease the costs of reducing emissions, which ultimately lowers the allowance price. The allowance price could thus be lower than anticipated, to the extent that the options of saving energy, using renewables and applying CCS become cheaper over time. Technological improvements and learning-by-doing will drive down their costs, supported by (possibly additional) subsidies. Nonetheless some governments aim to raise the carbon price, for instance by applying a national carbon tax which operates as a domestic price floor for allowances (as the Netherlands does) to stimulate investments in low-carbon technologies. Furthermore, some scholars and politicians are in favour of imposing a pre-announced minimum 'reserve' price when auctioning allowances in the EU to stimulate such low-carbon investments. At some point in time, therefore, the allowance price could become so high that enhanced energy savings or storing emissions underground is more cost-effective for emitters than trading allowances to comply with their emissions caps.

3.3   EMISSIONS TRADING DESIGN AND HYBRID CARBON PRICING SCHEMES

'Emissions trading' is an umbrella concept for multiple market-based environmental policy instruments. There are three basic design variants of emissions trading, all of which will be discussed in this section:

- Cap-and-trade;
- Performance standard rate trading;
- Project-based credit trading.

---

[6]   These cost ranges are indicative only. A few energy saving options even have negative costs (such as increasing the energy efficiency of trucks). However, some energy saving options (such as fiscal measures to stimulate zero-emission cars), as well as various CCS projects (for instance those with offshore CO_2 storage), have higher costs per tonne of avoided CO_2 than the cost ranges mentioned here. See for example: K Gillingham and JH Stock, 'The Cost of Reducing Greenhouse Gas Emissions' (2018) 32(4) *Journal of Economic Perspectives* 53–72; R Koelemeijer et al., *Kosten energie- en klimaattransitie in 2030 – update 2018* (2018) PBL Planbureau voor de Leefomgeving, Publicatienummer 3241; L Beck and L Temple-Smith, 'Is CCS Expensive? Decarbonisation Costs in the Net-Zero Context' (2020) Global CCS Institute Brief; and: S Budinis et al., 'An Assessment of CCS Costs, Barriers and Potential' (2018) 22 *Energy Strategy Reviews* 61–81.

All these design variants are relevant for understanding the EU ETS, which will be clarified in the next section. The EU ETS is primarily based on cap-and-trade, but industry lobbies have tried to push its design towards performance standard rate trading, with increasing success. Moreover, importing project-based credits into the EU ETS was possible until 31 December 2020, albeit severely restricted. Finally, some Member States have decided to raise the carbon price by supplementing EU emissions trading with a national carbon tax. We will first describe the design variants of emissions trading and explain their differences. Then we will explain what happens when a domestic carbon levy is added to an emissions trading system.

3.3.1   Cap-and-Trade

Cap-and-trade refers to the trading of emission entitlements under an emissions cap.[7] A cap-and-trade system imposes a cap on the annual emissions of a group of emitters, such as companies, for a certain period of time. Emission rights, called 'allowances', are allocated to established companies for this period. The allowances are allocated either for free or through an annual sale by auction (a combination is also possible). The allowances are tradable. Newcomers and companies seeking to expand production must purchase allowances from established companies or from a government reserve. A company closing down a plant can sell its allowances.

3.3.2   Performance Standard Rate Trading

Performance standard rate trading refers to the trading of emission entitlements based on a performance standard.[8] Such a system of tradable reduction credits is based on a government-mandated emissions standard adopted for a group of companies. The emissions standard dictates permitted emissions per unit of energy consumption (such as electricity) or per unit of production output (such as steel). In this system, emission reduction credits can be earned by emitting less than is prescribed by the emissions standard. These credits can then be sold to companies who can use them to compensate their emissions in excess of the emissions standard applying to them. If the economy grows, the supply of credits also increases because companies do not operate under an 'absolute' emissions ceiling but have to observe the 'relative' emissions standard. An energy-intensive company which expands production, or a newcomer entering the industry, therefore has a right to new emissions, as long as he obeys the emissions standard. This means that emissions will grow (in 'absolute' terms). To prevent that from happening, the emissions standard can be strengthened. A company closing down loses its credits.

---

[7]   Cap-and-trade (C&T or CAT) is also referred to as allowance trading or permit trading.

[8]   Performance standard rate trading (PSR) is also referred to as credit trading, dynamic allocation or output-based allocation (OBA).

3.3.3   Project-based Credit Trading

Project-based credit trading refers to the trading of emission entitlements based on individual projects. The most common examples are Joint Implementation (JI) projects mainly in Eastern European countries and Clean Development Mechanism (CDM) projects in developing countries under the 1997 Kyoto Protocol to the United Nations Framework Convention on Climate Change (UNFCCC). Comparable projects are likely to become possible under Article 6 of the 2015 Paris Agreement, informally referred to as Sustainable Development Mechanism (SDM) projects. Emission reduction credits are generated on the basis of the difference between baseline emissions and predicted (or actual) emissions at the project site. Take the example of a project in which a coal-fired power plant is rebuilt into a more climate-friendly, gas-fired one in a foreign country where such an investment is cheaper. The baseline is an estimation of future emissions at the project site in the absence of the project, in this case the relatively high greenhouse gas emissions of a coal-fired power plant. This baseline is a counterfactual that will never materialize, because the plant will turn into a gas-fired one which produces fewer emissions. The difference in tonnes of CO_2-equivalent between baseline emissions and actual emissions at the foreign project site can be used as credits for companies to comply with their own emission reduction obligations at home.[9]

3.3.4   Comparative Performance of These Design Variants

Although there is a lack of consensus in the emissions trading literature about which design variant is 'best', most authors agree that cap-and-trade outperforms both performance standard rate trading and project-based credit trading in terms of effectiveness and efficiency.[10] In terms of acceptance the situation depends, among other things, on the political preferences of the nation(s) involved and on the type of industry to which the scheme applies. We will now briefly discuss the effectiveness and efficiency of the aforementioned emissions trading design variants.

Effectiveness
Effectiveness is about reaching the emission targets of the emissions trading system. Cap-and-trade is environmentally effective because it imposes an absolute limit on total emissions, which also should become more stringent over the years. Effectiveness will only be a problem if emissions are not adequately monitored or if non-compliance measures are not enforced. As long as monitoring and enforcement are in order, cap-and-trade will be effective.

Performance standard rate trading, however, has an additional problem in reaching an absolute, predefined emission level for the industry to which the relative standard applies.

---

[9]   The national version of this is domestic offset projects (DOPs), where a domestic party invests in a domestic project to generate such credits.

[10]   A Nentjes and E Woerdman, 'Tradable Permits versus Tradable Credits: A Survey and Analysis' (2012) 6(1) *International Review of Environmental and Resource Economics* 1–78.

When industrial production increases, the emissions of companies rise as well. The only way to deal with this problem is to strengthen the emissions standard (ad hoc or perhaps automatically), but energy-intensive industries could lobby against a stricter emission requirement.

Project-based credit trading is even more problematic in terms of effectiveness. There may be several plausible ways to calculate the baseline for (SDM, CDM or JI) emission reduction projects. Effectiveness is undermined if future emissions without the project are overestimated by inflating the baseline to claim more credits. Moreover, emissions may still increase in a host country outside the project location, in particular in developing countries that do not have a nationwide emissions cap.

Efficiency

Efficiency is about reaching the environmental targets at the lowest possible costs. Cap-and-trade has 'superior' efficiency properties.[11] In an emissions trading system based on emissions caps, each unit of emissions has a price.

In the case of allowance auctioning, the emitter has to buy allowances and obviously pays for each unit of emissions. In the case of free allocation, allowances have 'opportunity costs', equal to the allowance price. By using the allowances to cover the emissions, the emitter foregoes the opportunity to sell the allowances and thus misses out on sales revenues. Since each unit of emissions has a price in a cap-and-trade scheme, there is an incentive to examine all emission reduction possibilities and apply the least costly option.

In a performance standard rate trading system, there is a credit price. The received amount of money for credits sold is equal to the sum paid to purchase the credits by companies that exceed the emissions standards. However, the emissions within the limits set by the emissions standard remain without a price. For the group of companies as a whole, the cost of the permitted emissions is nil. Free credits do not have opportunity costs: if a company stops emitting it has no credits to sell, because it will lose its credits. Credits can only be earned through reducing emissions per unit of energy or output. Economizing on fuel input or slowing production does not earn any credits. Total emission reduction costs are therefore higher compared to cap-and-trade.

Project-based credit trading improves the cost-effectiveness of reaching emission targets, but it is not a full-blown, 'top-down' scheme that prizes each unit of emissions in (some part of) the economy. It also suffers from relatively high transaction costs.[12] Unlike the other design variants, CDM and (some) JI projects require pre-approval to check the environmental integrity of the project baseline, thereby raising transaction costs. Governments that do not want to impose an emissions cap on their industries, for instance those in developing countries, are likely to prefer project-based emissions trading over performance standard rate trading:

---

[11]   For example, T Tietenberg, M Grubb, A Michaelowa, B Swift and ZX Zhang, *International Rules for Greenhouse Gas Emissions Trading*, UNCTAD/GDS/GFSB/Misc.6, Geneva: United Nations Conference on Trade and Development (UNCTAD, 1999, p.106).

[12]   Transaction costs are the costs of making an economic exchange, such as information costs, contract costs and the costs of monitoring and enforcement.

projects are then merely an option for firms, whereas a performance standard is mandatory and more demanding in terms of administrative infrastructure.

3.3.5   Adding a Domestic Carbon Levy to Emissions Trading

Emissions trading is a 'quantity-based' instrument: emissions are priced on the market by trading emission allowances. Emissions taxation is a 'price-based' instrument where the government sets a price on emissions. There is a large body of literature on their relative pros and cons (referred to as 'quantities *versus* prices'), but legislators may also combine both instruments (quantities *and* prices).[13] For instance, the United Kingdom (UK) (since 2013) and the Netherlands (since 2020) have legislated to combine carbon trading for electricity producers with a carbon levy operating as a domestic carbon price floor. The Netherlands even extended this floor price in 2021 to its ETS industries. The result is a hybrid carbon pricing scheme, with the potential to raise the carbon price in those countries. The minimum price in the UK was £18 (pounds) until 2021; in the Netherlands it was €18 in 2020 increasing stepwise to €43 in 2030 for the power sector. Dutch ETS industries have faced a price floor of €30 since 2021, to increase to €125 in 2030. If the EU allowance (market) price is lower than this (regulatory) price floor, companies pay the difference per ton of CO_2. A higher carbon price due to such a domestic price floor has the disadvantage of increasing emission reduction costs, but has the potential advantage of strengthening the incentive for a national transition to low-carbon power and industrial production.

3.4   THE EU ETS DIRECTIVE

The EU Emissions Trading System (EU ETS) is a hybrid of the three emissions trading design variants mentioned in the previous section.[14] Although cap-and-trade forms its basis, importing CDM and JI credits into the EU ETS was possible until 31 December 2020 (albeit with restrictions) to further enhance cost-effectiveness. Moreover, to boost acceptance, some elements of performance standard rate trading have been incorporated into the EU ETS, such as free allocation to expanding industries as well as emissions standards to determine the allocation of free allowances. In this section and those which follow we will explain that this could lead to some inefficiencies in the EU ETS.

The EU ETS started in 2005. The legal basis for this scheme is Directive 2003/87/EC (the 'EU ETS Directive'),[15] amended a few times, for instance in 2009 by Directive 2009/29/EC, and

---

[13]   P Wood and F Jotzo, 'Price Floors for Emissions Trading' (2011) 39(3) *Energy Policy* 1746–53.

[14]   E Woerdman and A Nentjes, 'Emissions Trading Hybrids: The Case of the EU ETS' (2019) 15(1) *Review of Law and Economics* 1–32.

[15]   Directive 2003/87/EC of the European Parliament and the Council of 13 October 2003 establishing a scheme for greenhouse gas emission allowance trading within the Community and amending Council Directive 96/61/EC. OJ 2009 L.275/32–46.

revised in 2018 by Directive 2018/410.[16] The EU ETS operates in 27 EU Member States plus Norway, Iceland and Liechtenstein. It includes more than 11,000 installations from large emitters, covering about 40 per cent of greenhouse gas emissions in the EU. The EU ETS mainly targets CO_2 emissions, but since 2013 it has also included N_2O emissions from the production of certain acids as well as PFC emissions from aluminium production.

For reasons of effectiveness and efficiency, the EU ETS is primarily based on cap-and-trade: the emissions caps and tradability of the emissions allowances are central to the scheme. Before emissions can be traded, however, the operator of an installation must obtain a greenhouse gas emissions permit. The competent authority grants such an authorization to emit only if the emitter is deemed capable of monitoring and reporting emissions. The next step is to allocate allowances to emitters, either via auction or free of charge. After that has been done, allowances can be traded between emitters. To be able to enforce the emissions caps at company level, emissions should be monitored and allowance transactions must be registered. If an emitter does not have enough allowances to cover its emissions, penalties will be imposed. This section discusses the following essential design elements of cap-and-trade, which can also be found in the EU ETS Directive:

- Emissions cap;
- Allowance allocation;
- Emissions monitoring;
- Allowance tradability;
- Compliance and enforcement.

3.4.1 Emissions Cap

In a cap-and-trade scheme, the emissions cap is crucial. The emissions target defines this cap: the maximum amount of greenhouse gases that the covered installations are allowed to emit. The goal of the EU for 2030, set out in its climate and energy framework, is to reduce greenhouse gas emissions by at least 40 per cent below 1990 levels.[17] To achieve this goal, allowances allocated to installations under the EU ETS should be 43 per cent below their 2005 emissions by 2030. The 2013 EU-wide emissions cap has been set at 2,084,301,856 (about 2 billion) allowances. Article 9 of the amended and revised EU ETS Directive determines that the emissions cap should be reduced by 1.74 per cent each year as of 2013 and by 2.2 per cent each year as of 2021 to ensure that total emissions continue to fall. This will be sufficient to reach the aforementioned 2030 goal, but a higher percentage – and thus a more stringent emissions cap – is needed to meet the 2 (preferably 1.5) degrees Celsius global warming goal of the Paris Agreement and the corresponding ambition of the European Commission to be

---

[16]   Directive 2009/29/EC of the European Parliament and of the Council of 23 April 2009 amending Directive 2003/87/EC (...). OJ 2009 L.140/63–87. And Directive (EU) 2018/410 of the European Parliament and of the Council of 14 March 2018 amending Directive 2003/87/EC (...). OJ 2018 L.76/3–27.

[17]   A Policy Framework for Climate and Energy in the Period from 2020 to 2030, COM(2014) 15 final, 22.1.2014, Brussels: European Commission.

climate-neutral by 2050.[18] At the end of 2020, therefore, the European Council agreed with the European Commission's Green Deal proposal to strengthen the EU's reduction target to 55 per cent.[19]

The EU ETS has multiple trading phases (or compliance periods): the first phase 2005–07 (a pre-Kyoto learning period); the second phase 2008–12 (which coincided with the first commitment period of the Kyoto Protocol, discussed in Chapter 2 of this book); the third phase 2013–20; and the fourth phase 2021–30. The first two trading phases were based on national emissions caps, with allowances being allocated at Member State level via National Allocation Plans (NAPs). Since the trading phase 2013–20, however, the allowances have been distributed at EU level based on a single, Community-wide emissions cap. The mandatory emissions caps are imposed on installations that are owned by certain power companies, energy-intensive industries and commercial airlines. Member States submit a list of installations under the EU ETS to the European Commission. These lists are referred to as National Implementation Measures (NIMs).

The amended EU ETS Directive (in Annex I) broadened the scope of covered industries, now including oil refineries and steel works as well as the production of iron, aluminium, metals, cement, lime, glass, ceramics, pulp, paper, cardboard, acids and bulk organic chemicals. Power companies have a central position, since they produce most of the emissions under the EU ETS. Capture, transport and underground storage of CO_2 were also added as activities covered by the EU ETS as of 2013, in conformity with Directive 2009/31/EC, which is the legal basis for CCS. Allowances do not have to be surrendered for CO_2 emissions that are permanently stored; they only need to be surrendered when emissions would leak from CO_2 capture installations, transport pipelines and storage sites.

Aviation was effectively added to the EU ETS in 2012, based on Directive 2008/101/EC.[20] Power stations and industrial installations have a separate emissions cap from aviation because different types of allowances are issued to them. Allowances for fixed installations are general allowances, whereas the aviation sector has aviation allowances. Airlines can use both types of allowances for compliance, but fixed installations cannot use aviation allowances. The Community-wide emissions cap for aviation in the trading period 2013–20 was set at 5 per cent below the average annual level of emissions in the years 2004–6. Unlike the declining cap for fixed installations, the aviation emissions cap remained the same in each year of this third trading period, namely 208,502,525 (about 200 million) aviation allowances per year. From 2021 onwards, however, the aviation emissions cap will start to decrease by 2.2 per cent each year. Various aircrafts and airlines are exempted from the EU ETS, such as police, state

---

[18]   A Clean Planet for All: A European Strategic Long-Term Vision for a Prosperous, Modern, Competitive and Climate Neutral Economy, COM(2018) 773 final, 28.11.2018, Brussels: European Commission.

[19]   The European Green Deal, COM(2019) 640 final, Brussels: European Commission.

[20]   Directive 2008/101/EC of the European Parliament and of the Council of 19 November 2008 amending Directive 2003/87/EC so as to include aviation activities in the scheme for greenhouse gas emission allowance trading within the Community. OJ 2009 L.8/3–21.

and rescue flights; airlines operating limited services within the EU; and airlines from least developed countries.

According to the Court of Justice of the EU (CJEU), the application of the EU ETS to aviation is compatible with international law. The CJEU drew this conclusion on 21 December 2011 as part of proceedings brought by a number of American airlines and their trade associations against the UK Secretary of State for Energy and Climate Change.[21] However, after political protest and legal action taken by various jurisdictions, including the US and China, the EU decided in early 2013 to 'stop the clock' by postponing enforcement of the EU ETS for flights from or to non-European countries. Initially non-enforcement only applied to the year 2012, but this was extended first until, and later also after, 2016.[22] In 2013 the International Civil Aviation Organization (ICAO) agreed to develop a global market-based mechanism (MBM) by 2016 to tackle international aviation emissions as of 2021. The ICAO came up with a Carbon Offsetting and Reduction Scheme for International Aviation (CORSIA). This scheme is voluntary until 2026 and aims to stabilize CO_2 emissions from international aviation at 2020 levels by allowing operators to buy offset credits from emission reduction projects in other sectors. The EU ETS will remain limited to flights within Europe, but will also be subject to a new review in the light of CORSIA.

To further enhance the cost-effectiveness of climate policy, importing (relatively cheap) project-based credits into the EU ETS was possible until 2020. To guard the effectiveness of the scheme, the import of CDM and JI credits after 2013 was (roughly) restricted for operators to 11 per cent (of their allocation in the period 2008–12) or for newcomers to 4.5 per cent (of their verified emissions during the period 2013–20).[23] An additional reason for these quantitative restrictions was that the environmental integrity of project-generated credits is considered to be weaker than that of allowances in a cap-and-trade system. There were also qualitative restrictions: international credits were not allowed to be imported from nuclear energy projects, afforestation or reforestation activities, or, since 2013, projects involving the destruction of industrial gases. As of 2021, however, the importation of project-based credits is no longer allowed, because it increases the overall emissions cap in the EU.

At the time of writing this chapter, shipping is not yet included in the EU ETS; the European Commission announced in its European Green Deal a proposal to do so as of 2023. In seeking to reduce greenhouse gas emissions from the maritime sector, the EU preferred a global MBM under the International Maritime Organization (IMO), but lack of international agreement triggered the EU to take action itself. Since 2018, large ships (more than 5,000 gross tons) are required to monitor, report and verify CO_2 emissions using EU ports. Also road transport

---

[21]   Judgment of the Court of Justice of 21 December 2011 in Case C-366/10.

[22]   Regulation (EU) 2017/2392 of the European Parliament and of the Council of 13 December 2017 amending Directive 2003/87/EC to continue current limitations of scope for aviation activities and to prepare to implement a global market-based measure from 2021. OJ 2017 L.350/7–14.

[23]   The exact rules are more complex: see Commission Regulation (EU) No 1123/2013 of 8 November 2013 on determining international credit entitlements pursuant to Directive 2003/87/EC of the European Parliament and of the Council. OJ 2013 L.299/32–33.

and buildings are not included in the EU ETS. Those sectors are now primarily subject to (non-tradable) emissions standards, direct regulation, taxation and labelling (see Chapter 4), but in its European Green Deal the Commission proposed to consider their inclusion into the EU ETS. Although it is theoretically possible to include individuals in a separate emissions trading scheme, its chances of political success are slim, for instance because of behavioural barriers due to CO_2 price uncertainty, fairness concerns regarding how many allowances each individual should get and double counting of the same emissions under such a personal carbon trading scheme and the EU ETS.[24]

3.4.2　Allowance Allocation

Article 3 of the EU ETS Directive defines an allowance as 'an allowance to emit one tonne of carbon dioxide equivalent during a specified period, which […] shall be transferable […]'. In the first and second trading phases, respectively 95 per cent and 90 per cent of the allowances were allocated free of charge to firms, based on Article 10 of the EU ETS Directive. Only a very small part of the allowances was auctioned. Since 2013, however, the default method for allocating allowances is through auctioning them. Power generators need to buy all of their allowances at auction. The reason for this is that power companies do not compete on an international product market, so they can pass on most of their carbon costs to electricity consumers.

Various energy-intensive industries, however, do operate on international markets, where they are exposed to competition with non-EU companies that are often not subject to any form of carbon pricing. This could result in carbon leakage: the undesirable moving of companies, and thus emissions, to countries without an emissions trading system.[25] To prevent such leakage, these 'exposed' industries appear on a so-called carbon leakage list and receive their allowances for free (if they satisfy certain criteria under Article 10a of the amended EU ETS Directive). The Commission is required to make a new list every five years: the latest list was adopted in 2019 and applies to the years 2021–30. For energy-intensive industries that are not on this list, free allocation will decline every year to reach 30 per cent in 2025 and zero per cent in 2030. The number of allowances available for industries receiving free allowances will, of course, decline in line with the total cap for all companies. Under certain conditions, power companies in Eastern European Member States could receive some allowances for free until

---

[24]　For example, A Brohé, 'Personal Carbon Trading in the Context of the EU Emissions Trading Scheme' (2010) 10(4) *Climate Policy* 462–76.

[25]　Carbon leakage may also occur via reduced demand for fossil fuels in the emissions trading system, which lowers their price and therefore actually stimulates their use (and thus emissions) in countries outside the scheme. If it occurs, carbon leakage undermines the effectiveness of the EU ETS since emissions on a global scale would not be reduced and could even rise. Carbon leakage is also inefficient, since the emissions that move outside the scheme would remain unpriced.

2020, following Article 10c of the amended EU ETS Directive. The European Council decided to continue this exception until 2030, albeit with some additional limitations.[26]

In the first two trading phases, free allowances were allocated based on historical emissions, which is called 'grandfathering' in the literature. This favours 'dirty' firms with relatively high emissions, which was considered necessary at the start of the EU ETS to gain acceptance by the industry. Since 2013, however, free allocation of allowances takes place on the basis of a carbon standard per unit of production multiplied by production in a certain base year. Such an emissions standard, referred to as an 'ex-ante benchmark' in Article 10a of the amended EU ETS Directive, is determined based on the average emissions of the 10 per cent of installations with the lowest carbon emissions per unit of product or energy output in an industrial sector. These complex rules essentially amount to handing out free allowances based on low carbon performance. This favours 'clean' firms with relatively few emissions per product.

By using an emissions standard ('benchmarking') to calculate the allocation of free allowances, an element of performance standard rate trading is brought into the EU ETS. In principle, in a cap-and-trade scheme, emissions can be (auctioned off or) allocated free of charge on the basis of any allocation rule that is politically desirable. Allocation rules only have distributional consequences, with no consequences for the emissions cap or for allowance trading, which is called the 'independence property' of emissions trading.[27] Historical emissions usually form the basis of free allocation in order to garner support from the 'dirtier', emissions-intensive industries. This was also applied at the start of the EU ETS in 2005, but since 2013 the EU ETS has used the aforementioned carbon standard per unit of production as the basis for calculating the yearly fixed (so-called ex-ante) allocation of free allowances. An advantage of 'benchmarking' is that it favours 'cleaner' industries with relatively low emissions levels, which most politicians consider to be fairer than favouring 'dirtier' firms. A potential disadvantage, however, is that such a 'relative' standard is one of the building blocks of a performance standard rate trading system. If industry lobbies succeed in incorporating more design elements of performance standard rate trading, which is less effective and less efficient than cap-and-trade, a sub-optimal design of the EU ETS would emerge. One potential disadvantage was already anticipated by the Commission: if the total number of free allowances calculated per firm based on those benchmarks were to exceed the EU-wide emissions cap, a so-called Cross-Sectoral Correction Factor (CSCF) would be applied. Each firm's free allocation would then be slightly reduced. However, the calculation of this factor was successfully challenged before the CJEU in 2016, after which the Commission had to calculate more stringent reduction percentages.[28] In a different case, the Court ruled that allowances should not be

---

[26] Free allowances for the energy sector should be no more than 40 per cent of the auctioned allowances in these low-income Member States. See: Conclusions on 2030 Climate and Energy Policy Framework, SN 79/14, 23.10.2014, Brussels: European Council, p.3.

[27] RW Hahn and RN Stavins, 'The Effect of Allowance Allocations on Cap-and-Trade System Performance' (2011) 54(4) *Journal of Law and Economics* 267–94.

[28] Judgment of the Court of Justice of 28 April 2016 in Joined Cases C-191/14, C-192/14, C-295/14, C-389/14 and C-391/14 to C-393/14.

considered property: Member States are entitled to claim back wrongly allocated allowances without compensation.[29] The CJEU also made clear that Member States are not allowed to tax allowances allocated free of charge which have not been used or which have been transferred.[30]

All allowances which are not allocated free of charge will be auctioned. Member States are responsible for ensuring that the allowances assigned to them are auctioned. Each Member State could choose between developing its own allowance auction or cooperating with other Member States in a regional or EU-wide auction. The UK, Poland and Germany wanted to have their own national auction platforms, but eventually all Member States decided to use the European Energy Exchange (EEX) in Leipzig as a common auction platform (selected via a procurement procedure).[31]

Article 10 of the revised EU ETS Directive determines that, rather than 100 per cent, 90 per cent of the allowances to be auctioned will be distributed to Member States, in shares identical to their share of verified emissions in 2005. For the purpose of solidarity, 10 per cent of the allowances to be auctioned will be distributed among the least wealthy Member States. In practice, this amounts to a redistribution of wealth from Western European to Eastern European Member States. For the period after 2020, the European Council decided to set aside (and auction) a new reserve of 2 per cent of the EU ETS allowances to modernize the energy systems in low-income Member States, now referred to as the 'Modernisation Fund' under Article 10d of the revised EU ETS Directive.[32]

Each auction of allowances is open to buyers from anywhere in the EU, Norway, Iceland and Liechtenstein. Allowance auctions are subject to Regulation 1031/2010/EU, referred to as the 'Auctioning Regulation', which has been amended several times.[33] This Regulation addresses a number of legal and economic issues, including access to the auctions, the minimum number of allowances in a bid, the timing of auctions and the prevention of market abuse. The Auctioning Regulation assigns various tasks to several entities that assist in containing market manipulation, including (a) the common auction platform of the European Energy Exchange (EEX) in Leipzig; (b) an auction monitor (still to be selected via a procurement procedure); (c) the European Commission; and (d) several national competent bodies supervising the financial sector.[34] Some inefficiencies may arise in this framework, for instance because of the

---

[29]   Judgment of the Court of Justice of 8 March 2017 in Case C321/15.

[30]   Judgment of the Court of Justice of 12 April 2018 in Case C-302/17.

[31]   The UK appointed ICE Futures Europe (ICE) in London as its auction platform, whereas Germany and Poland both selected the European Energy Exchange (EEX) in Leipzig.

[32]   Conclusions on 2030 Climate and Energy Policy Framework, SN 79/14, 23.10.2014, Brussels: European Council, pp.3–4.

[33]   Commission Regulation (EU) No 1031/2010 of 12 November 2010 on the timing, administration and other aspects of auctioning of greenhouse gas emission allowances pursuant to Directive 2003/87/EC of the European Parliament and of the Council establishing a scheme for greenhouse gas emission allowances trading within the Community. OJ 2010 L.302/1–41.

[34]   These are the competent national authorities charged with tasks under Directive 2014/65/EU on financial instruments (MiFID), Directive 2003/6/EC on insider dealing and market manipulation (MAD) and Directive (EU)

concurring competences of these enforcement agents, but coordination by Member States could alleviate this potential problem.[35]

According to Article 10(3) of the amended EU ETS Directive, at least 50 per cent of the allowance auction revenues should (not shall) be used for emission mitigation and adaptation measures and technologies, including renewable energy, energy efficiency and CCS. This is basically a non-mandatory call for governments to earmark part of the auction money for climate protection. Economists generally perceive the earmarking of money to be undesirable (as there could be more efficient ways of spending government finance), but in practice Member States spent around 80 per cent of their auction revenues on climate and energy measures.[36]

As of 2013, airline companies were required to buy 15 per cent of the aviation allowances at auction, while 82 per cent could be obtained free of charge based on a 'benchmark' (which stipulates that an airline receives 0.6422 allowances per 1,000 tonne-kilometres flown).[37] The remaining 3 per cent is put in a special reserve for fast-growing and new aircraft operators. All auction revenues from selling aviation allowances should (again: not shall) be used: (a) to fund emission mitigation and adaptation measures or (b) to cover the administrative costs of the scheme.

3.4.3 Emissions Monitoring

To ensure correct monitoring and reporting of emissions, the operator of an installation under the EU ETS needs a greenhouse gas emissions permit – which is reviewed at least every five years – from a national competent authority, according to Articles 4–8 of the EU ETS Directive. In order to receive such a permit the operator is required to submit a monitoring plan, which needs to be approved, based on Article 14 of this Directive.

Article 19 of the EU ETS Directive says the operator needs to apply for an account with a registry, comparable to an online banking account. In 2012 the EU merged the national registries into a Union registry, covering all 30 countries participating in the EU ETS. The Union registry is an online database that holds accounts for stationary installations and aircraft operators. The registry records: (a) a list of installations under the EU ETS, mentioned under the NIMs; (b) accounts of operators, companies or natural persons holding allowances; (c) allowance transactions by the account holders; (d) annual verified CO_2 emissions from installations or aircrafts; and (e) an annual reconciliation of allowances and verified emissions.

2018/843 on the prevention of the use of the financial system for the purpose of money laundering and terrorist financing.

[35] SE Weishaar and E Woerdman, 'Auctioning EU ETS Allowances: An Assessment of Market Manipulation from the Perspective of Law and Economics' (2012) 3(3) *Climate Law* 247–63.

[36] Report on the functioning of the European carbon market, COM/2019/557 final/2, Brussels: European Commission.

[37] Commission Decision of 26 September 2011 on benchmarks to allocate greenhouse gas emission allowances free of charge to aircraft operators pursuant to Article 3e of Directive 2003/87/EC of the European Parliament and of the Council. OJ 2011 L.252/20–21.

The European Union Transaction Log (EUTL) automatically checks, records and authorizes allowance transactions between accounts in the Union registry.

Based on Articles 15 and 16 of the EU ETS Directive, the operator needs to monitor the emissions of its installations or aircrafts and submits an annual emissions report, with a verification report, to the competent authority before 1 April. The verification report must be compiled by an independent accredited verifier (which is, in practice, an accountancy company such as KPMG or PwC). Each year the allowances must be surrendered by the operator, from his account in the registry to the competent authority, before 1 May. The number of surrendered allowances must be equal to the volume of emissions reported by the operator. The competent authority has the implicit power to correct the reported emission numbers in case of mistakes or in cases of cheating. This annual procedure of monitoring, reporting and verification (MRV) is referred to as the 'compliance cycle' of the EU ETS.

As of 2013, two new Regulations apply in this context. The Monitoring and Reporting Regulation (MRR) lays down rules for the monitoring and reporting of emissions and activity data.[38] The Accreditation and Verification Regulation (AVR) contains provisions for the verification of emission reports and for the accreditation, mutual recognition and supervision of verifiers.[39]

3.4.4   Allowance Tradability

The primary objective of the EU ETS Directive is to meet the emissions caps at the lowest possible costs. In the wording of Article 1 of this Directive, the emissions trading system is established 'to promote reductions of greenhouse gas emissions in a cost-effective and economically efficient manner'. To make that happen, Article 3 of the EU ETS Directive defines an emission right as 'an allowance to emit one tonne of carbon dioxide equivalent during a specified period, which […] shall be transferable […]'. As explained at the beginning of this chapter, the trading of allowances enables companies to realize the emission reductions in an efficient way. The tradability of the emission entitlements is therefore essential in an emissions trading system.

Next to meeting the emissions caps at low cost, some argue that the EU ETS Directive also has a secondary (or implied) objective, namely the stimulation of low-carbon technologies. Recital 20 of the Preamble, for instance, mentions that this Directive will 'encourage the use of more energy-efficient technologies'. This is done under Article 10, for instance, by using part of the allowances in the new entrants' reserve and part of the auctioning revenue to support CCS and CCU (carbon capture and utilization) as well as innovative renewable energy and energy storage technologies. It could be argued that the EU ETS Directive has even more secondary objectives, including the prevention of carbon leakage. As explained previously, Article

---

[38]   Commission Regulation (EU) No 601/2012 of 21 June 2012 on the monitoring and reporting of greenhouse gas emissions pursuant to Directive 2003/87/EC of the European Parliament and of the Council. OJ 2012 L.181/30–104.

[39]   Commission Regulation (EU) No 600/2012 of 21 June 2012 on the verification of greenhouse gas emission reports and tonne-kilometre reports and the accreditation of verifiers pursuant to Directive 2003/87/EC of the European Parliament and of the Council. OJ 2012 L.181/1–29.

10 attempts to prevent carbon leakage by providing free allowances to companies that compete with firms outside the EU.

Alongside trading allowances, there are some additional flexibility tools for firms operating under the EU ETS. If a company expects a higher carbon price in the future, it can bank relatively low-priced allowances for later use to make future compliance cheaper. In principle, banking thus enhances the efficiency of climate policy. Alongside banking allowances for later use within a compliance period, the banking of allowances from one compliance period (2013–20) to another (2021–30) is made possible under Article 57 of the 'Registries Regulation'.[40]

Borrowing allowances from future allocations could also improve efficiency, but at the same time increases the risk of future non-compliance. Borrowing effectively increases current emissions caps and lowers future emissions caps, which may then turn out to be too difficult to meet. For this reason, borrowing from the future is typically avoided in emissions trading systems. In the EU ETS, however, firms receive their yearly allocation of allowances no later than 28 February, prior to the compliance date of 30 April for emissions in the previous year. In practice, this allows firms to 'borrow' these allowances to cover the previous year's emissions. With a stock of allowances which is higher than the annual amount allocated, firms that are short on an annual basis can therefore postpone the abatement of their pollution to some extent.

3.4.5 Compliance and Enforcement

After each year, as explained above, a company must surrender enough allowances to cover the emissions of its installation(s). If the operator does not surrender sufficient allowances, Article 16 of the EU ETS Directive determines that the following penalties will be imposed:

- a fine of €100 per tonne;
- a requirement to repair the emissions deficit;
- naming and shaming.

First, such an operator must pay a fine of €100 for each tonne of CO_2-equivalent emitted that is not covered by allowances. Since 2013, this penalty has increased by the level of inflation in European consumer prices. Second, the operator must repair this emissions deficit by surrendering an amount of allowances equal to these excess emissions in the next calendar year. Third, the Member State publishes the names of the operators who are in breach of the requirement to surrender sufficient allowances under the EU ETS. According to the CJEU,

---

[40]   Commission Regulation (EU) No 389/2013 of 2 May 2013 establishing a union registry pursuant to Directive 2003/87/EC of the European Parliament and of the Council, Decisions No 280/2004/EC and No 406/2009/EC of the European Parliament and of the Council and repealing Commission Regulations (EU) No 920/2010 and No 1193/2011. OJ 2013 L. 122/1–59.

these penalties apply even if the operator who did not surrender its allowances on time had enough allowances available to cover its emissions for the past year on 30 April.[41]

This is a serious set of penalties, if only in view of the allowance price, which is about €30 at the time of writing this chapter. It is one of the reasons why compliance is nearly 100 per cent in the EU ETS (namely, 96 per cent in the period 2005–07, 100 per cent in the period 2008–12 and around 99 per cent in the period 2013–20).[42]

3.5   LINKING THE EU ETS TO NON-EU EMISSIONS TRADING SYSTEMS

Although the EU ETS was already linked to developing countries via the restricted import of credits from CDM projects until 2020, 'linking' typically refers to connecting the EU ETS with other cap-and-trade schemes around the globe. Article 25 of the amended EU ETS Directive makes it possible to link the EU ETS to compatible mandatory greenhouse gas emissions trading systems with absolute emissions caps established in other countries or regions. Linking enables participants in one scheme to use emission entitlements from another (linked) scheme for compliance purposes. A bigger emissions market increases market liquidity and reduces market power, while it also enhances cost-effectiveness if that other country or region has cheaper emission reduction possibilities. Moreover, linking emissions trading systems could help, as a 'bottom-up' approach, to build or expand international climate policy.

The relevance of greenhouse gas emissions trading is steadily growing, given its continuing spread around the globe. Emissions trading systems have emerged in North America, includ-ing the Regional Greenhouse Gas Initiative (RGGI) and the Québec Cap-and-Trade System in Canada. Eight pilots are being carried out in China, for instance in Beijing and Shanghai, and the roll-out of a national China ETS (launched in November 2017) was completed in early 2021. Other ETSs include the Korea ETS (K-ETS), Tokyo's Cap-and-Trade Program and an ETS in Kazakhstan that (after temporary suspension) was re-launched in 2018. The New Zealand Emissions Trading Scheme (NZ ETS) is up and running, but Australia's Carbon Pricing Mechanism (AUS CPM) came under attack in domestic Australian politics and was eventually abolished per 1 July 2014. Geographically closer to the EU is the Swiss Emissions Trading Scheme. Six jurisdictions (including Mexico and Colombia) have officially scheduled implementation of a carbon ETS, while 12 others (including Brazil, Russia and Indonesia) are considering adopting one.[43]

The abolishment of emissions trading plans in Australia was a major setback for global climate policy in general and for linking proponents in particular. In 2012 the European

---

[41]   Judgment of the Court of Justice of 17 October 2016 in Case C203/12.

[42]   This first two percentages have been calculated on the basis of www.eea.europa.eu/data-and-maps/data/data -viewers/emissions-trading-viewer. The latter percentage has been taken from: Report on the Functioning of the European Carbon Market, COM(2019) 557 Final, Brussels: European Commission.

[43]   A global, interactive overview of emissions trading systems is available online at: https://icapcarbonaction .com/ets-map.

Commission and Australia agreed on a pathway for linking the EU ETS and the AUS CPM.[44] A full two-way link between those cap-and-trade schemes was supposed to start no later than 2018. Companies would then be able to use carbon units from the Australian scheme or from the EU ETS for compliance under either system. The abolishment of the AUS CPM put an end to these plans. The EU and Switzerland signed an agreement in 2017 to link their emissions trading systems as of 2021. A future link between the EU ETS and the (post-Brexit) UK Emissions Trading Scheme (UK ETS) is not unimaginable. In North America, the California Cap-and-Trade Program has been linked to the Québec Cap-and-Trade System since 2014.

There are many potential legal problems when linking emissions trading systems, related to differences in sector coverage, cap definition, allowance allocation, allowance registration, monitoring and enforcement, carbon offset types and limits, allowance price controls, competitiveness safeguards, carbon leakage measures, treatment of new entrants and plant closures, allowance banking and borrowing, privacy regulation, tax liabilities and remedies in the event of loss or theft of the allowances.[45] Such legal differences could lead to allowances migrating to one particular registry, which requires some degree of harmonization when linking emissions trading schemes. Some aspects need to be harmonized, such as the nature of the emissions targets (absolute cap versus relative goal), while other aspects may differ after linkage, such as sector coverage (which would expand abatement options).[46] Linking also implies that the allowance price increases in the emissions trading system with the lower price and decreases in the system with the higher price. This is likely to trigger industry lobbying, especially by net buyers in the former system where the allowance price goes up. Linking is therefore subject to serious political debate and requires years of legal preparation.

3.6 IMPLEMENTATION PROBLEMS OF THE EU ETS AND SOLUTIONS

Thus far the EU ETS has been quite successful in realizing emission reductions at low costs, but it has also encountered many implementation problems. For example, the economic crisis of 2008 led to lower emissions and thus a lower allowance price, weakening the incentive to invest in carbon-friendly technologies. Subsidies for renewable energy and energy efficiency made those technologies available on the market in larger quantities, but also raised emission reduction costs and further undermined the allowance price. Industry lobbying led to some inefficient rules and exceptions. Identity fraud and allowance theft temporarily threatened the

---

[44] Australia and European Commission Agree on Pathway towards Fully Linking Emissions Trading Systems, European Commission, IP/12/916, 28 August 2012.

[45] For example: M Ranson and RN Stavins, 'Linkage of Greenhouse Gas Emissions Trading Systems: Learning from Experience' (2016) 16(3) *Climate Policy* 284–300.

[46] A Kachi et al., 'Linking Emissions Trading Systems: A Summary of Current Research' (2015) Berlin: ICAP.

EU ETS. This section analyses the functioning of the EU ETS by focusing on the following implementation problems and their solutions:

- Over-allocation;
- Windfall profits;
- New entrants and closures;
- Allowance prices and structural reform;
- Tax fraud and allowance theft.

3.6.1   Over-allocation

The pre-Kyoto phase 2005–7 of the EU ETS, a learning period prior to the first commitment period of the Kyoto Protocol, was intended to have a gradual start without overly stringent emissions caps. Public and private parties could then gain experience with emissions trading and industries subject to international competition would not face severe competitive disadvantages. Initially the allowance price rose from €10 in 2005 to €30 in 2006, but after the generosity of allowance allocations became known the allowance price fell to almost zero (less than €1 euro) in 2007. In the EU as a whole, the number of allowances allocated turned out to be 4 per cent more than actual emissions. This is referred to in the literature as allowance 'over-allocation'.[47]

What caused the allowance surplus? First, EU Member States allocated too many allowances to protect the competitiveness of their own industries, partly in response to national industry lobbying. Second, governments also faced the problem of incomplete information about the emission levels of companies, which they had to estimate in advance to set the cap. Business growth predictions were over-optimistic and turned out to be lower than expected, so that actual emissions were also lower than predicted. Third, the (albeit limited) import of CDM credits into the EU further increased the number of available carbon entitlements.

To prevent further 'over-allocation', the European Commission forced some Member States to set more stringent allowance allocation levels in the second trading period 2008–12.[48] Unfortunately, this attempt to increase carbon scarcity was undermined by the severe, unanticipated economic crisis that hit the EU in 2008. This was the main cause of the surplus increasing to around two billion allowances. For the third trading period 2013–20, the EU decided to distribute the allowances at EU level based on a single, Community-wide emissions cap. This was expected to prevent Member States from allocating too many allowances to their industries.

---

47   D Ellerman and B Buchner, 'Over-allocation or Abatement? A Preliminary Analysis of the EU ETS based on the 2005–06 Emissions Data' (2008) 41(2) *Environmental & Resource Economics* 267–87.

48   By doing so, the European Commission went beyond its legal powers, according to the CJEU. The Court concluded that the Commission had neither the right to impose an emissions ceiling for a NAP nor the right to replace NAP data with its own data. See: T-183/07 *Poland v Commission* [2009] ECR II-03395, paras 120 and 131. The appeal case C-504/09P *Commission v Poland* [2012] ECLI:EU:C:2012:178 was unsuccessful.

However, this was insufficient to prevent an increase in the emissions surplus in the third trading phase 2013–20. More was thus needed to reform the EU ETS after 2020; this will be discussed in section 3.6.4.

3.6.2 Windfall Profits

In the first two trading periods, large companies emitting CO_2 mainly received their allowances free of charge. Since 2013, however, allowances have been auctioned to power companies. Energy-intensive industries still receive their allowances for free if they are subject to a significant risk of carbon leakage, but for those that do not face this risk auctioning is continuously phased in. By the end of the third trading period, more than half of the total number of allowances was auctioned. This transition to allowance auctioning was triggered by politicians and the energy consumer lobby, including small households and major industries, in various Member States.[49] They contested the so-called windfall profits earned by power companies. These companies passed a substantial part of the market price of free allowances on to small and large consumers. Consumers' electricity bills increased, while power companies received the allowances for free. It was claimed that these companies earned 'windfall profits': additional profits for which they did nothing. Various politicians and consumer organizations called for a change.

From an economic point of view, however, these 'windfall profits' are not problematic at all. The external costs should be internalized: the damage of emitting greenhouse gases should be paid by the final energy consumer. Power companies are legally required to cover their emissions with a corresponding number of allowances. Prior to 2013, they obtained their allowances for free. They used the allowances to cover emissions when producing electricity. However, instead of using the allowances, the company could have sold them. As a consequence, the company wants to see the missed sales revenues back in the electricity price. Using the allowances thus comes at a cost, namely not being able to sell them. Free allowances have 'opportunity costs' when they are used, equal to the allowance price. Therefore, consumers should see their electricity bills increase, since the emission of greenhouse gases now has a price, irrespective of whether the allowances are auctioned or allocated for free.[50]

'Windfall profits' are thus not an economic but a political problem. Passing on the allowance price may be efficient, but it also feels unfair if the power companies got the allowances for free. Because free allowances have a market value, they make the shareholders of electricity firms richer. Consumers considered this to be unacceptable and declined to pay for something that power producers had obtained free of charge. Allowance auctioning indeed solves that political problem. Moreover, economists like auctions, because the emission allowances then accrue to those who value them the highest. For that reason, the switch to allowance auctioning was

---

[49]   See for example the discussions in the UK Parliament: *The Role of Carbon Markets in Preventing Dangerous Climate Change: Written Evidence*, Session 2009–10, House of Commons – Environmental Audit Committee.

[50]   For the same reason, free allowances in a cap-and-trade scheme are also in conformity with the polluter-pays principle: E Woerdman, A Arcuri and S Clò, 'Emissions Trading and the Polluter-Pays Principle: Do Polluters Pay under Grandfathering?' (2008) 4(2) *Review of Law and Economics* 565–90.

also supported by the European Commission. The end result of these political developments was an amended EU ETS Directive that prescribed a phased transition towards allowance auctioning.

Power companies also passed on their carbon costs to large industrial electricity consumers, such as chemical industries and steel manufacturers. Because this affects their international competitiveness, those industries have successfully lobbied for financial compensation of their increased electricity bill. This should improve the international level playing field for these industries. New state aid rules have been formulated to make this compensation possible.[51] The aid intensity amounted to up to 85 per cent starting in 2013 and subsequently fell to 75 per cent in 2020: this means that most of the increase of these manufacturers' electricity bills due to carbon pricing was paid by the government. However, some EU Member States make these compensation payments while others do not, which actually undermines the level playing field for those industries within Europe.[52] In addition, climate ambition clashes with international competitiveness and the pursuit of these multiple goals has led to inefficient policies. Where carbon costs are largely compensated by governments to protect competitiveness, environmental externalities are not fully being internalized by these protected industries. This compensation scheme weakens industries' incentives to use less and cleaner energy. Moreover, especially in the first years of the third trading period, it was unlikely that companies would run into international competition problems as a result of the low allowance price of €5 at that time, while they were still being compensated with tax-payers' money.

3.6.3   Expansions and Closures

In the textbook model of cap-and-trade, newcomers and companies that want to expand need to buy allowances from established companies or from a government reserve. In the third phase (2013–20) of the EU ETS, however, both newcomers and industries expanding their production capacity (in excess of 10 per cent) were allocated allowances for free. This was desired by the industry lobby.[53] A newcomers' reserve had been created to facilitate this, equalling 5 per cent of the total number of allowances. In addition, allowances in the textbook model of cap-and-trade do not expire when an installation closes down or when its capacity is reduced. In the third phase of the EU ETS, however, allowances needed to be surrendered

---

51   Communication from the Commission *Guidelines on Certain State Aid Measures in the Context of the Greenhouse Gas Emission Allowance Trading Scheme Post 2012*, 2012/ C 158/4 of 5.6.2012.

52   At the beginning of the third trading phase, only Germany, the UK, the Netherlands, Belgium and Norway provided such compensation, but by the end of this phase 12 Member States (including France, Spain, Finland and Slovakia) helped to pay the increased electricity bills of their industries.

53   See, for instance, the industry-funded report by: B Wesselink et al., *The IFIEC method for the allocation of CO2 allowances in the EU Emissions Trading Scheme: A review applied to the electricity sector* (2008) PECSNL074036, Utrecht: Ecofys. Or: *CO2 Emissions Trading: Progress EU ETS in 2007*, PowerPoint Presentation by Vianney Schyns of the Utility Support Group (including DSM and SABIC) at the 5th Congress of the European Chemical Regions Network (ECRN), 29–30 November 2007, Ludwigshafen.

in cases of plant closure or significant decline in production capacity.[54] This was desired by politicians who considered it unfair if firms kept their allowances when they no longer needed them to cover emissions. The consequence of these rules in the case of newcomers, expansions and closures – which resemble those of a performance standard rate trading system – is that they lead to the following two inefficiencies.[55]

First, if companies were able to keep their allowances in case of installation closure, it would be more attractive to shut down inefficient old plants since the allowances could then be sold on the carbon market. The rule in the EU ETS that companies have to surrender free allowances in case of closure undermines this desirable incentive. It makes the closure of dated, climate-unfriendly plants less attractive.

Second, when product prices are so low that (variable) production costs can no longer be covered, a company would normally shut down its installations and leave the market. As a result of the aforementioned expansion and closure rules of the EU ETS, however, companies that make losses maintain their production capacity in order to continue receiving free allowances which can be sold on the carbon market. In the (unlikely) case of a very high allowance price, it would even be profitable to invest in production capacity only to obtain allowances to sell. From a social welfare point of view, this is inefficient. Investing in or maintaining capacity that is not deployed for production purposes constitutes a social waste.

Two elements of performance standard rate trading were already incorporated in the EU ETS as of 2013 to boost industry acceptance, namely emissions standards ('benchmarking') used to calculate the allocation of free allowances, as described earlier, and the inefficient rules for expansions and closures, as just described. Industry lobbyists should plead for two more changes to realize a full-blown performance standard rate trading scheme: (a) flexibly (ex-post) correcting the yearly allocation of free allowances based on firms' actual production levels; and (b) reverting the absolute emissions caps for firms into 'intended' emissions caps.

This would be favourable for the industry but it would also be undesirable from a social welfare perspective: performance standard trading is less environmentally effective and less efficient than cap-and-trade, as explained before. The European Commission has never been willing to abandon the absolute nature of the emissions caps for firms, but after more than a decade of industry lobbying the Commission had to partly give up its resistance against sub-optimal ex-post corrections.[56] The European Council concluded in relation to the 2030 Climate and Energy Policy Framework: 'Future allocations will ensure better alignment with

---

[54]   This is not the case for aircraft operators that cease operations, which should continue to be issued with allowances until the end of the period for which free allowances have already been allocated.

[55]   E Woerdman and A Nentjes, 'Emissions Trading Hybrids: The Case of the EU ETS' (2019) 15(1) *Review of Law and Economics* 1–32. See also: D Ellerman, 'New Entrant and Closure Provisions: How Do They Distort?' (2007) 29 *Energy Journal* 63–76.

[56]   In the Netherlands, for instance, government and industry have agreed to a 'joint lobby in Brussels' in favour of 'a 100 per cent free allocation of rights based on realistic benchmarks and actual production': in the SER Energy Agreement for Sustainable Growth of 06.09.2013, p.9.

changing production levels in different sectors.'[57] This is also what happened in the revised EU ETS Directive. As of 2021, if production in an installation has increased (or decreased) by more than 15 per cent, the number of free allowances shall be adjusted upwards (or downwards) (Article 10a). This adjustment takes place every five years (Article 11). Regulators will see their administrative costs increase, because they now also have to process and check companies' production data next to these companies' emissions figures.

The aforementioned inefficiencies can be overcome by auctioning all allowances. In its Green Deal, the European Commission has indeed proposed to abolish free allocation and replace it with a carbon border adjustment mechanism to reduce the risk of carbon leakage.

3.6.4   Allowance Prices and Structural Reform

Governments allocated too many emission allowances at the start of the EU ETS in 2005, while the economic crisis in 2008 led to lower than anticipated emissions. The import of CDM credits further increased the available emissions space. When the cap is higher than actual emissions, an allowance surplus emerges. This decreased the allowance price, which weakened the incentive to invest in low-carbon technologies. Allowance banking aggravates this problem. The possibility to bank (relatively cheap) allowances from one trading phase to another (probably more expensive) trading phase improves cost-effectiveness, but also implies that over-allocated allowances can be used in the future. In 2008, policymakers and traders expected allowance prices of €25–35 in 2010 and of €35–50 in 2020.[58] At the start of the third trading phase 2013–20, however, the allowance price was about €5 (and stayed around that level for at least five years), while the emissions surplus was 2.1 billion allowances.

Some perceive this emissions surplus as an imbalance that undermines the orderly functioning of the EU ETS. Others stress the advantage of a low allowance price in times of economic recession: it does not weigh heavy on the shoulders of big and small consumers of energy and industrial products. When the economy starts to grow again, the allowance price will increase and consumers will be better able to bear it. Emissions trading therefore works 'countercyclically': allowance prices increase when economic times are good and will decrease when economic times are bad.

Because a low allowance price erodes the development of low-carbon technologies, in 2013 the European Commission launched a debate on structural reform of the EU ETS and formulated the following six options:[59]

● Increasing the EU emission reduction target (from 20 per cent) to 30 per cent in 2020;
● Retiring a number of allowances in the third trading phase permanently;
● Revising the annual linear reduction factor of 1.74 per cent to make it steeper;

---

[57] Conclusions on 2030 Climate and Energy Policy Framework, SN 79/14, 23.10.2014, Brussels: European Council, p.2.

[58] Point Carbon, *Carbon 2008: Post-2012 is Now* (Point Carbon, 2008, p.31).

[59] European Commission (2012), *The State of the European Carbon Market in 2012*, Brussels 14.11.2012, COM(2012) 652 final.

- Extending the scope of the EU ETS by including more sectors;
- Limiting access to international credits;
- Introducing discretionary price management mechanisms.

Each option has its pros (such as stimulating low-carbon technologies) and cons (such as increasing complexity or reducing cost-effectiveness). Rather than discussing all options, we will focus on the choices made: basically, a combined variation of all of them, to apply beyond 2020.

The Council had already approved the so-called back-loading proposal of the European Commission to postpone the auctioning of 900 million allowances for a number of years.[60] The auction volume was reduced by 400 million allowances in 2014, by 300 million in 2015 and by 200 million in 2016. It was then supposed to increase by 300 million allowances in 2019 and by 600 million in 2020. This was made possible under Article 29a of the amended EU ETS Directive and required a change in the Auctioning Regulation. 'Back-loading' reduced allowance supply, raising the allowance price a bit. The problem was that this effect would only be temporary, since 'back-loading' was supposed to increase allowance supply prior to 2020, which would then lower the allowance price again. Only a permanent retirement would help to tackle the emissions surplus. This was politically unacceptable, but as a compromise those 'back-loaded' allowances were put into a newly created reserve, which we will discuss below.

As a next step, the Commission proposed to lower the emissions cap for ETS sectors by 2.2 per cent each year from 2021 and to establish a so-called Market Stability Reserve (MSR) that reduces the allowance auction volume in case of an allowance surplus.[61] These choices were supported by the European Council.[62] Lowering the cap by increasing the annual linear reduction factor is an obvious choice: it reduces the allowance surplus, depending on the political decision as to how steep the yearly decrease of the emissions cap will be. Introducing discretionary price management mechanisms, however, is more controversial. Some argue that price management could support the allowance price and strengthens the incentive to invest in low-carbon technologies, which is good for climate policy in the long term.[63] However, others argue that price management is a dangerous option in the short and medium term, because it could undermine the essence of the EU ETS as a market mechanism in which allowance

---

[60] Commission Regulation (EU) No 176/2014 of 25 February 2014 amending Regulation (EU) No 1031/2010 in particular to determine the volumes of greenhouse gas emission allowances to be auctioned in 2013–20. OJ 2014 L.56/11–13.

[61] Proposal for a decision of the European Parliament and of the Council concerning the establishment and operation of a market stability reserve for the Union greenhouse gas emission trading scheme and amending Directive 2003/87/EC, COM(2014) 20/2, Brussels: European Commission.

[62] Conclusions on 2030 Climate and Energy Policy Framework, SN 79/14, 23.10.2014, Brussels: European Council, p.2.

[63] For example: C de Perthuis and R Trotignon, 'Why the European Emissions Trading Scheme Needs Reforming, and How This Can Be Done' (2013), Information and Debates Series No. 24, Paris: Climate Economics Chair.

supply and demand should efficiently determine the allowance price.[64] Besides increasing the complexity of the EU ETS, there is also a risk of regulatory failure in price management due to incomplete information on what the 'right price' would have to be. Incentives to invest in climate-friendly technologies could be harmed if price management is unpredictable or if its rules are unclear to investors.

Therefore, instead of discretionary price-based interventions, the EU opted in 2015 for non-discretionary quantity-based interventions to apply as of 2019 in the form of the afore-mentioned MSR.[65]

The MSR automatically reduces the allowance auction volume in the case of an allowance surplus. If the allowance surplus is 'big' in a certain year (more than 833 million allowances) – which can only be calculated in the next year – 12 per cent of this surplus will be placed in the reserve by reducing the auction volume by a corresponding amount of allowances in the year thereafter. If the allowance surplus is 'small' (fewer than 400 million allowances), 100 million allowances will be released from the reserve. Allowances will also be re-injected if the allowance price is relatively high for more than six months. This proposal was criticized for putting surplus allowances in a reserve rather than permanently retiring them, and for the two-year time lag of auction volume adjustments, which may increase allowance price volatility.[66] As a response to this criticism, the EU decided to double the yearly intake of surplus allowances (from 12 to 24 per cent) between 2019 and 2023, and as of 2023 to cancel those allowances held in the MSR that are above the previous year's auction volume.[67] The latter is crucial: it means that the MSR may only hold as many allowances as were auctioned in the previous year. The rest will be cancelled.

There is a lot of discussion in the literature about the consequences of this new cancellation policy. Some are mildly optimistic and predict that overlapping climate policies will now finally start to work. For instance, if a government were to impose a regulatory shutdown of its coal-fired power plants, more allowances will become available on the market, which would cause the MSR to grow, as a result of which some of these allowances will be automatically deleted.[68] Other examples are the Directives for renewable energy and energy efficiency that help to lower emissions. Prior to the MSR, if those policies were successful they would also lower demand for allowances, which undermines the allowance price and ironically weakens the incentive to invest in renewable energy and energy efficiency. Subsidies for renewable

---

[64]   *IETA's Vision for a Comprehensive EU ETS Reform*, 28.02.2013, Geneva: International Emissions Trading Association (IETA).

[65]   Decision (EU) 2015/1814 concerning the establishment and operation of a market stability reserve for the Union greenhouse gas emission trading scheme and amending Directive 2003/87/EC. OJ 2015 L 264/1-5.

[66]   For example: R Trotignon, F Gonand and C de Perthuis, 'EU ETS Reform in the Climate-Energy Package 2030: First Lessons from the ZEPHYR Model' (2014) Policy Brief No. 2014-01, Paris: Climate Economics Chair.

[67]   These decisions were made in 2018, as part of the revised EU ETS Directive 2018/410 (in Article 2, Amendments to Decision EU 2015/1814).

[68]   G Perino, 'New EU ETS Phase 4 Rules Temporarily Puncture Waterbed' (2018) 8 *Nature Climate Change* 260–71.

energy and energy efficiency would imply spending tax-payers' money to achieve emission reductions that free up allowances which could be sold and partly used to cover emissions elsewhere in the EU ETS. Fortunately, this undesirable 'waterbed effect' has now been partly undone by the MSR and its cancellation policy. However, others are more pessimistic, because scarcity is only temporarily increased as the MSR will spit out allowances later on. The MSR's complexity could also increase price volatility due to greater uncertainty about the future allowance stock. Some even argue that the design of the MSR could create a perverse incentive to postpone abatement, because fewer allowances will be available in the future if firms reduce more now (as this would lead to more allowances in the reserve that will be partly cancelled).[69]

Discussions about the design of the MSR are therefore likely to continue, but so far the initial effect of the MSR is that the allowance price increased to around €30 and the emissions surplus fell to around 1.65 billion allowances at the end of the third trading phase.

3.6.5 Tax Fraud and Allowance Theft

Value-added tax (VAT) fraud in the EU ETS was discovered by Europol in 2009. It can also be described as missing-trader fraud or carousel fraud.[70] Suppose that X buys allowances from Y in another Member State, a transaction which is exempt from VAT. X then sells to Z in its own Member State and charges VAT. X obtains the VAT by disappearing or by declaring itself insolvent without paying dues to the tax authorities. Z then applies to deduct the VAT, which implies a loss to the tax authorities. This loss was substantial: about 5 billion euros in total; sometimes hundreds of millions of euros per Member State (for example, 850 million euros for Germany alone). To combat this type of fraud, the VAT Directive 2006/112/EC was amended by Directive 2010/23/EU to make the buyer (instead of the seller) responsible for paying VAT on allowance transactions. This so-called reverse-charge VAT mechanism is implemented in most but not all Member States, although VAT fraud has become less likely since allowances gained the legal status of financial instruments in 2014.[71] A number of people in different Member States (including Germany and the UK) have been sentenced to jail, with sentences ranging between a few months and several years, for evasion of taxes on EU ETS allowances.[72]

In addition, computer hackers stole tens of million euros' worth of allowances from the national registries of several Member States, especially Romania and the Czech Republic. In 2010 and 2011 hackers entered companies' allowance accounts, after which they transferred the allowances to themselves and sold them. Once this was discovered, spot trading was

---

[69] K Bruninx et al., 'The Unintended Consequences of the EU ETS Cancellation Policy' (2019), MPRA Paper 96437.

[70] For example: K Nield and R Pereira, 'Fraud on the European Union Emissions Trading Scheme: Effects, Vulnerabilities and Regulatory Reform' (2011) 20(6) *European Energy and Environmental Law Review* 255–89.

[71] This happened thanks to Directive 2014/65/EU on financial instruments (MiFID II). See: K Nield and R Pereira, 'Financial Crimes in the European Carbon Markets', in: SE Weishaar (ed.), *Research Handbook on Emissions Trading*, Cheltenham: Edward Elgar Publishing, pp.195–231.

[72] For example: 'Carbon Credit VAT Fraudster to Pay Back £13 Million', HM Revenue & Customs (GOV.UK News Stories), 17 October 2013.

suspended and national registries closed until additional security measures had been implemented. Policymakers quickly prepared legal changes. First, the introduction of a single Union registry in 2012 reduced the potential for fraud and theft. Second, a Registries Regulation was adopted which strengthened 'know-your-customer' (KYC) inspections (by checking bank account details, VAT details and criminal records), prescribes a 26-hour delay of allowance transfers and allows national administrators to suspend access to accounts under certain conditions.[73] Although these security improvements are significant, Interpol and the European Court of Auditors consider the EU ETS still to be vulnerable to carbon crime – especially if the system links to emissions trading systems in non-EU countries, such as Switzerland, the UK or perhaps the US.[74]

3.7   CONCLUSION

The primary objective of any cap-and-trade scheme is to meet emissions caps (effectiveness) at the lowest possible cost by allowing emissions to be traded (efficiency). Cost-effectiveness is also the basic rationale of the EU ETS, by allocating tradable emission allowances to large emitters in order to achieve low-cost emission reductions. Solidarity requires that a part of the allowances to be auctioned is redistributed from Western to Eastern European Member States, which does not affect the cost-effectiveness of the EU ETS.

Cost-effectiveness is compromised, however, by the political desire that the EU ETS should meet other adjacent goals as well, such as stimulating investments in low-carbon technologies, protecting the competitiveness of internationally operating industries and preventing carbon leakage where companies could move to countries without carbon pricing. These additional objectives have led to some inefficient measures in the EU ETS, such as free allowances for industries that expand production and partial financial compensation for energy-intensive companies' increased electricity bills.

Cost-effectiveness is also threatened by some EU ETS reforms. Largely due to the economic crisis, an over-allocation of allowances led to a low allowance price, weakening investments in low-carbon technology. A complex MSR reduces this allowance surplus, which raises the allowance price but could also increase price volatility. Moreover, some Member States impose a domestic carbon price floor on electricity producers, which could increase their abatement costs.

Nevertheless, the EU ETS operates quite successfully in meeting the greenhouse gas emissions caps at low cost. It also inspires countries around the globe to build comparable systems

---

[73]   Commission Regulation (EU) No 389/2013 of 2 May 2013 establishing a union registry pursuant to Directive 2003/87/EC of the European Parliament and of the Council, Decisions No 280/2004/EC and No 406/2009/EC of the European Parliament and of the Council and repealing Commission Regulations (EU) No 920/2010 and No 1193/2011, OJ 2013 L. 122/1–59.

[74]   Interpol, *Guide to Carbon Trading Crime*, Lyon June 2013 (International Criminal Police Organisation 2013). See also: European Court of Auditors, *The Integrity and Implementation of the EU ETS*, Special Report No 06 (2015), Luxembourg: ECA.

of their own, including in Asia and in (parts of) North America, which could eventually be linked to the EU ETS. As a result of the multiple goals pursued, however, the EU ETS has become a complex, sub-optimal instrument of climate law with some inefficient design elements.

CLASSROOM QUESTIONS

1. Which legal objectives does the EU ETS have and are they compatible with each other?
2. What would you prefer to change in the design of the EU ETS to improve its functioning?
3. What do you consider to be the two biggest legal issues when linking the EU ETS to a non-EU emissions trading system?

SUGGESTED READING

Books

Weishaar SE (ed.), *Research Handbook on Emissions Trading (Edward Elgar Publishing 2016)*.

Articles and chapters

Kotzampasakis M and Woerdman E, 'Linking the EU ETS with California's Cap-and-Trade Program: A Law and Economics Assessment' (2020) 4(4) *Central European Review of Economics and Management* 9–45.

Narassimhan E, Gallagher KS, Koester S and Rivera Alejo J, 'Carbon Pricing in Practice: A Review of Existing Emissions Trading Systems' (2018) 18(8) *Climate Policy* 967–91.

Woerdman E and Nentjes A, 'Emissions Trading Hybrids: The Case of the EU ETS' (2019) 15(1) *Review of Law and Economics* 1–32.

Policy documents

European Commission website on the EU ETS https://ec.europa.eu/clima/policies/ets_en.

European Commission, *Report on the Functioning of the European Carbon Market*, e.g. 2020: https://ec.europa.eu/clima/sites/clima/files/news/docs/com_2020_740_en.pdf.

International Carbon Action Partnership website https://icapcarbonaction.com/en/.

4

Regulation of emissions from non-ETS sectors

ABSTRACT

- Regulation 2018/842 (Effort Sharing Regulation) is the main instrument to regulate greenhouse gas emissions from the non-ETS sectors, such as transport, agriculture, buildings and waste;
- The Effort Sharing Regulation is an additional measure to the ETS Directive, extending the EU's greenhouse gas emission reduction commitments to non-ETS sectors;
- Solidarity is pursued under the Effort Sharing Regulation, for instance because greenhouse gas reduction requirements are linked to Gross Domestic Product (GDP) per capita of each Member State;
- The Effort Sharing Regulation only establishes annual emission allocations in the period 2021–30 for each Member State, while leaving great flexibility on how to achieve these targets;
- The measurement and compliance system of the Effort Sharing Regulation can erode the effectiveness of the regulatory framework, because the system for reporting and monitoring basically relies on self-incrimination and non-exact measurements;
- From a cost-effectiveness point of view, flexibility in how to achieve the emission targets for non-ETS sectors has both positive effects as tailor-made regulation and negative effects by further undermining an already weak compliance system.

4.1 INTRODUCTION

This chapter focuses on EU regulation of greenhouse gas emissions resulting from activities not covered by the Emissions Trading System (ETS), hence called 'non-ETS' sectors. As discussed in Chapter 3, an important step for the EU in combating climate change is to reduce greenhouse gas emissions, most notably carbon dioxide (CO_2), from power companies and large industrial installations. However, the sectors covered by the ETS only count for about 40 per cent of the EU's greenhouse gas emissions. This explains why, based on Article 192(1) TFEU thus following the ordinary legislative procedure, Regulation (EU) 2018/842 was adopted in

2018,[1] replacing Decision No 406/2009/EC.[2] Besides forcing EU Member States to take climate action in the non-ETS sectors, this so-called Effort Sharing Regulation can be used as a basis for the adoption of measures to accelerate energy efficiency improvements in the non-ETS sectors.[3] Measures to stimulate energy efficiency and to limit fluorinated gases are dealt with in Chapters 6 and 8 of this book, respectively. Here we examine Regulation 2019/631/EU,[4] repealing Regulations 443/2009/EC and 510/2011/EU,[5] on emissions performance standards for new passenger cars and light commercial vehicles. In addition we will discuss the newly introduced Land Use, Land Use Change and Forestry (LULUCF) Regulation.[6]

We will examine the EU regulatory framework for greenhouse gas emissions reduction in the non-ETS sectors from the perspective of the overarching EU climate policy principles of cost-effectiveness and solidarity. Section 4.2 provides an explanation of the main concepts in the context of the regulation of greenhouse gas emissions under the non-ETS sectors. Sections 4.3 and 4.4 provide a description of the Effort Sharing Regulation, the LULUCF Regulation and Regulation 2019/631/EU, by paying particular attention to cost-effectiveness and solidarity. Finally, section 4.5 answers the question to what extent the overarching principles of EU climate policy are honoured in these measures.

---

[1]   Regulation (EU) 2018/842 of the European Parliament and of the Council of 30 May 2018 on binding annual greenhouse gas emission reductions by Member States from 2021 to 2030 contributing to climate action to meet commitments under the Paris Agreement and amending Regulation (EU) No 525/2013 (Text with EEA relevance), [2018] OJ L156/26.

[2]   Decision No 406/2009/EC of the European Parliament and of the Council of 23 April 2009 on the effort of Member States to reduce their greenhouse gas emissions to meet the Community's greenhouse gas emission reduction commitments up to 2020, [2009] OJ L140/136.

[3]   See the Commission webpage on the Effort Sharing Decision for an overview of those measures, http://ec .europa.eu/clima/policies/effort/index_en.htm last accessed February 2020.

[4]   Regulation (EU) 2019/631 of the European Parliament and of the Council of 17 April 2019 setting CO2 emission performance standards for new passenger cars and for new light commercial vehicles, and repealing Regulations (EC) No 443/2009 and (EU) No 510/2011, [2019] OJ L 111/13.

[5]   Respectively, Regulation (EC) No 443/2009 of the European Parliament and of the Council of 23 April 2009 setting emission performance standards for new passenger cars as part of the Community's integrated approach to reduce CO2 emissions from light-duty vehicles, [2009] OJ L140/1; and Regulation (EU) No 510/2011 of the European Parliament and of the Council of 11 May 2011 setting emission performance standards for new light commercial vehicles as part of the EU's integrated approach to reduce CO2 emissions from light-duty vehicles, [2001] OJ L145/1.

[6]   Regulation (EU) 2018/841 of the European Parliament and of the Council of 30 May 2018 on the inclusion of greenhouse gas emissions and removals from land use, land use change and forestry in the 2030 climate and energy framework, and amending Regulation (EU) No 525/2013 and Decision No 529/2013/EU (Text with EEA relevance) [2018] OJ L156/1.

4.2   BASICS OF NON-ETS SECTOR GREENHOUSE GAS EMISSIONS

4.2.1   Non-ETS Sectors

Just like the ETS Directive, the Effort Sharing Regulation applies only to emissions coming from certain sectors. However, the Effort Sharing Regulation merely has residual value, in the sense that it applies to greenhouse gas emissions from certain activities as long as they are not covered under the ETS Directive. Moreover, a conjunctive reading of both the Effort Sharing Regulation and an external source, the IPCC Guidelines for National Greenhouse Gas Inventories,[7] is necessary in order to understand what sectors are truly covered by the Effort Sharing Regulation. Indeed, Article 2 of the Effort Sharing Regulation states that it applies to four of the five categories of activities defined by the IPCC guidelines. More precisely, it applies to:

1. Energy;
2. Industrial processes and product use;
3. Agriculture, but not to forestry and other land use;
4. Waste.

Land use, land use change and forestry (LULUCF) is not covered by the Effort Sharing Regulation but by a specific regulation, discussed in section 4.3.1. Under the IPCC Guideline the four categories covered by the Effort Sharing Regulation are further sub-categorised. For example, the category 'Fuel combustion' under 'Energy' covers, inter alia, 'Transport', which is in turn sub-divided into 'Civil aviation', 'Road transportation', 'Railways', 'Waterborne Navigation' and 'Other Transportation'. As stated before, the Effort Sharing Regulation applies to those sectors as long as they are not covered by the ETS Directive.[8]

The above shows that a two-step approach is necessary to establish whether a certain activity is covered by the Effort Sharing Regulation. Here we make use of the aviation sector as an example of how to establish the scope of application of the Effort Sharing Regulation. The first step requires a conjunctive reading of both the Effort Sharing Regulation and the IPCC guidelines. The IPCC guidelines shows that only civil aviation is covered under the sub-category 'Transport'. The second step is to look at the ETS Directive. Annex I of the EU ETS Directive clarifies that emissions from flights performed by aircrafts with a certified maximum take-off mass of less than 5700 kg fall outside the scope of application of the ETS Directive. Therefore, it is possible to conclude that emissions from civil flights performed by aircraft with a certified maximum take-off mass of less than 5700 kg are covered by the Effort Sharing Regulation.[9]

---

[7]   Available at www.ipcc-nggip.iges.or.jp/public/2006gl/ last accessed January 2019.

[8]   Article 2(1) of the Effort Sharing Regulation.

[9]   However, for CO_2 emissions coming from the civil aviation sector covered by IPCC guidelines, the Effort Sharing Regulation states that such emissions shall be treated as zero (Article 2(3) of the Effort Sharing Regulation).

4.2.2   National Emission Targets and Annual Emission Allocations

The Effort Sharing Regulation uses two special kinds of emission limit values to achieve its goal:

(a)   national emission targets;
(b)   annual emission allocations.

First, it uses the so-called national emission targets. Under Article 3(1) each Member State shall, by 2030, limit its greenhouse gas emissions by at least the percentage set for that Member State by the Effort Sharing Regulation. Hence, the Effort Sharing Regulation does not impose an emission limit value upon the emitters of greenhouse gases, as is the case under the Industrial Emissions Directive,[10] but it establishes an emission limit value upon the Member States. Moreover, differently from the ETS Directive, the Effort Sharing Regulation does not impose an EU-wide national emission target, but rather 28 (27 after Brexit) national emission targets. For example, Germany has to reduce its greenhouse gas emissions by, at least, 38 per cent in comparison to its emissions in 2005.

National emission targets under the Effort Sharing Regulation are calculated based on the Gross Domestic Product (GDP) per capita of each Member State. As further explained below, a Member State with a high GDP, such as Germany, has to reduce its emissions by at least 38 per cent, while a Member State with a low GDP, such as Bulgaria, can actually maintain its emissions covered by the Effort Sharing Regulation at the 2005 level. Therefore, a national emission target does not always mean that greenhouse gas emissions must decrease. Still, all annual emission targets under the Effort Sharing Regulation represent a tightening of the regime, even for Bulgaria: under the Effort Sharing Decision, Bulgaria was allowed to increase its emissions up to 20 per cent by 2020 in comparison with 2005.

Second, alongside national emission targets, the Effort Sharing Regulation uses the so-called annual emission allocations or AEAs.[11] The national emission targets to be met by 2030 must be reached by means of a linear decrease of emissions in the years 2021–30. In practice the EU legislator will establish for each Member State ten AEAs for the period 2021–30.[12] This means that there will be 280 (270 after Brexit) AEAs under the Effort Sharing Regulation for the period 2021–30.

This means that these emissions are covered but in practice do not impact the national emission targets and annual emission allocations of the Member States, introduced in the next section.

[10]   Directive 2010/75/EU of the European Parliament and of the Council of 24 November 2010 on industrial emissions (integrated pollution prevention and control), [2010] OJ L334/17.

[11]   Article 3(2) of the Effort Sharing Regulation.

[12]   For the 2013–20 period the yearly CAPs for each Member States had been established with Decision 2013/162/EU, OJ 2013 L 90/106. For example, the 2013 CAP for Belgium in 2013 was 81,206,753 tons, the 2014 CAP was 79,635,010 tons, the 2015 CAP was 78,063,267 tons, and so on until 2020, showing that Belgium had to reduce the amount of greenhouse gas emissions covered by the Effort Sharing Decisions by 1,571,743 tons per year, so that in 2020 Belgium would not emit more than 70,204,550 tons, which is 15 per cent less than in 2005.

4.2.3   Banking, Borrowing, Transfer and Safety Reserve

Article 5 of the Effort Sharing Regulation refers to the concepts of 'banking', 'borrowing' and 'transfer'. These are instruments used by the Effort Sharing Regulation to allow flexibility in how Member States can fulfil their obligations. *Banking* means that Member States which achieve a higher reduction target for a given year can carry over the surplus to subsequent years until 2030, up to a maximum of 30 per cent of their annual emissions allocation in the years 2022–29.[13] *Borrowing* means that a Member State wishing to exceed its AEA for a given year can carry forward some emissions allocation from the following year, up to a maximum of 10 per cent in the years 2021–5 and up to a maximum of 5 per cent in the years 2026–29.[14] *Transfer* means that a Member State can give a part of its AEA to another Member State, up to a maximum of 5 per cent in the years 2021–25 and up to 10 per cent in the years 2026–30.[15]

Besides banking, borrowing and transferring emission allocations, the Effort Sharing Regulation introduces a Safety Reserve with up to 105 million tonnes of CO_2 aiming at helping Member States with a GDP lower than the EU average should they have difficulty reaching their 2030 targets, setting a maximum of 20 per cent of the overall overachievement of the state requesting to make use of this safety net.[16]

4.3   THE EFFORT SHARING REGULATION

The Effort Sharing Regulation cannot be appropriately understood without taking into consideration its legal and political international background. Indeed, the Effort Sharing Regulation derives from the intention of the EU to meet its commitments under the United Nations Framework Convention on Climate Change (UNFCCC). Recital 3 to the Regulation clearly shows the relationship between the Effort Sharing Regulation and its international legal background. It states that it forms part of the implementation of the Union's contribution under the Paris Agreement.[17] Moreover, the scope of application of the Regulation is shaped by the content of the IPCC guidelines under the UNFCCC.

Besides, the Effort Sharing Regulation derives from the political will of the European Council to transform Europe into a highly energy-efficient and low-carbon economy, which means that the EU should reduce its greenhouse gas emissions by at least 40 per cent by 2030 as compared to 1990. By the end of 2020, the Council had agreed to strengthen the EU's 2030 reduction target to 55 per cent.[18] The Effort Sharing Regulation is designed to help the EU to

---

[13]   Article 5(3) of the Effort Sharing Regulation.

[14]   Article 5(1 and 2) of the Effort Sharing Regulation.

[15]   Article 3(4 and 5) of the Effort Sharing Decision.

[16]   Article 11 of the Effort Sharing Regulation.

[17]   Paris Agreement, [2016] OJ L282/4.

[18]   European Council meeting (10 and 11 December 2020) – Conclusions, EUCO 22/20, Brussels, 11 December 2020.

achieve these goals in a cost-effective way, by supplementing the climate action pursued under the ETS Directive.

We will first look at the scope of application of the Effort Sharing Regulation (section 4.3.1) and the way in which the national emission targets have been established (section 4.3.2). We will then describe the instruments envisaged under the Effort Sharing Regulation to achieve its goals, either by the Member States (section 4.3.3) or, as discussed in section 4.4, by the European institutions, with particular attention to the LULUCF Regulation.

4.3.1 Scope of the Effort Sharing Regulation

Recital 2 of the Effort Sharing Regulation states that all sectors of the economy should contribute to the achievement of the emission reduction targets set by the EU. The Effort Sharing Regulation represents an important step in this direction. The material scope of application of the Effort Sharing Regulation is limited to greenhouse gases, most notably carbon dioxide (CO_2), methane (CH_4), nitrous oxide (N_2O), hydrofluorocarbons (HFCs), perfluorocarbons (PFCs), nitrogen trifluoride (NF_3) and sulphur hexafluoride (SF_6).

It should be noted that the Effort Sharing Regulation applies only to certain sectors, namely (1) energy; (2) industrial processes and product use; (3) agriculture, with the exception of forestry and other land use; and (4) waste, as defined under the IPCC Guideline. Activities falling under the category 'Land use, Land use change and Forestry' are covered by the LULUCF Regulation, to be discussed in section 4.4.1 below. Moreover, the Effort Sharing Regulation applies to emissions coming from the categories mentioned in the IPCC Guideline only if they are not already covered by the ETS Directive. As we can see, there is a clear link between the scope of application of the ETS Directive and that of the Effort Sharing Regulation. The Effort Sharing Regulation should be considered as an additional measure to the ETS Directive, extending the EU's greenhouse gas reduction commitments to sectors of the economy which were previously left unregulated. With the Effort Sharing Decision, the predecessor of the Effort Sharing Regulation, the EU legislator redressed a situation which was considered by some parties as unfair.[19] However, others argued that comparable polluting activities are regulated under two different legal regimes.[20] Nevertheless, thanks to the legally binding target for the non-ETS sectors, the EU and its Member States do not need to put more pressure on the ETS sectors to achieve the 40 per cent target by 2030.[21] Now, nearly all sectors have to contribute towards the reduction of greenhouse gas emissions.

---

[19] See most notably on this point the Arcelor case, ECJ Case C-127/07, *Société Arcelor Atlantique et Lorraine and Others v Premier ministre, Ministre de l'Écologie et du Développement durable and Ministre de l'Économie, des Finances et de l'Industrie*, [2008] ECR I-9895.

[20] M. Peeters and M. Stallworthy, 'Legal Consequences of the Effort Sharing Decision for Member States Action', in M. Peeters, M. Stallworthy and J. de Cendra de Larragán (eds), *Climate Law in the EU Member States – Towards National Legislation for Climate Protection* (Edward Elgar Publishing 2012), 15–38, at 33.

[21] Cf. N. Lacasta, S. Oberthür, E. Santos and P. Barata, 'From Sharing the Burdens to Sharing the Effort: Decision 406/2009/EC on the Member States Emissions Targets for Non-ETS Sectors', in Sebastian Oberthür and Marc Pallamaerts (eds), *The New Climate Policies of the European Union* (Brussels University Press 2010), 93–116, at 102.

The residual nature of the Effort Sharing Regulation also implies that the greater the number of activities which will be covered by the ETS Directive, the fewer the number of activities which will be covered by the Effort Sharing Regulation. This relation is reflected in Article 10 of the Effort Sharing Regulation, stating that the maximum quantity of emissions for each Member State under the Effort Sharing Regulation shall be adjusted in accordance with amendments to the scope of application of the ETS Directive.

4.3.2   Understanding the Functioning of National Emission Targets and AEAs

The linkage between the cap under the ETS Directive and the annual emissions allocations (AEAs) under the Effort Sharing Regulation should lead to an increase of cost-effectiveness. On average, marginal abatement costs are higher for non-ETS sectors than for ETS sectors.[22] This is because the ETS Directive covers only large sources of greenhouse gas emissions, while the Effort Sharing Regulation covers a heterogeneous multitude of small emitters, ranging from buildings to cars and waste. The cost-effectiveness argument therefore justifies the fact that the ETS sectors have to achieve relatively higher abatements targets than the non-ETS sectors. More precisely, under the Effort Sharing Regulation, Member States have to achieve a reduction of greenhouse gas emissions of only 30 per cent in comparison with 2005, while under the ETS Directive they have to achieve a reduction of 43 per cent in comparison with 2005. Taking the reduction of the ETS- and non-ETS sectors together, the EU will reduce its emissions by at least 40 (probably 55) per cent in 2030 in comparison with 1990.[23] Apparently, the EU legislator considers that such a differential treatment contrasts neither with the principle of equality nor with the polluter-pays principle.

In order to achieve a 30 per cent reduction target in the non-ETS sectors, including the LULUCF sectors, the EU legislator has established 28 (27 after Brexit) national emission targets that Member States must achieve by 2030. Due to differences in economic development between the Member States, an equitable share of burdens under the Effort Sharing Regulation was deemed to be necessary.[24] Accordingly, the European Council indicated in 2007 that a differentiated approach to the contributions of the Member States was needed. This approach had to reflect fairness and transparency and take into account national circumstances as well

---

[22]   C. Böhringer, T. Hoffmann, A. Lange, A. Löschel and U. Moslener 2005 'Assessing Emission Regulation in Europe: An Interactive Simulation Approach', *Energy Journal* 26, 1–22. See also S. Peterson 2006, 'Efficient Abatement in Separated Carbon Markets: A Theoretical and Quantitative Analysis of the EU Emissions Trading Scheme', Kiel Working Paper 1271, as confirmed in, for example, M. Jalard and others 2015, 'The EU ETS Emissions Reduction Target and Interactions with Energy and Climate Policies' in *Exploring the EU ETS beyond 2020: First Assessment of the EU Commission's Proposal* (I4CE 2014) and B. Bye, T. Fæhn and O. Rosnes 2019, *Marginal Abatement Costs under EU's Effort Sharing Regulation: A CGE analysis* (Statistics Norway 2019).

[23]   European Commission, Impact Assessment from the Commission to the European Parliament, the Council, European Economic and Social Committee and the Committee of the Regions. 20 20 by 2020. Europe's Climate Change Opportunity up to 2020, COM (2008) 17 final.

[24]   In general see Lacasta et al. (n 21), at 93–116.

as the relevant base years for the first commitment period of the Kyoto Protocol.[25] Following this indication, the European Commission decided to link the national reduction targets to Gross Domestic Product per Capita (GDP/Capita) of the Member States and to use 2005 as the reference year for the reductions, since this was the first year in which reliable data under the ETS were available.[26]

Each national reduction target is formulated as an amount of emissions reduction that has to be achieved by 2030, expressed in a percentage. As shown in Table 4.1, Member States with a relatively high GDP/Capita, such as Luxembourg, have to achieve a higher level of reduction than Member States with a low GDP/Capita, such as Greece. Bulgaria does not have to reduce emissions as its target is a 0 per cent reduction. Given the connection between the national emissions targets and the GDP of the Member States, each Member State contributes to the achievement of the EU target in accordance with its possibilities. This should ensure solidarity among Member States.[27] However, solidarity has its limits. At the time of writing, the official maximum reduction target has been set at 40 per cent. If there were no limitation, Denmark, Sweden and Luxembourg would have to comply with targets going beyond 40 per cent, with Luxembourg having to comply with a target as high as 61 per cent, based on the GDP/Capita criterion.[28] This implies that Luxembourg is a relative winner, which was justified by the fact that the GDP/Capita criterion would have run against the cost-effectiveness criterion for these states.

Procrastination is a problem that affects not only those students who wait until the very last moment to start studying for their exams, but also Member States. The Effort Sharing Regulation uses AEAs to avoid Member States adopting measures when it is already too late to achieve their 2030 target. Each year, greenhouse gas emissions covered by the Effort Sharing Regulation must therefore decrease in a linear manner. This means that the difference between the target to be achieved by 2030 and the level of emissions in 2021 has been divided by ten. Hence, each year an equal amount of emission reduction must be achieved.

---

[25]   European Council 2007, Presidency conclusions 7224/1/07 REV 1, at 12.

[26]   Lacasta et al (n 21) at 104.

[27]   See L. Saikku and S. Soimokallio 2008 'Top-Down Approaches for Sharing GHG Emission Reductions: Uncertainties and Sensitivities in the 27 European Union Member States' *Environmental Science and Policy* 11(8), 723–34 and Robert Harmsen, Wolfgang Eichhammer and Bart Wesselink 2011 'Imbalance in Europe's Effort Sharing Decision: Scope for Strengthening Incentives for Energy Saving in the Non-ETS Sectors' *Energy Policy* 39, 6636–49 for a discussion of whether the GDP/Capita criterion was the best one to ensure solidarity among Member States.

[28]   Commission Staff Working Document Impact Assessment Accompanying the Document Proposal for a Regulation of the European Parliament and of the Council on binding annual greenhouse gas emission reductions by Member States from 2021 to 2030 for a resilient Energy Union and to meet commitments under the Paris Agreement and amending Regulation No 525/2013 of the European Parliament and the Council on a mechanism for monitoring and reporting greenhouse gas emissions and other information relevant to climate change, SWD/2016/0247 final – 2016/0231 (COD), pp.38–41.

Table 4.1   National targets and flexibility arrangements

2030 target compared to 2005		Maximum annual flexibility (as a per cent of 2005 effort sharing sectors emissions)	
		Flexibility from ETS to Effort Sharing Regulation	Flexibility from land use sector to Effort Sharing Regulation
LU	−40%	4%	0.2%
SE	−40%	2%	1.1%
DK	−39%	2%	3.6%
FI	−39%	2%	1.4%
DE	−38%		0.5%
FR	−37%		1.5%
UK	−37%		0.4%
NL	−36%	2%	1.0%
AT	−36%	2%	0.4%
BE	−35%	2%	0.5%
IT	−33%		0.3%
IE	−30%	4%	5.7%
ES	−26%		1.2%
CY	−24%		1.4%
MT	−19%	2%	0.3%
PT	−17%		1.1%
EL	−16%		1.1%
SI	−15%		1.1%
CZ	−14%		0.4%
EE	−13%		1.7%
SK	−12%		0.5%
LT	−9%		4.9%
PL	−7%		1.2%
HR	−7%		0.5%
HU	−7%		0.4%
LV	−6%		3.6%
RO	−2%		1.7%
BG	−0%		1.5%

Source: Based on: European Commission, Factsheet on the Commission's proposal on binding greenhouse gas emission reductions for Member States (2021–2030), at https://ec.europa.eu/clima/policies/effort/proposal_en#tab-0-0 last accessed March 2020.

It should be noted that the Effort Sharing Regulation allows for a certain amount of flexibility regarding the AEAs. As explained above, by means of borrowing, Member States can carry forward some allocated emissions from the following year, up to a maximum of 10 per cent.[29]

Moreover, by means of banking, they can carry over to the following years that part of the AEA that has remained unused in a given year. Besides, the Safety Reserve can be invoked by those Member States with a GDP lower than the EU average if they have difficulties reaching their 2030 targets.[30] Furthermore, under Article 6 of the Effort Sharing Regulation, nine Member States may achieve their national targets by covering up to 2 per cent,[31] or 4 per cent,[32] of their emissions with (a given maximum of) EU ETS allowances, which would normally have been auctioned under the ETS Directive. EU-wide, this cannot amount to more than 100 million EU ETS allowances over the period 2021–30. These states were required to notify to the Commission by 2020 whether they wanted to make use of this option.

Overachievement under the LULUCF sectors is another source of flexibility to achieve fairness and cost-effectiveness under the Effort Sharing Regulation. Under Article 7, Member States can use up to 280 million credits over the entire period 2021–30 to comply with their national targets. All Member States are eligible to make use of this flexibility if needed to achieve their target, while access is higher for Member States with a larger share of emissions from agriculture. This recognises that there is a lower mitigation potential for the agriculture sector.

Table 4.1 provides an overview per Member State of their overall targets as well as the room for flexibility based on the ETS sector and the LULUCF sector.

The various flexibility instruments mean that the AEAs of each Member State can be less stringent than indicated under the EU framework. The European Union's multilateral assessment presented in the context of the UNFCCC COP 24 at Katowice in 2018 shows that the flexibility instruments under the Effort Sharing Regulation can lead to different targets than indicated under the Effort Sharing Regulation, where Bulgaria is allowed an increase of 1.5 per cent in emissions from the non-ETS sectors by 2030 in comparison with 2020.[33]

4.3.3   Member State Instruments to Achieve the Targets

The Effort Sharing Regulation grants considerable discretion to the Member States on how to achieve the various AEAs and the 2030 national emission target. Most importantly, the Effort Sharing Regulation does not specify what measures Member States should adopt to achieve their targets. Member States' action may, accordingly, range from command-and-control measures to economic instruments and other initiatives, such as a campaign to stimulate a low-carbon lifestyle.

---

[29]   Article 5(1) of the Effort Sharing Regulation.

[30]   Article 11 of the Effort Sharing Regulation.

[31]   These Member States are Belgium, Denmark, Malta, the Netherlands, Austria, Finland and Sweden.

[32]   These Member States are Ireland and Luxemburg.

[33]   See the documents about the European Union's multilateral assessment available on the UNCCC site, https://unfccc.int/MA/European_Union#eq-1 last accessed March 2020.

The absence of prescribed measures to reach environmental targets is a legislative technique common to other regulated sectors. The Air Framework Directive, Directive 1996/62/EC, established, for example, a limit value to be respected by the Member States, but it did not prescribe any specific measure on how the limit values must be achieved.[34] This approach, however, did not lead to cost-effectiveness – at least not immediately. In 2006, a study comparing how certain Member States had complied with the Air Framework Directive showed considerable differences.[35] Germany, for example, had a more flexible system than the Netherlands to achieve the targets set out by the Directive. Dutch industry reacted by strongly criticising the instruments chosen by the Dutch legislator for implementation. Today, the Netherlands has aligned its implementation methods to those adopted in Germany.[36] Without going into the details of this particular example, it shows that leaving flexibility on how to achieve an EU target to the Member States does not ensure cost-effectiveness. It only transfers the responsibility to achieve cost-effectiveness from the EU legislator to the national legislator.

The Effort Sharing Regulation, however, does provide some instruments facilitating Member States to achieve cost-effectiveness. As mentioned before, next to banking and borrowing, the Effort Sharing Regulation allows each Member State to transfer part of its Annual Emission Allocation (AEA) to another Member State. The only limitation is that a Member State cannot transfer to another Member State more than 10 per cent of its overachievement. Such transfers may be carried out in a manner that ought to be convenient for both Member States, including auctioning, the use of market intermediaries acting on an agency basis or bilateral agreements.[37] Although in a rather unclear manner, the Effort Sharing Regulation seems to indicate that a transferring Member State must request to make a payment to a receiving Member State. Indeed, what does 'ought to be mutually convenient' mean? The lack of clarity comes from the fact that Article 5(4) of the Regulation, regulating the transfer of annual emission allocations from one state to another, is silent about this issue. Yet, in Article 5(7) of the Regulation, it is written that transfer *may be* the result from an economic transaction between a selling state and a receiving state. The 'may be' formula used in this action does not actually exclude other forms of transfer. Similarly, this provision does not speak about a *buying* state, but only about a *receiving* state. In the light of the polluter-pays principle, it could be argued that a receiving Member State should not be able to receive extra allocations without having to pay for them. Transfer without payment could also affect the cost-effectiveness of the Effort Sharing Regulation, at least over time. A receiving Member State would have fewer incentives to reduce its own emissions, and therefore to stimulate low-carbon technological developments, if it does not have to pay to achieve its reduction target under the Effort Sharing

---

[34]   Council Directive 96/62/EC of 27 September 1996 on ambient air quality assessment and management, [1996] OJ L296/55, now recast by Directive 2008/50/EC, which applies a similar approach.

[35]   Ch. W. Backes, Internationale vergelijking implementatie EU-richtlijnen luchtkwaliteit, Centrum voor Omgevingsrecht en beleid/NILOS, (Utrecht 2006).

[36]   L. Squintani, *Beyond Minimum Harmonisation: Green-Plating and Gold-Plating of European Environmental Law* (Cambridge University Press 2019), 13–71.

[37]   Recital 20 of the preamble to the Effort Sharing Regulation.

Regulation. Since technological development increases the cost-effectiveness of reducing emissions, transfer without payment could affect cost-effectiveness under the Effort Sharing Regulation.

Member States' freedom regarding how to achieve the targets is controlled by the EU in two different ways. First, the Effort Sharing Regulation establishes monitoring and reporting duties upon the Member States and a specific set of enforcement provisions. Second, these *internal* compliance mechanisms are supplemented by *external* ones, such as the infringement procedure under Article 258 TFEU and a series of sector-specific measures requiring the Member States to adopt specific measures in order to reduce greenhouse gas emissions, such as the LULUCF Regulation and Regulation 2019/631, analysed later.

As regards monitoring, the Effort Sharing Regulation amended the Monitoring Mechanism Regulation,[38] which sets detailed provisions for monitoring and reporting greenhouse gas emissions at national and EU levels. Under the Monitoring Mechanism Regulation, Member States have to report on: (a) their annual greenhouse gas emissions; (b) the use, geographical distribution and types of – and the qualitative criteria applied to – flexibility instruments; and (c) projected progress towards meeting their obligations under the Effort Sharing Regulation and about more stringent protective measures, if any.[39] Under the Monitoring Mechanism Regulation, Member States must submit to the Commission, on an annual basis, their inventories on the amount of greenhouse gas emitted two years earlier.[40] For example, in 2023, Member States will submit their national inventories on the emission that took place in 2021. The methodology applied to fill in the national inventories must comply with the requirements established under the UNFCCC.[41] This requirement is understandable because the EU uses the data included in the national inventories to write the report that it submits to the UNFCCC. The national reports form the basis for the United Nations' assessment of whether the EU complies with its international commitments. If the methodology used for the national inventories differed from the one required by the UNFCCC, it would be impossible to use the data from the national inventories to prove compliance with the Convention. The methodology required under the Effort Sharing Regulation is considered complicated,[42] as it is based on estimates, averages and statistical data. In a nutshell, the reporting duty under the Effort Sharing Regulation is based on self-reporting and estimations, and not on an exact

---

[38]   Regulation (EU) No 525/2013 of the European Parliament and of the Council of 21 May 2013 on a mechanism for monitoring and reporting greenhouse gas emissions and for reporting other information at national and Union level relevant to climate change and repealing Decision No 280/2004/EC, [2013] OJ L165/13.

[39]   Articles 7 and 13 of the Monitoring Mechanism Regulation.

[40]   Article 7(3) of the Monitoring Mechanism Regulation.

[41]   Article 3(2) and Articles 7 and 13 of the Monitoring Mechanism Regulation, with reference to Decision 19/CMP.1 or other relevant decisions of UNFCCC or Kyoto Protocol bodies.

[42]   M. Peeters and M. Stallworthy (n 20), at 24, with reference to the Effort Sharing Decision, which used the same system as the Effort Sharing Regulation.

measurement of the greenhouse gases emitted by each Member State.[43] For example, as further discussed later, emissions from passenger cars in Europe are based on estimations calculated on the basis of the so-called Worldwide Harmonized Light Vehicles Test Procedure (WLTP test). This testing methodology replaced the obsolete New European Driving Cycle (NEDC) test, as real-life tests proved that emissions were between 30 and 50 per cent higher than those indicated under the NEDC test.[44]

If a Member State does not meet its AEA for a given year, two different enforcement systems apply, namely (a) a set of penalties and (b) the infringement procedure. Both systems take the national inventories provided by the Member States as their basis, meaning that they are based on self-incrimination. The first enforcement system provides three different penalties, which take place automatically, meaning that no one has to wait for any further decision of the EU.[45] First, the AEA for the following year is reduced by a factor of 1.08 multiplied for the amount of greenhouse gas emissions in excess. For example, an excess of 5 tonnes in 2021 leads to a 5.4-tonne reduction of the AEA for 2022. Second, the Member State concerned is not allowed to transfer its emission allocations to other Member States for that year. Third, the Member State concerned must submit a corrective action plan showing how the requirements of the Regulation will be fulfilled.[46] The second enforcement system is that provided for by Article 258 TFEU, namely the infringement procedure. This enforcement system does not apply automatically since it requires the adoption of additional measures by the European Commission. Moreover, it should be noticed that the European Commission has the *competence* but not the *obligation* to start an infringement procedure.

Considering the above described regime of monitoring and sanctioning, it could be argued that the Effort Sharing Regulation has put in place a system which could stimulate the reporting of incorrect or misleading data. Moreover, it is not clear whether the European Commission will have the means and the political will to check compliance. Artificially 'improved' reports from the Member States are beneficial to the EU as a whole in the context of the UNFCCC.

4.4   EU INSTRUMENTS TO SUPPORT THE EFFORT SHARING REGULATION

Thus far, the EU legislator has adopted four sets of measures to help Member States achieve their reduction targets. As stated previously, measures concerning energy efficiency and fluorinated gases are dealt with in Chapters 6 and 8 of this book, respectively. Here, we examine the LULUCF Regulation and the legal framework regulating the emissions from vehicles, namely the Directive on the quality of fuels and the Regulation setting emissions requirements for cars and light commercial vehicles.

---

[43]   It should be noted that the European Environmental Agency, Eurostat and Joint Research Centre perform initial checks on the submitted data.

[44]   TNO, TNO-rapport, TNO 2013 R10703, Praktijkverbruik van zakelijk personenauto's en plug-in voertuigen.

[45]   Article 9 of the Effort Sharing Regulation.

[46]   Article 8 of the Effort Sharing Regulation.

One of the reasons why land use and land use change, on the one hand, and transport, on the other, have attracted particular attention from the EU legislator is that these two sectors are major contributors to greenhouse gas emissions. Emissions from agriculture represented about 10 per cent of the EU's total GHG emissions in 2017.[47] In addition, land use and land use change was unaccounted for under the EU GHG emissions framework until the adoption of the LULUCF Regulation. Farming activities such as the use of fertilisers and livestock are covered by the Effort Sharing Regulation, but this does not cover the emissions coming from, stored and captured in the land and forests.[48] This lacuna was particularly problematic in the context of biomass production and use. Emissions from biomass are indeed considered to be zero in the energy sector, and were not accounted for in the non-ETS sectors. The LULUCF Regulation partially solves this lacuna, as further discussed in section 4.4.1.

As regards transport, the Commission indicates that one fifth of carbon dioxide emissions in Europe come from road transport.[49] Moreover, emissions from road transport continued to increase until 2007[50] and again since 2014,[51] simply because more people drive cars and more goods are transported around Europe.[52] Only in the period between 2008 and 2014 did emissions from road transport show a decrease, partially due to increased oil prices and the economic recession of the time.[53] EU measures in the this field can be divided into measures focusing on fuel quality (discussed in section 4.4.2) and measures focusing on emission performance of vehicles (discussed in section 4.4.3).

---

[47] See Eurostat, 'Agri-environmental Indicator – Greenhouse Gas Emissions', available at https://ec.europa.eu/eurostat/statistics-explained/index.php?title=Agri-environmental_indicator_-_greenhouse_gas_emissions last accessed January 2019.

[48] Commission Staff Working Document Impact Assessment Accompanying the Document Proposal for a Regulation of the European Parliament and of the Council on the inclusion of greenhouse gas emissions and removals from land use, land use change and forestry into the 2030 climate and energy framework and amending Regulation No 525/2013 of the European Parliament and the Council on a mechanism for monitoring and reporting greenhouse gas emissions and other information relevant to climate change, SWD(2016) 249 final, pp.4–6.

[49] Several studies performed on this topic are accessible on the website of the Commission at https://ec.europa.eu/clima/policies/transport/vehicles_en last accessed March 2020. See in particular European Commission, 'EU Transport in figures 2011' (no longer available on the site).

[50] European Commission 2014, Communication from the Commission to the European Parliament, the Council, the European Economic and Social Committee and the Committee of the Regions, A policy framework for climate and energy in the period from 2020 to 2030, COM(2014) 15 final, at 14.

[51] See the information available on the site of the European Environment Agency: www.eea.europa.eu/data-and-maps/indicators/greenhouse-gas-emission-trends-6/assessment-3 last accessed March 2020.

[52] Patrick ten Brink, 'Mitigating CO_2 Emissions from Cars in the EU (Regulation (EC) No 443/2009)', in Sebastian Oberthür and Marc Pallamaerts (eds), *The New Climate Policies of the European Union* (Brussels University Press 2010), 179–210, at 180.

[53] European Commission 2014 (n 50), at 14; European Environmental Agency (2013), Annual European Union Greenhouse Gas Inventory 1990–2011 and inventory report 2013, executive summary, at 8 with the exact data over the period 2009–11. See also the source mentioned at n 51 for more recent figures.

4.4.1   LULUCF Regulation

Vegetation is a net remover of carbon dioxide: it absorbs more carbon dioxide than it emits under normal circumstances. Human intervention can affect this natural phenomenon. For example, by cutting down trees and producing biomass, carbon dioxide removal potentials are lost and the carbon dioxide stored in trees is emitted. After long and complex negotiations,[54] the LULUCF Regulation has prescribed that national emissions from land, land use and forestry should not exceed removals, calculated as the sum of total emissions and total removals on a Member State territory in the periods 2021–5 and 2026–30.[55] This means that emissions from the LULUCF sectors do not need to diminish; it is sufficient that they are halted (the so-called no-debit rule).

The existence of two subsequent time periods is linked to the kinds of land use that are covered by the regulation. The LULUCF Regulation applies to emissions and removals coming from three kinds of land use activities during the period 2021–30:

(a) 'managed forest land'[56] – land use reported as:

- forest land remaining forest land;
- cropland, grassland, wetlands, settlements or other land, converted to forest land ('afforested land');
- forest land converted to cropland, grassland, wetlands, settlements or other land ('deforested land');

(b) 'managed cropland' – land use reported as:

- cropland remaining cropland;
- grassland, wetland, settlement or other land, converted to cropland; or
- cropland converted to wetland, settlement or other land;

(c) 'managed grassland' – land use reported as:

- grassland remaining grassland;
- cropland, wetland, settlement or other land, converted to grassland; or
- grassland converted to wetland, settlement or other land.

For the period 2026–30, a fourth LULUCF sector is added to the scope of application of the LULUCF Regulation: 'managed wetland', which is land use reported as wetland remaining wetland, settlement or other land, converted to wetland, or wetland converted to settlement or other land.

---

[54]   A. Savaresi and L. Perugini 2019 'The Land Sector in the 2030 EU Climate Change Policy Framework: A Look at the Future' *Journal for European Environmental and Planning Law* 16, 164.

[55]   Article 4 of the LULUCF Regulation.

[56]   Cf. the LULUCF Regulation, which mentions afforested land and deforested land separately from managed forest land, which applies to forest land remaining forest land.

With regard to afforested and deforested land, though, not all forests are covered by the Regulation. Annex II to the Regulation indicates three parameters for determining the forest area relevant for calculating emissions and removals: (a) the size of the area (ranging from a minimum of 0.05 ha in, for example, Austria to a minimum of 1.0 ha in, for example, Portugal); (b) the tree crown coverage within this area (ranging from a minimum of 10 per cent in, for example, Bulgaria to a minimum of 30 per cent in, for example, Estonia); and (c) the tree height (ranging from a minimum of 2 metres in, for example, Greece to a minimum of 5 metres in, for example, Ireland).[57] This means than only a portion of trees in a country is covered by the concept of forest under the LULUCF Regulation. Moreover, it means that the portion of trees covered in each country takes into account the characteristics of that country. Activities concerning trees outside the covered areas do not account for the emissions of that country – neither under the LULUCF Regulation nor under the Effort Sharing Regulation. There are thus still emissions from land use that are not accounted for under the EU regulatory framework.

To prove compliance, Member States must apply accounting and reporting rules set out under the LULUCF Regulation, in conformity with the IPCCC methodology.

Some flexibilities are included in the Regulation to lower implementation burdens. As already indicated above, Member States can use overachievements under the LULUCF Regulation to achieve their targets under the Effort Sharing Regulation (so-called inter-pillar flexibility). Other two flexibility mechanisms are known from the Effort Sharing Regulation, namely banking and transfer (so-called intra-account flexibility).[58] Finally, there is a specific flexibility mechanism within the LULUCF Regulation, called 'Managed Forest Land Flexibility'.[59] It is indeed possible to allow compensation for net emissions in a Member State. To be able to use this exception, two main conditions must be met.[60] First, the Member States concerned must envisage ongoing or planned specific measures ensuring the conservation or enhancement of forest sinks and reservoirs. Second, the EU as a whole must ensure that its total emissions do not exceed the acceptable level. Use of this flexibility mechanism is possible only up to a given threshold (set out under Article 13 of, and in conjunction with Annex VII to, the LULUCF Regulation). Each country has its own maximum amount of compensation. For example, the EU country with the highest compensation threshold is France, with a compensation limit of 61.5 million tons of CO_2 equivalent for the period 2021–30.

---

[57] It includes also areas with trees, including groups of growing young natural trees, or plantations that have yet to reach the minimum values for tree crown cover or equivalent stocking level or minimum tree height, including any area that normally forms part of the forest area but on which there are temporarily no trees as a result of human intervention, such as harvesting, or as a result of natural causes, but which area can be expected to revert to forest (Article 3(6) of the LULUCF Regulation).

[58] Article 12(3) of the LULUCF Regulation.

[59] Article 13 of the LULUCF Regulation.

[60] Article 13(2) of the LULUCF Regulation.

4.4.2   Fuel Quality Directive

The EU has been regulating fuel quality for many years in an attempt to improve air quality. Since 2009, fuel quality regulation has also contributed to the fight against climate change. To this extent, Directive 2009/30/EC has amended Directive 98/70/EC (the Fuel Quality Directive)[61] in order to add provisions ensuring the reduction of greenhouse gas emissions related to fuel and its supply. As of 2021, the decarbonisation of transport fuels is addressed under the Renewable Energy Directive, which also sets out the criteria for classifying fuels as biofuel, as discussed in Chapter 5 of this book.

4.4.3   Cars and Vehicles Regulation

In the 1990s, emissions from passenger cars were regulated by means of voluntary agreements, improvements in consumer information and the promotion of fuel-efficient cars by means of fiscal measures.[62] However, despite the adoption of voluntary agreements by auto manufacturers from Europe, Japan and Korea, the reductions accomplished in the period 2000–7 were considered insufficient to achieve the target of 120 grams of carbon dioxide emissions per kilometre as the average for a new car in Europe.[63] Accordingly, mandatory targets have been established by the EU legislator regarding passenger cars and light commercial vehicles. In 2018, the Commission presented a proposal to regulate emissions from heavy-duty vehicles.[64] This should help the EU to move towards a reduction of 60 per cent of its greenhouse gas emissions from transport by 2050 compared to 1990, as set out in the White Paper on Transport.[65]

Regulation 2019/631 sets mandatory targets for manufacturers of 'passenger cars'[66] registered in the Community for the first time and which have not previously been registered outside the Community, that is, 'new passenger cars'. From 2020 onwards the average of the emissions from new passenger cars in the EU was not permitted to exceed 95 grams.[67] Regulation 2019/631 sets the average for carbon dioxide emissions for new light commercial

---

[61]   Directive 98/70/EC of the European Parliament and of the Council of 13 October 1998 relating to the quality of petrol and diesel fuels and amending Council Directive 93/12/EEC, [1998] OJ L350/58.

[62]   European Commission, 2007 Proposal for a Regulation of the European Parliament and of the Council setting emission performance standards for new passenger cars as part of the Community's integrated approach to reduce CO_2 emissions from light-duty vehicles, COM(2007) 856 final, at 3.

[63]   For an overview of the development of the discussion on voluntary agreements see ten Brink 2009 (n 52), at 185–8.

[64]   European Commission 2018, Proposal for a regulation setting CO_2 emission performance standards for new heavy-duty vehicles, COM(2018) 284.

[65]   European Commission 2011, Report from the Commission to the European Council, Trade and Investment Barriers Report 2011 Engaging our strategic economic partners on improved market access: Priorities for action on breaking down barriers to trade SEC(2011) 298 final, COM(2011) 144 final.

[66]   Passenger cars are defined as motor vehicles of category M1 as defined in Annex II to Directive 2007/46/EC.

[67]   Article 1 of Regulation 2019/631.

vehicles at 147 grams per kilometre as from 2020. Both targets will be further sharpened as from 2025 and 2030. These targets must be achieved by means of technological development of vehicles' engines, and will be complemented by means of technological fixes capable of inducing European drivers to adopt an environmentally friendly driving style. The latter instrument represents the so-called integrated approach.[68]

Technological development of engines is assessed by means of the standards set out in Regulation 715/2007 and its implementing measures,[69] which are based on the real driving emissions (RDE test) and the WLTP test, replacing the obsolete NEDC methodology.[70] The regulation sets binding targets upon those manufacturers registering their motor vehicles in the EU.

The new testing methodology was necessary as, under the NEDC methodology, there was too much discrepancy between official emissions and real-life emissions, which could be up to 1.5 higher than declared.[71] Discrepancies were highest for those vehicles with low fuel

---

[68]   European Commission 2007 (n 62).

[69]   Regulation (EC) No 715/2007 of the European Parliament and of the Council of 20 June 2007 on type approval of motor vehicles with respect to emissions from light passenger and commercial vehicles (Euro 5 and Euro 6) and on access to vehicle repair and maintenance information, [2007] OJ L171/1. See also Commission Regulation (EC) No 692/2008 of 18 July 2008 implementing and amending Regulation (EC) No 715/2007 of the European Parliament and of the Council on type-approval of motor vehicles with respect to emissions from light passenger and commercial vehicles (Euro 5 and Euro 6) and on access to vehicle repair and maintenance information, [2008] OJ L199/1; and Commission Regulation (EU) 2016/646 of 20 April 2016 amending Regulation (EC) No 692/2008 as regards emissions from light passenger and commercial vehicles (Euro 6) (Text with EEA relevance), [2016] OJ L109/1.

[70]   The RDE and WLTP tests are an EU reaction to what became known as 'Dieselgate'. Regulation (EC) No 715/2007 of the European Parliament and of the Council of 20 June 2007 on type approval of motor vehicles with respect to emissions from light passenger and commercial vehicles (Euro 5 and Euro 6) and on access to vehicle repair and maintenance information, [2007] OJ L171/1. See also Commission Regulation (EC) No 692/2008 of 18 July 2008 implementing and amending Regulation (EC) No 715/2007 of the European Parliament and of the Council on type-approval of motor vehicles with respect to emissions from light passenger and commercial vehicles (Euro 5 and Euro 6) and on access to vehicle repair and maintenance information, [2008] OJ L199/1. Both more times amended. In particular, the WLTP test was introduced with Commission Regulation (EU) 2017/1151 of 1 June 2017 supplementing Regulation No 715/2007, amending Directive 2007/46, Regulation No 692/2008 and Commission Regulation (EU) No 1230/2012 and repealing Regulation No 692/2008, [2017] OJ L175/1. The RDE test was introduced with Commission Regulation (EU) 2016/427 of 10 March 2016 amending Regulation No 692/2008 as regards emissions from light passenger and commercial vehicles (Euro 6), [2016] OJ L82/1. As from 2021, measurements will take place based on Commission Regulation (EU) 2017/1151 of 1 June 2017 supplementing Regulation (EC) No 715/2007 of the European Parliament and of the Council on type-approval of motor vehicles with respect to emissions from light passenger and commercial vehicles (Euro 5 and Euro 6) and on access to vehicle repair and maintenance information, amending Directive 2007/46/EC of the European Parliament and of the Council, Commission Regulation (EC) No 692/2008 and Commission Regulation (EU) No 1230/2012 and repealing Commission Regulation (EC) No 692/2008, [2017] OJ L175/1.

[71]   TNO-rapport (n 44).

consumption and emissions.[72] In general, the reason for higher real-life emissions was the flexibility present in the NEDC test, which manufacturers exploited to the maximum.[73]

Regulation 2019/631 and Regulation 692/2008 take the shortcomings under the NEDC into consideration, most notably by requiring compliance not only with the laboratory tests based on the WLTP methodology, but also with the RDE test, hence a methodology reproducing real-life conditions. Moreover, Article 12 of Regulation 2019/631 requires the Commission to monitor the representativeness of the tests and develop a strategy to prevent the gap between the results of the RDE and WLTP tests and real-world emissions from growing.

Manufacturers thus enjoy less flexibility under the testing regulatory framework than was the case in the past. However, they still enjoy considerable flexibility under the EU regulatory framework. First, the binding target for each manufacturer is calculated by means of the so-called limit curve value. This means that the mass of a vehicle is taken into consideration when establishing the amount of emissions per kilometre that can be expected from that vehicle. For example, a passenger car weighing 2000 kilograms is expected to emit 115.64 grams of carbon dioxide per kilometre and a light commercial vehicle of the same mass is expected to emit 169.43 grams of carbon dioxide per kilometre.[74] Each manufacturer must, for each calendar year, ensure that the expected average emissions from their passenger-car fleet and from their light commercial vehicle fleet do not exceed the average calculated for them under Article 4 of Regulation 2019/631.

This comparison shows that the expectations from passenger cars are higher than those from light commercial vehicles and, more generally, that heavier vehicles are allowed higher emissions than lighter vehicles, while preserving the overall fleet average. Moreover, it shows that the emission targets are tailor-made to meet the characteristics of the various manufacturers, at least to a certain extent. Standardisation is achieved by manipulating three different parameters of the formula to set the emission limit value of each manufacturer. Basically, the formula moves from the assumption that a passenger car with a mass of 1372 kilograms produced in accordance with the actual technological state of the art should not emit more than 95 grams of carbon dioxide per kilometre. A state of the art light commercial vehicle with a mass of 1706 kilograms should not emit more than 147 grams of carbon dioxide per kilometre. Every extra kilogram of mass is multiplied by a coefficient of 0.0333 for passenger cars and 0.096 for light commercial vehicles and then added to 95 and 147 (grams), respectively. This means that the EU legislator can standardise the reduction targets: (a) by lowering the emission level used as the reference point (that is, 95 grams and 147 grams of carbon dioxide per kilometre); (b) by lowering the coefficients; or (c) by increasing the mass of the vehicles which is expected

---

[72] Ibid, p.34 for petrol and diesel engines and p.36 for plug-in hybrid engines. The data on plug-in engines are mainly based on lease autos. Owners of such autos have less incentives to exploit the electric engines of those autos, given that they do not themselves pay for the petrol or diesel.

[73] TNO-rapport (n 44), at 35 and 38.

[74] This examples are based on the formulas established under Article 4 of the Regulation in conjunction with Annex I (Part A for passenger cars and Part B for light commercial vehicles) to Regulation 2019/631. This formula is discussed further below.

to comply with the emission value used as a reference point. It is possible to combine one or more of these actions. As a matter of fact, Regulation 2019/631 reduced the reference point and increased the vehicle mass used as a reference point in comparison with Regulation 443/2009 and Regulation 510/2011. As indicated above, the emission levels used as a reference point for passenger cars and light commercial vehicles will be lowered first in 2025 and again in 2030. Moreover, Article 14 of Regulation 2019/631 tasks the European Commission to adjust the mass used as a reference point in light of the parameters set out under Regulation 2019/631. Basically, the EU legislator will sharpen the emission targets in relation to technological development, assuming that there will be technical innovation in the interim.

Another means by which the EU legislator offers flexibility to manufacturers is the so-called phased-in approach, which means that the emission targets are applied only to a percentage of the fleet, which increases with time. As regards the 2020 target of 95 grams of carbon dioxide per kilometre, the phased-in approach means that only 95 per cent of the passenger-car fleet was taken into account in assessing compliance.[75] From 2021 onwards, 100 per cent of the fleet will be taken into consideration.

It should be noted that the Regulation allows for derogations for manufacturers registering less than a certain amount of vehicles in the EU.[76] A manufacturer registering fewer than 10,000 passenger cars each year or fewer than 22,000 light commercial vehicles might apply for derogation to the European Commission. If accepted, instead of the binding target set unilaterally by the EU legislator, the manufacturer has to comply with a self-established emission target. The European Commission can accept the proposal if it is consistent with the reduction potential of the manufacturer applying for the derogation, including the economic and technological potential for reducing its specific emissions of carbon dioxide, and taking into consideration the characteristics of the market for the type of vehicle manufactured. Manufacturers registering more than 10,000 but fewer than 300,000 passenger cars can also apply for a relaxation of the emission target set by the EU legislator. Also in this case, the European Commission can accept the request for derogation only if it is consistent with the reduction potential of the manufacturer applying for the derogation. For light commercial vehicles, however, there is not a similar possibility. By allowing for derogations, the EU legislator seems to have paid attention to the fact that the costs of meeting the EU's standards are relatively more burdensome for manufacturers registering few vehicles than for those registering many vehicles, mainly due to economies of scale. It could be argued that there are substantive equality benefits from this.

Moreover, the Regulation focuses on the average of emissions from a manufacturer's fleet, which also allows flexibility to manufacturers of passenger cars and light commercial vehicles. Indeed, manufacturers are free to manufacture passenger cars and light commercial vehicles emitting more carbon dioxide than the average, as long as they also produce enough vehicles emitting less carbon dioxide than the average. In 2020, when the 95 gram/kilometre target for passenger cars entered into force, cars emitting less than 50 grams of carbon dioxide per kilo-

---

[75] Article 4 of Regulation 2019/631.

[76] Article 10 Regulation 2019/631.

metre were counted as two cars for calculating the average of the fleet.[77] This is the so-called super-credit approach, which aims at stimulating the introduction of climate-friendly technologies, such as those used in electric cars. This 'super credit' will disappear in 2023, following a progressive decrease.[78]

Another instrument allowing manufacturers flexibility as to how to achieve their targets is 'pooling'.[79] This refers to the possibility for two or more manufacturers to combine their fleets for the purpose of complying with the requirements set out by the Regulation. They basically form a pool of manufacturers, as long as competition rules are complied with. For example, in 2019 Fiat Chrysler Automobiles (FCA) pooled with Tesla in order to benefit from Tesla's credits and comply with the EU regime.[80]

As well as technological development of engines, both regulations focus on the deployment of innovative technologies aiming at stimulating a low-emission driving style. For example, low tyre pressure increases fuel consumption and emissions by about 4 per cent.[81] Alarm systems advising the driver to increase tyre pressure could therefore lead to a reduction of carbon dioxide emissions. Article 1 of Regulation 2019/631 indicates that additional measures based on the integrated approach shall lead to a reduction of 10 grams of carbon dioxide per kilometre. The first step in the implementation of the integrated approach is to give credits to manufacturers equipping their fleets with innovative technologies, such as stop-and-start technology.[82] Under the Regulation, manufacturers can be granted emission credits equivalent to a maximum emissions saving of 7 grams per kilometre per year for their fleet. To be eligible for the credits, manufacturers have to equip their vehicles with such innovative technologies. Moreover, the benefits associated with the innovative technology proposed by a manufacturer must be verified by independent data showing that this technology makes a verified contribution to the reduction of carbon dioxide.

The EU legislator can draw inspiration from the innovative technologies notified to the Commission in order to be eligible for credits and can then decide to take the second step in the integrated approach, which is making that technology compulsory for all manufacturers. For example, Article 9 of Regulation 661/2009 requires that all new passenger cars must be equipped with 'tyre-pressure monitoring systems', indicating to the driver when to increase

---

[77]   Article 5 of Regulation 2019/631.

[78]   Article 5 of Regulation 443/2009, which limits the use of super credits to a maximum of 7.5 grams/kilometres per manufacturer.

[79]   Article 6 of Regulation 2019/631.

[80]   E.g. Peter Campbell, 'Fiat Chrysler to spend €1.8bn on CO2 credits from Tesla', *Financial Times online*, 3 May 2019 last accessed June 2019.

[81]   European Commission 2010, Report from the Commission to the European Parliament, the Council, and the European Economic and Social Committee, Progress report on implementation of the Community's integrated approach to reduce CO2 emissions from light-duty vehicles, COM(2010) 656 final, at 6.

[82]   Article 11 of Regulation 2019/631.

tyre pressure.[83] Similarly, Article 9 requires all new car models to be equipped with 'low rolling resistance tyres'. Once an innovative technology has been made mandatory, manufacturers are no longer eligible for credits for deploying that technology. Innovative technologies have been adopted as regards, inter alia, gear shifting indications and air condition performances.[84] Clearly, such technologies' contribution to the reduction of carbon dioxide emissions from cars depends on the actual shift in driving styles.

Monitoring and penalties are also among the instruments used by the regulations to achieve their goals.[85] Crucially, the Regulation establishes a penalty for those manufacturers that do not comply with their emission targets. For the first gram/kilometre of exceedance, manufacturers have to pay a so-called excess emissions premium of €95 per gram/kilometre, multiplied by the number of vehicles registered by the manufacturer concerned. As in other fields, the level of the penalty will determine the degree of compliance by the manufacturers. Rational manufacturers might actually consider it more advantageous to trespass the limits set out under the Regulation, for example because they estimate they will sell more cars so that the extra revenues exceed the penalties. It is clear that the penalty systems envisaged by the regulations are an important instrument to achieve the reduction targets set out by the EU. Although it makes sense from a public finance perspective, it could be seen as regrettable that the money coming from the penalties will not be reserved to counterbalance eventual flaws in the system envisaged by the regulation, but will become part of the general budget of the EU.[86]

It should be noted that the Regulation has been adopted on the basis of Article 192 TFEU, which triggers the possibility to maintain or adopt more stringent protective measures. Member States could thus choose to adopt stricter measures. However, Recital 46 of the preamble to Regulation 2019/631 states that Member States shall not impose additional or more stringent penalties on manufacturers who fail to meet their targets under the regulation.[87]

---

[83] Regulation (EC) No 661/2009 of the European Parliament and Council of 13 July 2009 concerning type-approval requirements for the general safety of motor vehicles, their trailers and systems, components and separate technical units intended therefore, [2009] OJ L200/1.

[84] European Commission 2010 (n 81), section 3.

[85] Monitoring requirements can be found in Article 7 of Regulation 2019/631.

[86] Article 8(4) of Regulation 2019/631.

[87] It would go beyond the didactic purposes of this book to enter into a detailed discussion on whether secondary law adopted on the basis of Article 192 TFEU can prohibit the adoption of more stringent protective measures under Article 193 TFEU and, more generally, on whether a recital from the preamble of a EU secondary measure can set aside a provision of primary law. Here, we limit ourselves in pointing out that the legislative practise used while drafting these two regulations seems quite awkward. Those readers interested in a more extensive discussion on this topic are invited to read more specific publications. such as Squintani (n 36), ch. 1, with further references. As regards the discussion on the use of a similar legislative technique concerning the insertion of emission limit values for greenhouse gas emission covered by the ETS Directive in permits granted under national law implementing the Industrial Emissions Directive, see L. Squintani, J.M. Holwerda and K.J. De Graaf, 'Regulating Greenhouse Gas Emissions from ETS Installations: What Room Is Left for the Member States?' in M. Peeters, M. Stallworthy and J. de Cendra de Larragán (eds), *Climate Law in the EU Member States: Towards National Legislation for Climate Protection* (Edward Elgar Publishing 2012), 67–88, with further references.

4.5   CONCLUSION

It will be interesting to see what degree of success is achieved by the EU regulatory framework for reducing greenhouse gas emissions from the non-ETS sectors by 30 per cent by 2030. With respect to cost-effectiveness, the Effort Sharing Regulation allows for considerable flexibility in methods used to achieve the AEAs and facilitates a tailor-made approach, which can reduce compliance costs. The positive attitude of the EU towards a tailor-made approach was also evident in the context of the Regulation on emissions from cars and light commercial vehicles. The limit values imposed upon manufacturers take the needs of the various manufacturers into consideration by allowing higher emissions from heavier and commercial vehicles. Solidarity was largely seen in the Effort Sharing Regulation, where Member States' capacities to reduce emissions were reflected in the AEAs.

The main shadow cast over the EU package regulating the non-ETS sectors concerns the functioning of the monitoring and reporting system. Since the enforcement system is mainly based on self-incrimination and exact measurements are lacking, doubts can be raised about the degree to which the data provided by Member States on the level of their emission reductions may be trusted. The fact that the European Commission, the watchdog controlling those data, requires positive figures in order to show compliance with its international commitments does not improve that situation.

CLASSROOM QUESTIONS

1. Does the EU non-ETS legal framework ensure enough legal certainty that the EU will achieve its objectives under the Paris Agreement?
2. Has the EU non-ETS legal framework sufficiently accounted for the early claims that EU climate law breaches the equality principle?
3. Does the EU non-ETS legal framework regulate the relationship between EU action and Member States' action in a manner which helps to pursue a high level of environmental protection?

SUGGESTED READING

Books

No specific books available (to our knowledge).

Articles and chapters

Romppanen S, 'The EU Effort Sharing and LULUCF Regulations: The Complementary yet Crucial Components of the EU's Climate Policy Beyond 2030' in M Peeters and M Eliantonio (eds), *Research Handbook on EU Environmental Law* (Edward Elgar Publishing 2020).

de Sadeleer N, 'Light-vehicles Emissions Standards under EU Law in the wake of the "Dieselgate"' in M Peeters and M Eliantonio (eds), *Research Handbook on EU Environmental Law* (Edward Elgar Publishing 2020).

Savaresi A and Perugini L, 'The Land Sector in the 2030 EU Climate Change Policy Framework: A Look at the Future' (2019) 16(2) Journal for European Environmental and Planning Law 148–64.

Policy documents

European Academies' Science Advisory Council (EASAC), Negative Emission Technologies: What Role in Meeting Paris Agreement Targets? EASAC policy report (2018) 1.
Erbach G, 'Effort Sharing Regulation, 2021–2030: Limiting Member States' Carbon Emissions', European Parliamentary Research Service briefing (2018) 11.
Romppanen S, 'How does the LULUCF Regulation affect EU Member States' forest management?' Institute for European Studies, Policy Briefs (2019) 4.

5
Renewable energy consumption

ABSTRACT

- Promoting the consumption of renewable energy is one of the key means to reduce carbon emissions from electricity production, heating and transportation, while it also improves energy security and contributes to economic growth;
- Repealing the early Directives 2001/77/EC and 2003/30/EC that aimed to promote renewable sources, Directive 2009/28/EC (2009 RES Directive) established a legal framework for increasing the consumption of energy from renewable sources in the EU to 20 per cent of overall energy consumption by 2020. At that time, Member States were given legally binding national targets for achieving specified levels of renewable energy growth;
- The current Directive EU/2018/2001 (2018 RES Directive) requires States to act in solidarity to achieve collectively a target of 32 per cent renewable energy in energy consumption by 2030, but replaces binding national targets with detailed governance arrangements for scrutinising individual contributions of Member States to the Union target;
- The 2018 RES Directive places greater emphasis on cost-effectiveness than the 2009 RES Directive, for instance by requiring support schemes for renewable electricity to give producers market exposure in most circumstances. It also encourages Member States to seek out lower cost options for increasing renewable energy consumption by allowing them to collaborate through cooperation mechanisms;
- The 2019 Electricity Directive (2019/944) and the 2019 Electricity Regulation require Member States and sectoral actors to enable renewable energy integration in networks and markets by reforming rules on their operation and development;
- The 2018 RES Directive requires Member States to promote the use of renewables in energy consumed for transport, while preventing this and other sources of demand for bioenergy from giving rise to unsustainable fuel production practices;
- Member States employ support schemes to encourage investment in RES production capacity, but legal questions arise over their compatibility with TFEU provisions concerning the free movement of goods and state aid.

5.1   INTRODUCTION

This chapter examines the legal framework established by the EU to promote renewable energy consumption. Its main focus is on the Renewable Energy Directive of 2018 (2018 RES Directive) that was adopted to implement the European Union's target of increasing the proportion of energy from renewable sources in overall energy consumption to at least 32 per cent by 2030.[1]

Section 5.2 introduces the topic of renewable energy by explaining what this is and why the replacement of fossil fuel energy with energy from renewable sources is promoted by the EU and by many of its Member States. Section 5.3 begins by considering the EU's early development of a policy on renewable energy, in the 1990s, and the 2001 Directive on Renewable Electricity and the 2003 Directive on Biofuels that this policy spawned.[2] It examines the limitations of these early Directives for promoting renewables, as well as considering their successor, the Renewable Energy Directive of 2009 (2009 RES Directive), which was enacted to provide stronger support for the sector's growth.[3] Section 5.4 provides an overview of the various respects in which the 2018 RES Directive that replaces it seeks to promote renewable energy consumption and to address both perceived barriers to this and the potential environmental consequences of a major increase in the consumption of biomass-based energy.

Section 5.5 examines legal issues associated with Member States' provision of financial support for renewable energy. Sections 5.5.1 and 5.5.2 explain why financial support, whether provided directly by the public sector or by private sector actors acting in accordance with legal direction, is often needed to secure investment in renewable energy. Sections 5.5.3 and 5.5.4 describe the types of schemes that Member States have typically used to provide support. Section 5.5.5 considers the influence that aspects of EU Treaties have had on the take-up and design of support schemes by Member States. Section 5.6 concludes.[4]

5.2   BASICS OF RENEWABLE ENERGY

5.2.1   What Is Renewable Energy?

The term 'renewable' is used to describe energy derived from sources that are replenished at the same rate as they are used. This is in contrast to fossil fuels, the consumption of which

---

[1]   Directive 2001/2018/EU of the European Parliament and of the Council of 11 December 2018 on the promotion of the use of energy from renewable sources [2018] OJ L 328/82 (the 2018 RES Directive).

[2]   Council Directive 2001/77/EC of 27 September 2001 on the promotion of electricity produced from renewable energy sources in the internal electricity market [2001] OJ L283/33 (the 2001 Directive); Council Directive 2003/30/EC of 8 May 2003 on the promotion of the use of biofuels or other renewable fuels for transport [2003] OJ L 123/42 (the Biofuels Directive).

[3]   Council Directive 2009/28/EC of 23 April 2009 on the promotion of the use of energy from renewable sources and amending and subsequently repealing Directive 2001/77/EC and 2003/30/EC [2009] OJ L 140/16 (the 2009 RES Directive).

[4]   Separate laws have been enacted by the EU on the simultaneous production of electricity and heat from renewable energy sources through 'cogeneration', which is considered in Chapter 6 of this book. The legal provision that has been made in EU law to address grid access challenges for renewable electricity is considered in Chapter 9. Chapter 3 examines the extent to which the overarching principles of EU climate policy are honoured in the EU ETS Directive and what balance has been struck between them.

reduces the stock available for future generations. The principal sources of renewable energy are the sun, the wind, waves, tides, tidal currents, geothermal energy and organic matter (biomass). The majority of these sources are the product, either directly or indirectly, of energy from the sun. The exceptions to this are tidal and geothermal energy, which are derived respectively from the gravitational effect of the moon and from the heat of the Earth's interior. Most of these sources are fully renewable, but biomass and geothermal are only renewable to the extent that consumption does not exceed the capacity of the Earth and its interior to replace them. Technologies have been developed to produce energy from all of these sources. Some of the technologies are well established and widely used for commercial energy production (for example, wind and solar energy) while others are at an earlier stage of development (for example, wave and tidal current energy).[5]

Renewable sources can be used to meet demands for energy for electricity production, heating and transportation. Electricity can be generated from solar energy (including through photovoltaic (PV) units), through the release of water stored behind dams (hydropower and tidal barriers) and through turbines driven by wind, wave and tidal currents and by the burning of biomass. In addition to meeting current demand for services such as lighting, renewable electricity is expected to have a growing role in providing energy for heating and transportation if fossil fuel consumed for these purposes is to be replaced by lower-carbon alternatives. Energy for heating can be attained directly from the sun, including through its heating of the air and water, from the burning of biomass and gases derived from them in boilers, from the capture of heat produced as a by-product of electricity generated from renewable sources in combined heat and power units and through tapping geothermal energy. Fuels derived from a wide variety of biomass feedstocks and from organic waste can be used to power road, marine and air transportation.

A common characteristic of most renewable sources is that carbon dioxide is not emitted during the production of energy from them. The exception to this is biomass, which is described as a 'carbon neutral' energy source because the carbon dioxide that organic matter absorbs during its growth is released into the atmosphere when it is burnt. In practice, there will be some carbon emission associated with all renewable energy production when this is calculated on a life-cycle basis. For example, emissions will be produced through the consumption of electricity during the manufacture of wind turbines or by the production of fertilisers to cultivate biomass. Even so, it is possible to say that carbon emissions associated with renewable energy production tend to be much lower than those produced by fossil fuel combustion.

5.2.2   Why Do We Need Renewable Energy?

The growth of renewable energy production has been actively supported by the EU and by many of its Member States since the 1990s. The principal reason for this support is the need to mitigate climate change by securing rapid reductions of greenhouse gas emissions in general,

---

[5]   For a fuller explanation of renewable sources see G. Boyle, B. Everett and G. Alexander, 'Introducing Renewable Energy' in G. Boyle (ed.) *Renewable Energy: Power for a Sustainable Future* (Oxford: Oxford University Press, 2012, 3rd ed.), pp.14–17.

and particularly those associated with energy consumption. The energy production industry, because of its current dependence on fossil fuels, is the largest sectoral contributor to the EU's greenhouse gas emissions (26 per cent in 2018).[6] Energy consumption as a whole, including for terrestrial transportation and industrial, commercial and domestic consumption, was responsible for a staggering 74.68 per cent of the EU's emissions in 2018.[7] Promoting renewable energy therefore forms a key part – alongside increasing energy efficiency – of the EU's strategy for meeting its political commitments to cut greenhouse gas emissions by at least 40 per cent of 1990 levels by 2030.

The mitigation of climate change provides reason enough for a switch from fossil fuels to lower-carbon renewable energies, but there are also other factors that make the growth of renewable energy consumption a desirable policy objective for the EU. The first is to improve energy security by reducing the EU's dependence on imported fossil fuel energy. The EU imported 58.2 per cent of its energy consumption in 2018, but with reliance on imported petroleum and products and on natural gas at 94.6 per cent and 83.2 per cent respectively.[8] In contrast, renewable energy sources (save for imported feedstocks for some biomass-based energies) are largely indigenous. Their exploitation may also provide a more stable economic base over the long term than relying on fuel sources whose prices are prone to volatility.

Second, the development of specialisations in renewable energy technologies can contribute to job and wealth creation in the EU and enable it to become a world leader in the export of technologies and expertise that are required worldwide in connection with global efforts to mitigate and adapt to climate change. The Commission advised in 2012 that strong renewables growth to 2030 could generate more than three million jobs, and emphasised the value for the EU's global competitiveness of maintaining its leadership in renewable technologies as 'clean tech' industries become increasingly important around the world.[9] Investment in renewable energy during the 2020s is also seen as a key aspect of the EU's COVID-19 crisis recovery plan.[10]

---

[6]   Commission, *EU Energy in Figures: Statistical Pocketbook 2020* (Publications Office of the European Union, 2020), pp.164–7.

[7]   Ibid.

[8]   Ibid., p.24.

[9]   Commission, 'Renewable Energy: A Major Player in the European Energy Market' COM (2012) 271 final, 2.

[10]   Commission, '2020 Report on the State of the Energy Union pursuant to Regulation (EU) 2018/1999 on Governance of the Energy Union and Climate Action' COM (2020) 950 final, 17–22.

5.3 THE DEVELOPMENT OF RENEWABLE ENERGY LEGISLATION

5.3.1 The Electricity and Biofuels Directives

The Commission made its first formal statement on renewable energy in a green paper of 1996,[11] and followed this in 1997 with a white paper setting out a Community Strategy and Action Plan for renewables.[12] The white paper and following Commission policy documents on renewable energy use the three reasons that are given in section 5.2.2 above to justify intervention in this area at the European level.[13]

The process initiated by the white paper bore fruit in the adoption, in 2001, of the Renewable Electricity Directive.[14] The Directive's goal was to increase the share of electricity from renewable sources to 22.1 per cent of total EU electricity consumption by 2010. Article 3(1) placed an obligation on Member States to 'take appropriate steps to encourage greater consumption of electricity produced from renewable energy sources'. It advised them that this should be done in conformity with national indicative targets which they were required by Article 3(2) to set. Member States had to take account of reference values stated in the Annex to the Directive for renewable electricity consumption when setting national indicative targets, but were not legally obliged under European law to achieve either their reference values or the indicative targets set by them. The 2003 Directive on promoting the consumption of biofuels in transportation took a similar approach, with States being expected to ensure that a minimum proportion of biofuels and other renewable fuels would be placed on the market.[15] Article 3(1)(b)(ii) requires Member States to set national targets for this minimum proportion using a reference value of 5.75 per cent of all petrol and diesel for transport purposes placed on their markets by 31 December 2010, but the Directive does not place an obligation on them under European law to achieve that level of renewable fuels availability.

Member States' performance in relation to their non-binding targets was somewhat patchy. The Commission's 2009 progress report (capturing the position in 2007) records that two Member States had already reached their targets for electricity, but that several States had either made no progress towards them or had seen declines in renewable energy shares due to increasing energy consumption.[16] The 2013 report advises that 15 Member States failed to

---

[11]   Commission, 'Energy for the future: renewable sources of energy – Green Paper for a Community Strategy and Action Plan' COM (96) 576 final.

[12]   Commission, 'Energy for the future: renewable sources of energy – White Paper for a Community Strategy and Action Plan' COM (97) 599 final.

[13]   Ibid., 4–6. For example, the justifications are repeated in Commission, COM (2012) 271 final (n 9).

[14]   The 2001 Directive (n 2).

[15]   The Biofuels Directive (n 2).

[16]   Commission, 'The Renewable Energy Progress Report: Commission Report in accordance with Article 3 of Directive 2001/77/EC, Article 4(2) of Directive 2003/30/EC and on the implementation of the EU Biomass Action Plan, COM (2005) 628', COM (2009) 192 final, 3.

reach their indicative 2010 target for the share of renewables in the electricity mix, and that 22 Member States had not achieved the 5.75 per cent target under the Biofuels Directive.[17] This situation of some States free-riding on others that were exceeding their targets led to the view that more intrusive legal measures would be required to ensure that all Member States should contribute, in a spirit of solidarity, to increasing energy consumption in the EU.[18]

5.3.2   Renewable Energy Directive 2009

In March 2007, the European Council agreed to increase renewable energy consumption to 20 per cent of overall energy consumption in the EU by 2020.[19] This agreement formed part of the 20/20/20 strategy that sought to reduce the EU's greenhouse gas emissions to 80 per cent of 1990 levels by 2020. The other key plank of this strategy was to increase the efficiency of energy consumption by 20 per cent by 2020. Improving energy efficiency is valuable in itself for reducing emissions from energy production and use. It will also make it easier to achieve the commitment on renewables, as growth in this sector will take place against a backdrop of shrinking demand for energy.

The Commission was invited by the European Council to prepare a legal instrument to implement its political commitments. A new Directive for renewable energy was proposed in January 2008,[20] and was adopted in April 2009 after passing through the European legislative process. It entered into force in June 2009 and Article 27(1) required Member States to transpose it into their national legislations by 5 December 2010. The 2009 RES Directive repealed certain provisions of the 2001 Directive on 1 April 2010, with the Directive and the Biofuels Directive being repealed and replaced in their entirety by the 2009 RES Directive with effect from 1 January 2012. Those parts of the 2009 RES Directive concerned with the promotion of biofuels and bioliquids in energy consumption were amended in 2015, with transposition required by September 2017, in view of difficulties with controlling indirect land use as a consequence of biomass-based energy production (see section 5.4.5 below).[21]

There were two key differences between the 2009 RES Directive and its predecessors. The first was that it imposed upon Member States national targets to achieve increases in renewable energy consumption that were legally binding at the EU level. This was felt to be necessary because of the poor performance by a majority of Member States under the 2001 Directive and the Biofuels Directive.[22] All Member States were thus obliged to contribute to realising the EU's 2020 renewable energy objective by achieving national targets set out in Annex I to

[17]   Commission, 'Renewable Energy Progress Report' COM (2013) 175 final, 4.

[18]   Commission, 'Renewable Energy: Progressing towards the 2020 target', COM (2011) 31 final, 3–4.

[19]   Council of the European Union, 'Presidency Conclusions: 8/9 March 2007', Ref.7224/1/07/Rev.1.

[20]   Commission, 'Proposal for a Directive of the European Parliament and of the Council on the promotion of the use of energy from renewable sources' COM (2008) 19 final.

[21]   Directive (EU) 2015/1513 of the European Parliament and of the Council of 9 September 2015 amending Directive 98/70/EC relating to the quality of petrol and diesel fuels and amending Directive 2009/28/EC on the promotion of the use of energy from renewable sources [2015] OJ L 239/1.

[22]   Commission, COM (2011) 31 final (n 18), 3–4.

the Directive. Second, the targets under the 2009 RES Directive addressed increasing the con-tribution of renewable sources to energy sources for heating and cooling and transportation in addition to electricity. As noted above, the 2001 Directive was concerned with electricity only. It was for Member States to decide on how to reach their targets. Greater weight could be placed on decarbonising electricity than on heating and cooling and transportation, or vice versa. However, all Member States also had a separate legally binding target of achieving 10 per cent of energy from renewable sources in energy consumption for transportation.[23] The Commission's view in proposing a common target was that a concerted effort from Member States would be required to reduce emissions from a sector that is largely dependent on fossil fuel energy.[24] It also felt that the need to reflect national differences in possession of renewable resources for electricity, heating, and cooling was not present for transport, as biofuels are more easily traded than electricity.[25]

Growth of the EU renewables sector under the 2009 RES Directive improved significantly on the progress seen under the 2001 Directive and the Biofuels Directive. In contrast to the failure to meet EU targets under these Directives, the Commission's modelling predicted a renewable energy share in 2020 of 22.8 per cent to 23.1 per cent in total energy consumption for the European Union and of 12 per cent in energy for transport.[26] In 2018, five Member States were assessed as being at moderate (Luxembourg and the Netherlands) or severe (Belgium, France and Poland) risk of failing to meet national renewable energy targets.[27] In all, 11 Member States were expected to fall short of the 10 per cent transport target, but with three missing it by only a very small margin.[28] The figures compare favourably with the 15 and 22 Member States who fell short of indicative national targets under the 2009 RES Directive's predecessors.[29] Member States that fail to meet binding targets under Articles 3(1) and 3(4) of the 2009 RES Directive will be in breach of EU law and infraction proceedings against them can be initiated by the European Commission, potentially leading to the imposition of fines by the European Court of Justice.[30] The Directive gives Member States the opportunity to make up for shortfalls in their national efforts by arranging a statistical transfer (see section 5.4.6) with a State whose proportion of renewable energy in total energy consumption is on track to exceed its national target. Statistical transfers can also be arranged under the 2009 RES Directive of renewable energy consumed in energy for transportation alone in connection with

---

[23] 2009 RES Directive (n 3), Article 3(4).

[24] Commission, '2020 by 2020: The Community's Climate Change Opportunity' COM (2008) 30 final, 7–8; Commission, COM (2008) 19 final (n 20), 8.

[25] Ibid.

[26] Commission, 'Renewable Energy Progress Report' COM (2020) 952 final, 9–12.

[27] Ibid., 9.

[28] Ibid., 12.

[29] Commission, COM (2013) 175 final (n 17), 4.

[30] D. Benink, H. Croezen and M. van Valkengoed, *The Accountability of European Renewable Energy and Climate Policy* (CE Delft, April 2011), available at www.cedelft.eu/en/publications/1143/the-accountability-of-european-renewable-energy-and-climate-policy accessed 17 June 2021.

the 10 per cent goal.[31] The Commission recorded in its 2020 renewable energy progress report that four such agreements had been made, but that more may be concluded to enable Member States that had fallen behind required rates to make up for their shortfall.[32]

5.3.3 Replacing the 2009 RES Directive

The Commission initiated discussion of the policy and legal framework for climate and energy in the period from 2021 to 2030 in January 2014.[33] This led to agreement by the European Council in October 2014 on an EU-wide target of at least 27 per cent energy from renewable sources in overall energy consumption by 2030.[34] The Commission's proposal, endorsed by the Council, called for a different approach from the previous regime. It records the Commission's views that European and national targets may have driven 'strong action by the Member States and growth in emerging industries', but that they did not always fit well with EU policy goals for undistorted competitive energy markets.[35] In addition, the proposal expresses concern over the affordability of energy for consumers and businesses, over the effect of energy costs on the competitiveness of the EU's economy, and at the possibility that binding targets may have been responsible for impairing the cost-effectiveness of national efforts to implement Union climate and energy policy, including by requiring States to develop renewable energy to a specified level even when this was not the most cost-effective means open to them for reducing greenhouse gas emissions.[36]

The Commission's proposal in order to reconcile these concerns with ensuring the further growth of renewable energy was to replace legally binding national targets for Member States with an overall Union target for renewable energy, thereby allowing Member States more flexibility in deciding on how to meet greenhouse gas reduction targets in the most cost-effective way while imposing a collective responsibility for ensuring growth of renewables consumption. The overall Union target would be backed up by rigorous European-level governance arrangements to keep individual and collective progress by Member States towards its achievement under review.

The 27 per cent target was widely criticised as lacking in ambition. The Commission, the Parliament and the Council reached political agreement in June 2018 on a higher target of at least 32 per cent energy from renewable sources in overall energy consumption by 2030.[37]

---

[31]   Directive (EU) 2015/1513 (n 21), Article 2(4).

[32]   Commission, COM (2020) 952 final (n 26), 68.

[33]   Commission, 'A policy framework for climate and energy in the period from 2020 to 2030' COM (2014) 15 final.

[34]   European Council, 'Conclusion on 2030 Climate and Energy Policy', SN 79/14, 23 and 24 October 2014, available at www.consilium.europa.eu/uedocs/cms_data/docs/pressdata/en/ec/145356.pdf accessed 17 June 2021.

[35]   COM (2014) 15 final (n 33), 1–7.

[36]   Ibid.

[37]   Commission, 'Europe leads the global clean energy transition: Commission welcomes ambitious agreement on further renewable energy deployment in the EU' (14 June 2018, Press Release), available at https://ec.europa.eu/commission/presscorner/detail/de/STATEMENT_18_4155.

A proposal for a new renewable energy directive was published by the Commission in November 2016 as part of a wider package of laws to implement climate and energy policy for 2021–30 and beyond under the Clean Energy for All Europeans programme.[38]

5.4   THE RENEWABLE ENERGY DIRECTIVE

The 2018 RES Directive received legislative approval in December 2018.[39] Member States are required to have transposed it into national laws by 30 June 2021 (Art. 36). The Directive enshrines the 32 per cent target, but takes note of views that changing circumstances could render it inadequate. It provides that the Commission will review the target (and others set under it) 'with a view to submitting a legislative proposal by 2023 to increase it' where change in the cost of renewable energy production, change in the Union's commitments regarding decarbonisation under international law or decline in the Union's energy consumption would justify an increase.[40]

5.4.1   Core Obligations

The most notable difference between the 2018 RES Directive and its predecessor is that it does not place legally binding national targets on Member States at the European level. Instead, they each have obligations to 'set national contributions to meet collectively' the Union target for 2030 and to 'collectively ensure' its achievement (Arts 3(1) and 3(2)). The expectation that Member States will achieve a Union target without setting out their responsibilities for ensuring that it is met in European law raises questions about how they can be held to account for any perceived inadequacy of their contributions to its realisation. What prevents failure to achieve the overall EU target or the poor performance of individual States if there are no national legally binding targets at the European level?

The Commission proposed in its policy statement of 2014 to plug the gap left by the absence of national targets by establishing overarching Union governance for all policy areas contributing to its goal of a 40 per cent reduction of greenhouse gas emissions compared to 1990 levels by 2030.[41] A Governance Regulation was proposed as part of the Clean Energy for All Europeans legislative programme,[42] and was adopted in December 2018.[43]

---

[38]   Commission, 'Proposal for a Directive of the European Parliament and of the Council on the promotion of the use of energy from renewable sources', COM (2016) 767 final.

[39]   2018 RES Directive (n 1).

[40]   Ibid., Articles 3(1) and 25.

[41]   COM (2014) 15 final (n 33).

[42]   Commission, 'Proposal for a Regulation on the Governance of the Energy Union', COM (2016) 759 final.

[43]   Regulation (EU) 2018/1999 of the European Parliament and of the Council of 11 December 2018 on the governance of the Energy Union and Climate Action, amending Regulations (EC) No 663/2009 and (EC) No 715/2009 of the European Parliament and of the Council, Directives 94/22/EC, 98/70/EC, 2009/31/EC, 2009/73/EC, 2010/31/EU, 2012/27/EU and 2013/30/EU of the European Parliament and of the Council, Council Directives 2009/119/EC and

As a regulation, it had almost immediate direct legal effect. It required each Member State to prepare and submit a draft integrated national energy and climate plan for 2021 to 2030 to the Commission by the end of 2018, with the final plan to be notified to the Commission and published by the end of 2019 (Arts 3 and 9). The Regulation envisages that this process will be repeated every ten years with draft integrated plans and final plans for 2031–40 to be submitted by 1 January 2028 and 1 January 2029 respectively. The plan should state the contribution that the Member State will make to achieving the Union target on renewable energy together with the interim trajectory that will be followed to reach it in line with milestones specified in the Regulation (Arts 4(a)(2) and 5). Detailed prescribed information on national policies and measures that will be pursued and taken to effect the contribution should also be provided (Art. 3 and Annex I). Member States have extensive biennial reporting obligations during the period covered by the plan, including on progress made on increasing renewable energy consumption (Arts 17–28).

The Commission may issue country-specific recommendations for revising draft plans (Arts 9 and 31). If collective ambition is assessed to be inadequate for the achievement of Energy Union goals, the Commission may make recommendations calling on States whose contributions are deemed to be insufficient to increase their ambition (Art. 31). States falling below their expected contribution are to be identified by application of a formula set out in Annex II to the Regulation. Member States must take due account of recommendations when finalising their plans, but are not obliged to follow them (Art. 9(3)). The Commission reviewed and issued recommendations on draft Member State plans for 2021–30 during 2019.[44] Revisions made by Member States and included in their final plans for the period, submitted by the end of 2019, lifted their collective ambition from the 30.4–31.9 per cent shown in the draft plans to 33.15–33.7 per cent.[45]

Recommendations may also be made to a Member State if the Commission concludes through interim review that it is making insufficient progress towards implementing its climate plan, and to all States if it concludes that the Union is at risk of not meeting its target based on an aggregate interim assessment of performance (Art. 32). In addition, the Commission has an obligation, if collective national measures on renewable energy are assessed to be insufficient to achieve the 2030 target, to 'propose measures and exercise its power at the Union level' to ensure the target's achievement (Art. 32(2)).

These extensive requirements for Member State explanation of their climate and energy goals and policies will enable close scrutiny of their performance on renewable energy development. The resulting transparency may, if coupled with pressure from the Commission, from

---

(EU) 2015/652 and repealing Regulation (EU) No 525/2013 of the European Parliament and of the Council [2018] OJ L 328/1.

[44]   Commission, 'National Energy and Climate Plans (NECPs), available at National energy and climate plans (NECPs) | Energy (europa.eu).

[45]   Commission, 'United in delivering the Energy Union and Climate Action – setting the foundations for a successful clean energy transition', COM (2019) 285 final, 3; Commission, 'An EU-wide assessment of National Energy and Climate Plans', COM (2020) 564 final, 2.

other Member States and from civil society, force Member States to improve the ambition of their plans for renewable energy growth when they are found to be wanting, despite there being no legal obligation for them to do so. The Commission's powers when combined performance falls below the trajectory needed for achievement of 32 per cent by 2030 will help with promoting growth of renewable energy in the Union when this is flagging. Even so, there is no clear legal basis under the Directive and the Governance Regulation for compelling Member States to increase the proportion of renewable energy in national energy systems at a rate that will ensure achievement of the Union renewable energy target or for initiating infraction proceedings against them in the event that the 2030 Union target or a staging post towards this are not met. It remains to be seen whether thorough policing of Member State actions and enhanced scope for scrutiny from peers and civil society will be sufficient to avoid the poor performance experienced with the pre-2009 Directives.

Two features of the 2018 RES Directive seek to ensure that some progress is made by Member States beyond the levels of renewable energy development reached by the end of 2020, notwithstanding the lack of national targets. First, Member States commit not to allow the percentage of renewable energy in energy consumption to fall below their targets under the 2009 RES Directive and to take remediatory steps if such a decline occurs (Art. 3(4)). Second, the Commission takes on an obligation to 'support the high ambition of Member States through an enabling framework comprising the enhanced use of Union funds' (Art. 3(5)).

The Union target and Member States' contributions to this focus on the overall level of renewable energy in national consumption. Accordingly, there is no repeat of the separate legally binding target for renewable energy in transport under the 2009 RES Directive, perhaps in view of the many difficulties the Union has encountered in terms of meeting this (see 5.3.8 below). Member States are required, however, to provide separate details of estimated sectoral trajectories in their plans. With regard to transport, they must also set an obligation on fuel suppliers to ensure a minimum level of 14 per cent renewable energy in fuel supplies by 2030 (Art. 25(1)). In addition, the 2018 RES Directive makes specific provision, albeit through an obligation of conduct rather than result, for mainstreaming renewable energy in heating and cooling (Art. 23(1)). Member States must endeavour to increase the share of renewable energy in that sector by an indicative 1.3 per cent as an annual average for the periods 2021–25 and 2026–30. They must also lay down the necessary measures to ensure that district heating and cooling systems contribute to this increase (Art. 24).

5.4.2   Guarantees of Origin

Guarantees of Origin (GOs) are used to confirm that energy was produced from renewable sources. Their main role under the 2018 RES Directive (as it was under the 2001 Directive and the 2009 RES Directive) is to support the establishment of a 'green' energy market among environmentally conscious consumers by providing officially recognised backing for information given to them about the sources of their energy supplies. Electricity suppliers must provide information in bills and other documents on the contribution of each energy source to the fuel mix supplied over the preceding year.[46]

Article 19 of the 2018 RES Directive requires Member States to establish a formal procedure for generating documentary proof to back up the claims made by electricity suppliers about the sources from which their supplies are derived. Its sub-clauses give direction on the administrative arrangements that they should put in place to enable the issue and cancellation of GOs, and on the information that they should contain. Time limits and other criteria for the validity of GOs are also specified to prevent their misuse.

The Commission's first draft of the 2009 RES Directive proposed that GOs should also be used to demonstrate Member State compliance with targets for increasing renewable energy consumption and in connection with an EU-wide GO trading scheme. This suggestion was rejected and replaced, following Member State opposition, by mechanisms for interstate cooperation in which GOs serve no purpose. This remains the position under the 2018 RES Directive. Article 18(2) of the 2018 RES Directive, repeating the wording of Article 15(2) of the 2009 RES Directive, advises that 'the Guarantee of Origin shall have no function in term of a Member State's compliance with Article 3'.

It remains possible to trade GOs, but for the limited purpose of proving the inclusion of renewable energy within energy supplies to consumers.[47] In this regard, Member States must recognise GOs issued by each other except in the circumstances stated in Article 18(9). However, national laws that require actors to include a certain amount of renewable energy in energy produced, supplied or consumed by them tend to be associated with national schemes for the issue and trade of certificates that are distinct from GOs. In this regard, Recital 55 advises that it is important to distinguish between green certificates used for support schemes and GOs.

5.4.3 Reforming Administrative Procedures

Complexity and duplication in the administrative and regulatory regimes of the Member States has long been seen by the Commission as a constraining factor on the growth of renewable energy production.[48] Article 6 of the 2001 Directive sought to initiate a process that would lead to the eradication of administrative barriers by requiring Member States both to evaluate their existing legislative and regulatory frameworks for authorising renewable energy, and to publish reports by October 2003 stating actions that would be taken in light of the evaluation. It was clear however by the time of the Commission's 2006 review of progress on renewable

---

[46] Directive (EU) 2019/944 of 5 June 2019 on common rules for the internal market for electricity and amending Directive 2012/27/EU [2019] OJ L158/125, Article 18(6) and Annex I, paragraph 5.

[47] H.L. Raadal, E. Dotzauer, O.J. Hanssen and H.P. Kildal, 'The Interaction between Electricity Disclosure and Tradable Green Certificates' (2012) 42 *Energy Policy* 419.

[48] Commission, COM (2000) 279 final, 'Proposal for a Directive of the European Parliament and of the Council on the promotion of electricity from renewable energy sources in the internal electricity market', Explanatory Memorandum, Section 2.3, describes 'administrative and planning procedures' as a 'major barrier to the further development of RES electricity in the EU'.

energy that inappropriate or unnecessarily complicated administrative barriers were continuing to limit its expansion.[49] Article 13(1) of the 2009 RES Directive took a stronger line than its predecessor by placing detailed requirements on Member States to improve regulations governing renewable energy development. Its sub-clauses instruct Member States to modify rules for authorising renewable energy developments and related infrastructure with a view to making authorisation processes quicker, less complex and more transparent.

Article 15(1) of the 2018 RES Directive maintains Article 13(1)'s strong line on removing administrative barriers by requiring Member States to ensure that 'any national rules concerning the authorization, certification and licensing procedures' that are applied to plant for renewable energy production, for transforming biomass into energy products and for producing non-organic transport fuels are 'proportionate and necessary and contribute to the implementation of the energy efficiency first principle'. The remainder of Article 15 directs Member States to use regulations in ways that promote renewable energy consumption, particularly in the design and construction of new, and refurbishment of existing, building stock. For example, national buildings and codes should include measures that increase consumption of energy from renewable sources, including by requiring the use of minimum levels of renewable energy in new buildings and those subject to major renovation (Article 15(4)). In particular, they should promote renewable energy heating and cooling systems that achieve significant reductions in energy consumption (Article 15(6)). Article 18 also seeks to promote the diffusion of renewable energy technologies through its various requirements for Member States to ensure that information about renewable energy technologies and public support for their installation is widely available, and to establish arrangements that increase public confidence in their use, including certification schemes for installers.

The 2009 Directive encouraged Member States to streamline permitting regimes by establishing 'one-stop-shop' regimes, with one authority being responsible for awarding the permit.[50] The Commission complained in following progress reports that Member States were being slow to introduce this approach.[51] In view of this and other perceived inadequacies of Member State actions to streamline administrative processes, the 2018 RES Directive places additional obligations on Member States for reforming administrative processes. These are: a requirement to establish designated contact points that will, on an applicant's request, guide and facilitate the entire permit application (Art. 16); a time limitation of permit-granting processes to a maximum of two years for power plants and one year for small-scale electricity generation installations (capacities of less than 150kW) and repowering existing renewable energy plants, all unless 'extraordinary circumstances' justify a longer time-scale (Art. 16(4)–(6)); and simpler grid connection notification processes for low-capacity generating facilities (such as rooftop solar PV panels) (Art. 17).

---

[49]   Commission, 'Green Paper follow-up action: Report on progress in renewable electricity', COM (2006) 849 final, 17–19.

[50]   RES Directive 2009 (n 3), Article 22(3)(a).

[51]   Commission, 'Renewable Energy Progress Report', COM (2017) 57 final, 10.

5.4.4   Grid Access[52]

The 2001 Directive and the 2009 RES Directive placed obligations on Member States to address barriers to the expansion of renewable energy related to the operation and development of transmission and distribution systems for electricity in their territories.[53] The 2009 RES Directive also requires them to take steps supporting the integration of renewable gases into pipeline networks and district heating and cooling using renewable sources.[54]

Provisions on grid access in the 2018 RES Directive address the integration of renewable gases into existing infrastructure alone (Art. 20). There are four main reasons for its lack of provision on integrating renewable electricity in networks. First, the Electricity Regulation component of the Clean Energy for All Europeans legislative programme requires Member States to reform existing market structures which make it difficult for electricity systems to accommodate renewable electricity and for renewable generators to participate in system balancing.[55] For example, it mandates the introduction of markets operating in as close to real time as possible across the European Union (Art. 7(2)(c)). Second, the obligation under the 2009 RES Directive for Member States to afford priority or guaranteed access for renewable electricity to networks is, with regard to the former, incompatible with the move away from subsidies which guarantee legally that electricity produced by supported generating plants will be purchased at a set price (see 5.4.2 above) and, with regard to the latter, made unnecessary by duties for transmission and distribution system operators and market operators under the 2019 Electricity Regulation to enable the integration of renewables into electricity networks and trading platforms.[56] Third, provisions under the Electricity Regulation maintain obligations for Member States to prevent dispatch and curtailment by system operators in ways that disadvantage renewable generators unnecessarily (Arts 12 and 16). Fourth, the 2018 RES Directive obliges Member States to ensure that persons who produce and consume their own electricity and communities of those persons are able to do so and to sell their excess production by feeding it into the grid (Arts 21 and 22). These provisions form part of a wider package of measures under the legislative programme, with provisions under the Electricity Regulation and Directive also requiring adaptation of national laws to enable self-consumers and renewable energy communities to take part in energy systems including by selling excess electricity in markets directly or through aggregators.[57] Endorsement of possibilities for the democratisation of energy supplies afforded by renewable energy technologies responds to the problems that this presents for energy systems and related legal frameworks designed around centralised production facilities. It does this by recognising the potential for participation in

---

[52]   Subject matter covered in this section is also examined in Chapter 10 of this book, 'EU Climate Regulation and Energy Network Management'.

[53]   The 2001 Directive (n 2), Article 7; 2009 RES Directive (n 3), Article 16.

[54]   2009 RES Directive (n 3), Articles 16(9) to (11).

[55]   Regulation 2019/943 of 5 June 2019 on the internal market for electricity [2019] L158/54

[56]   Ibid.

[57]   Directive (EU) 2019/944 (n 46).

energy production and system management to increase public acceptance of the shift away from high-carbon energy supplies in general, and of renewable energy developments such as onshore wind farms, so often a source of public opposition, in particular.

5.4.5   Sustainability Criteria

The European Commission sees biofuels as the main contributor to the decarbonisation of energy consumed for transportation until other alternatives to petrol such as gas, electricity and hydrogen become more widely available.[58] However, it also recognises that a major increase in the consumption of biofuels, particularly where these are sourced from feedstocks that can also be used as food crops, could have unsustainable outcomes.[59] Particular fears are that this could lead to the degradation of environmentally valuable lands, increases in food prices as land used previously to grow food crops is turned over to the production of transportation crops, and growth of land in cultivation to enable biomass for transport fuels to be produced alongside food. Some biofuels could even be responsible for higher carbon emissions than fossil fuels when the full effects of their production, including direct and indirect changes in land use, are taken into account.

In view of these concerns, the 2009 RES Directive laid down certain 'sustainability' criteria that apply both to EU-produced and imported biofuels.[60] The Directive defines biofuels as 'liquid or gaseous fuel for transport produced from biomass'.[61] Biofuels that failed to meet the criteria could still be imported into and sold in the EU, but Article 17(1) advised that only those which satisfied them would be counted toward the 10 per cent target for renewables in energy consumed for transportation and be eligible for financial support. The criteria also applied to bioliquids, defined in Article 2(h) of the Directive as liquid fuels produced from biomass but used for other purposes than to provide energy for transportation (in other words, electricity, heating and cooling). The Commission felt that this was necessary to prevent biofuels that did not meet the sustainability criteria from receiving subsidies and contributing to Member State targets through the back door by being used as bioliquids instead.[62]

The 2018 RES Directive adopts the regime created by the 2009 RES Directive, which seeks to discourage the production of 'unsustainable' biofuels and bioliquids, and expands it to include 'biomass fuels', a defined term meaning 'solid and gaseous fuels produced from biomass' (Arts 2(27) and 29)). As a result, the sustainability of all biomass-based energy will be considered when assessing whether renewable energy consumption should count towards targets and be eligible to receive subsidies.

---

[58]   Commission, 'An EU Strategy for Biofuels' COM (2006) 34 final.

[59]   Ibid., 10.

[60]   L. Eremeichvili, 'Greening the Electricity Sector – Developing Markets for Trading Biomass' in M. Roggenkamp and H. Bjørnebye, *European Energy Law Report X* (Cambridge: Intersentia, 2014), pp.211–57 offers a full account of EU law and policy on sustainability concerns over biomass-based energy.

[61]   2009 RES Directive (n 3), Article 2(i).

[62]   2009 RES Directive (n 3), Recital 67.

Article 29 of the 2018 RES Directive specifies standards that biofuels, bioliquids and biomass fuels should achieve to be able to count towards targets and be eligible for national financial support. First, their consumption should achieve or exceed minimum levels of greenhouse gas emissions savings over fossil fuels. The minimum saving for biofuels, biogas when consumed in the transportation sector and bioliquids starts at 50 per cent where they are produced in installations starting operation on or before 5 October 2015, rising to 60 per cent where they are produced in installations starting operation between 6 October 2015 and 31 December 2020 and then to 65 per cent for installations starting operation on or after 1 January 2021.[63] The Directive also adds to restrictions under its predecessor by laying down minimum savings for electricity, heating and cooling produced from biomass fuels starting operation from 1 January 2021.[64] The electricity savings requirement applies only to installations meeting criteria set out at Article 29(11).

A mechanism for calculating emissions from biofuels, bioliquids and biomass fuels is set out in Article 30. A life-cycle approach is taken in which emissions from cultivation (including from direct land conversion and fertiliser use), from transportation of biomass and from the production, transportation and distribution of biomass, bioliquids and biomass fuels are taken into account.

Second, raw materials for biofuels, bioliquids and biomass fuels made from agricultural biomass must not be obtained from land possessing one or more of the statuses listed in Article 29(3)–(5) as at 1 January 2008 if they are to count towards the Directive's targets or be entitled to support. The listed statuses are collectively described as belonging to: land with high biodiversity value, including primary forests, legally designated areas for nature protection and highly biodiverse grassland (Art. 29(3)); land with high carbon stock, including wetlands, continuously forested areas and other forested areas meeting certain criteria (Art. 29(4)); and peat land (Art. 29(5)). Separate criteria apply for biofuels, bioliquids and biomass fuels produced from forest biomass (Arts 29(6) and 29(7)). Separate criteria and also relaxations of criteria apply for biofuels, bioliquids and biomass fuels and electricity, heating and cooling produced from waste and residues (Arts 29(1) and (2)).

Member States must require that 'economic operators […] show that the sustainability criteria […] have been fulfilled', using a methodology set out in Article 30(1). Article 30(4) allows them to prove compliance by participating in voluntary national and international schemes whose validity has been recognised by the EU. The Commission has recognised a number of voluntary schemes for biofuels compliance to date.[65]

Challenges with identifying emissions and environmental harm caused by indirect land use change due to already cultivated land being used for biomass energy crops bedevilled the European Union's attempts to dissuade the production of unsustainable biofuels and

---

[63]   2018 RES Directive (n 1), Article 29(10).

[64]   Ibid., Article 29(10)(d).

[65]   Commission, 'Voluntary Schemes', available at https://ec.europa.eu/energy/topics/renewable-energy/biofuels/voluntary-schemes_en?redir=1 accessed 17 June 2021.

bioliquids by applying sustainability criteria under the 2009 RES Directive.[66] Amendments to the 2009 RES Directive adopted in October 2015 responded to these challenges by restricting the use of biofuels derived from sources to which were attached high risks of unsustainable cultivation and low levels of emission reduction.[67] An amendment to Article 3(4) limited the contribution that biofuels from 'cereal and other starch-rich crops, sugars and oil crops' could make to meeting the 10 per cent target to 7 per cent of overall energy consumption for transportation in 2020.

The 2018 RES Directive goes further in discouraging unsustainable biomass energy production due to indirect land use change. The cap on the extent to which biofuels, bioliquids and biomass fuels produced from food and feed crops can count towards targets is set at 1 per cent higher than the share of such fuels in energy consumption for road and rail transport in 2020 in the Member State concerned, with a maximum of 7 per cent (Art. 26(1)). The contribution of biofuels, bioliquids and biomass fuels posing a high risk of indirect land use change for which a significant expansion of the production area into land with high carbon stocks is observed is also limited to the level of their consumption in 2019, declining to 0 per cent by 2030 (Art. 26(2)).[68]

The amendments to the 2009 RES Directive also placed further emphasis on promoting biofuels derived from waste, which are viewed as more sustainable because they do not create an additional demand for land. This was done by doubling their energy content when calculating their contribution to the 2020 target for 10 per cent of energy for transport from renewable sources (but not the 20 per cent target for renewable energy as a whole).[69] Member States were also required to set and endeavour to achieve a national target with a reference value of 0.5 per cent of energy content of the share of renewable energy in all forms of transport in 2020 for biofuels from the non-food, waste and algae feedstocks listed in Annex IX of the 2009 RES Directive (as amended) (Art. 3(4)(e)). The 2018 RES Directive maintains its predecessor's promotion of advanced biofuels by requiring that their share of final consumption of energy in the transportation sector should be at least 0.2 per cent in 2021, 1 per cent in 2025 and 3.5 per cent in 2030 (Article 25(1)).

5.4.6   Cooperation Mechanisms

One of the core objectives of the Commission in drafting what became the 2009 RES Directive was to promote cost-effectiveness in the growth of renewable energy production.[70] The

---

[66]   Commission, 'Report on indirect land-use change related to biofuels and bioliquids' COM (2010) 811 final, 14.

[67]   Directive (EU) 2015/1513 (n 21).

[68]   The high-risk feedstocks affected by this provision are defined by Commission Delegated Regulation (EU) 2019/807 of 13 March 2019 supplementing Directive (EU) 2018/2001 of the European Parliament and of the Council as regards the determination of high indirect land-use change-risk feedstock for which a significant expansion of the production area into land with high carbon stock is observed and the certification of low indirect land-use change-risk biofuels, bioliquids and biomass fuels [2019] OJ L 133/1.

[69]   RES Directive 2009 (n 3), Annex IX.

[70]   Commission, COM (2008) 30 final (n 24) 5.

Commission's view was that the overall cost of meeting the 2020 target for renewable energy could be reduced if Member States were to cease concentrating on national renewable energy resources and to look instead at possibilities for producing renewable energy in other Member States where this could be done at lower cost.[71] Articles 6–11 of the 2009 RES Directive sought to engender a 'European' mindset towards exploiting renewable energy resources by creating mechanisms that would enable Member State cooperation on meeting their national targets. The 2018 RES Directive also permits Member States to use the four mechanisms established by the 2009 RES Directive, but with some minor modifications to their design. The mechanisms are described briefly in the following sections.

Statistical transfers

A Member State may agree to transfer to another Member State a specified amount of the renewable energy that counts towards its target under the 2018 RES Directive.[72] This is described as a 'statistical' rather than a physical transfer of energy produced in one State to another State. Statistical transfers agreed independently by Member States become effective after all of the States involved have notified the transfer to the Commission.[73] The Commission is also given a duty to 'facilitate statistical transfers' by establishing a 'Union renewable development platform'.[74] The platform's purpose is to provide a marketplace for statistical transfers of national renewable energy consumption. Member States may, on a voluntary basis, submit annual data on their national contributions to the Union target or benchmarks set for monitoring progress together with statements on the amount by which they expect to exceed or fall short of contributions and the price at which they would accept a transfer of excess renewable energy production from or to another Member State, together with other conditions for transfer. The platform will establish a mechanism for matching requests and offers for transfer. The Commission is empowered to establish the platform and to set conditions for finalising transfers through it.[75] The platform had not yet been established at the time of writing.

Joint projects

This mechanism, provision for which is made by Articles 9 and 10 of the 2018 RES Directive, involves a statistical transfer of energy produced by a joint project 'relating to the production of electricity, heating or cooling from renewable energy sources' from the compliance account of one Member State to that of another. The project must have become operational after 25 June 2009 or have increased the capacity of an already existing facility after that date through refurbishment. Article 9(1) advises that the cooperation under this mechanism 'may involve private operators'. This means in practice that a private entity may seek support from a Member State for development in another State that is unable to provide sufficient backing

---

71 Commission, COM (2011) 31 final (n 22) 10–11.

72 2018 RES Directive (n 1), Article 8.

73 Ibid., Article 8(5).

74 Ibid., Article 8(2).

75 Ibid., Article 8(3).

for a proposed project to be pursued.[76] Control over the use of the mechanism in such circumstances remains exclusively with the States concerned, who must agree on the terms of support and the basis on which the renewable energy is to be shared between them.

Joint projects with third parties

The EU's strategy for decarbonising energy supplies looks not only to renewable energy production by Member States, but also to collaboration with third countries on generating electricity from their renewable sources. The Directive reflects this by including a mechanism under Articles 11 and 12 for cooperation between Member States and third countries. This allows electricity production in the latter to be counted towards Member States' targets. It is more prescriptive as to the circumstances in which it may be used than is the case for other mechanisms. Electricity can only be counted from installations that entered into operation or whose capacity was upgraded after 25 June 2009 and that were developed or upgraded as joint projects between the Member States concerned and the third country.[77] This is necessary to prevent existing renewable energy capacity being diverted from host States who may then make up the deficit through fossil fuel consumption.[78] The Directive also suggests in places that the new development should produce electricity for consumption domestically as well as in the EU, although this is not an express requirement for their output to count towards Member State targets.[79]

Member States may apply for electricity produced in third countries but not consumed in the EU to be counted towards their targets in the limited circumstances set out in Article 11(3). These describe a scenario in which renewable energy production from a joint project has commenced, but the energy cannot be transported to the EU because it is not possible for the interconnector that will be used to transport it to commence operation by the end of 2030. The exception is further limited by requirements that construction of the interconnector has begun by the end of 2026 and that it must be possible at least for it to become operational before the end of 2032.

Joint support schemes

The final mechanism, provision for which is made by Article 13 of the 2018 RES Directive, allows Member States that join or partly coordinate their national support schemes to determine how energy supported under the joint scheme is shared between them for the purposes of calculating their contributions to the European Union's 2030 goal. The States concerned may make statistical transfers or may set up a distributional rule for allocating supported energy to their compliance accounts. Distribution rules should be notified to the Commission within three months after the end of the month in which they take effect. Annual notifications of

---

[76]   Commission, 'Review of European and national financing of renewable energy in accordance with Article 23(7) of Directive 2009/28/EC', SEC (2011) 131 final, 8.

[77]   2018 RES Directive (n 1), Article 11(2)(b).

[78]   Ibid., Recital 41.

[79]   Ibid., Article 11(5)(d) and Recital 42.

energy produced during the preceding year that is allocated to them under the rule should also be provided by each of the participating States.

This mechanism comes closest to the Commission's original vision of a European-wide trading scheme as it allows the States concerned to focus on improving the cost-efficiency of renewable energy production within the extended territory that the joint scheme covers. It is also likely to be the most difficult mechanism to apply because of the many legal complexities associated with harmonising national approaches sufficiently to provide a common support scheme.[80]

The use and future of cooperation mechanisms

At the time of writing, limited use had been made of the mechanisms for achieving the target under the 2009 RES Directive. Statistical transfers of national renewable energy production from Lithuania and Estonia to Luxembourg were agreed in 2017.[81] Two additional statistical transfers had been agreed by October 2020, between the Netherlands and Denmark and Malta and Estonia.[82] The Commission expected further transfers to be agreed in view of likely failures by some States to achieve their 2020 targets for renewable energy and for renewables in energy for transportation.[83] Sweden and Norway adopted a joint green certificate scheme for promoting renewable energy in their territories.[84] Norway is not a Member State of the EU but applies the 2009 RES Directive along with Iceland, due to its incorporation in December 2011 into the corpus of EU law that States party to the European Economic Area Agreement agree to apply in their own territories.[85] In addition, Germany and Denmark agreed to a partial opening of their national support schemes to solar PV projects constructed in each other's territories.[86]

---

[80]   C. Klessmann, P. Lamers, M. Ragwitz and G. Resch, 'Design Options for Cooperation Mechanisms under the New European Renewable Energy Directive' (2010) 38 *Energy Policy* 4687–90.

[81]   N. Caldes, P. del Rio, Y. Lechon and A. Gerbeti, 'Renewable Energy Cooperation in Europe: What Next? Drivers and Barriers to the Use of Cooperation Mechanisms' (2019) 12 *Energies* 1–22.

[82]   Commission, COM (2020) 952 final (n 26), 6–7.

[83]   Ibid., 6–12.

[84]   Agreement on a Common Market for Electricity Certificates, Stockholm, 29 June 2011. An unofficial translation of the Agreement is available at www.regjeringen.no/globalassets/upload/oed/pdf_filer_2/elsertifikater/ agreement_on_a_common_market_for_electricity_certificates.pdf accessed 17 June 2021. See O. Boge, 'The Norwegian-Swedish Electricity Certificates Market' in M.M. Roggenkamp and H. Bjørnebye (eds) *European Energy and Law Report X* (Cambridge: Intersentia, 2014), pp.199–210.

[85]   European Free Trade Association, 'Directive on the promotion of renewable energy incorporated', 20 December 2011, www.efta.int/EEA/news/Directive-promotion-renewable-energy-incorporated-1086 accessed 17 June 2021. Liechtenstein was exempt from applying the Directive.

[86]   Governments of the Federal Republic of Germany and of the Kingdom of Denmark, Agreement on the Establishment of a Framework for the Partial Opening of National Support Schemes to support the generation of energy from solar photovoltaic projects and for the cross-border administration of such projects in the context of a single pilot run in 2016, 20 July 2016, www.bmwi.de/Redaktion/EN/Downloads/agreement-between-germany-and -denmark.pdf?__blob=publicationFile&v=4 accessed 17 June 2021. D. Dmitruk, 'Danish-German Cooperation on

The Commission speculated, after only one use had been made of the mechanisms by 2013, that Member State reluctance to enter into agreements was due to uncertainty over how to use them.[87] To address this, in November 2013 it produced guidelines for employing mechanisms, including methodologies for price setting, legal and institutional framework conditions and model agreements.[88] However, it also recognised that Member States prefer to concentrate on exploiting resources nationally because of the associated benefits for national economies and employment.[89]

The continued low use of mechanisms since 2013 seems to indicate that such considerations have continued to dominate Member States' thinking on how to go about meeting targets for renewable energy consumption. Despite this, the mechanisms have been given an on-going role in promoting interstate collaboration on renewable energy development under the 2018 RES Directive. The Commission sees particular scope for using them in connection with anticipated significant growth in the offshore wind energy capacity of Member States during the 2020s.[90]

5.5   FINANCIAL SUPPORT FOR RENEWABLE ENERGY

5.5.1   Why Are Subsidies Necessary?

The enormous investment in technologies for producing and consuming renewable energy that is required if EU targets are to be met is not likely to be stimulated by market prices alone. One reason for this is that prices in markets which remain dominated by long-established fossil fuel incumbents may not be high enough to create confidence that monies invested in facilities for producing electricity and fuels for heating and transport, as well as technologies required for their consumption, will be recovered together with an attractive profit margin through sales of these commodities and products.

Declining costs for electricity produced by better established renewable technologies such as onshore wind and solar PV often, although not always, enable projects using them to compete with fossil fuel generation at market prices.[91] However, newer forms of renewable

---

the first Cross-Border Tenders for Renewable Energy: A Blueprint for Future Cross-Border RES Projects?' in M.M. Roggenkamp and C. Banet (eds), *European Energy Law Report XII* (Cambridge: Intersentia, 2018, 1st ed.), pp.113–32.

[87]   Commission, 'Staff Working Document accompanying Renewable Energy: A major player in the European Energy Market', SWD (2012) 164 final, 16–17.

[88]   Commission, Guidance on the use of renewable energy cooperation mechanisms accompanying the document Communication from the Commission: Delivering the internal energy market and making the most of public intervention', SWD (2013) 440 final.

[89]   Commission, SWD (2012) 164 final (n 87), 16–17.

[90]   Commission, 'An EU Strategy to harness the potential of offshore renewable energy for a climate neutral future', COM (2020) 741 final, 16.

[91]   International Renewable Energy Agency, 'Renewable Power Generation Costs in 2017' (January 2018, report) www.irena.org/publications/2018/Jan/Renewable-power-generation-costs-in-2017 accessed 17 June 2021.

electricity production may struggle to compete with the prices attainable by established power-generating companies employing technologies which have become highly efficient through long experience with their use, and which benefit from economies of scale through centralisation. Investor concern that energy sales may not be sufficient to enable cost and profit recovery may also be heightened for all renewable electricity projects, however established the technology may be, by factors such as: the significant proportion of overall project capital costs incurred at the development stage, meaning that substantial confidence in their recovery is needed before development proceeds; the comparatively small scale of many renewable energy developments, again meaning that high confidence is needed in cost recovery up-front; and the difficulties and costs associated with connecting to networks and integrating into energy systems that were designed to transmit and distribute energy from centralised fossil fuel production facilities.

Means of producing renewable energy other than electricity for use in industry, heating, cooling and transportation are also disadvantaged by higher production costs and lower availability compared to fossil fuels such as petroleum and natural gas, which have benefited from several decades of experience with technologies for their production and consumption.[92]

5.5.2   Subsidies in EU Law

Member States of the EU have attempted to create investor confidence in renewable energy since they first began promoting its production in the 1980s, either by providing monies themselves from public funds or by placing legal obligations on private actors to support the growth of a renewables sector. The 2009 and 2018 RES Directives recognise that Member State provision of financial backing for renewable energy is required if targets for growth in its consumption are to be met. Both of them advise that Member States may apply 'support schemes' in order to reach their targets, defining 'Support Scheme' in Articles 2(k) and 2(5) respectively as

> any instrument, scheme or mechanism applied by a Member State or a group of Member States, that promotes the use of energy from renewable sources by reducing the cost of that energy, increasing the price at which it can be sold, or increasing, by means of a renewable energy obligation or otherwise, the volume of such energy purchased.

The definitions then go on to provide non-exclusive lists of ways in which support for renewable energy can be provided:

> The European Commission accepts that the provision of support is made necessary by the EU's renewable energy policy goals, but regards national subsidies as an undesirable depar-

---

[92]   International Renewable Energy Agency, 'Roadmap for a Renewable Future 2016 Edition', pp.106–20 (March 2016, report) www.irena.org/publications/2016/Mar/REmap-Roadmap-for-A-Renewable-Energy-Future-2016 -Edition accessed 17 June 2021.

ture from a preferred status quo of European-wide internal energy markets free of measures that distort competition including by advancing national preferences.[93]

What follows considers the main approaches that Member States have employed to support renewable energy development. These are grouped under measures used to enhance access to funding for development ('investment support') and those used to enable the recovery of development costs once the funded energy production plant is operational ('operating support'). The following text examines how State aid guidelines and provision under the 2018 RES Directive have been used to promote the integration of renewable energy into markets. It also considers legal questions to which Member States' use of support schemes has given rise concerning their compatibility with provisions of the Treaty on the Functioning of the European Union (TFEU) on the free movement of goods and state aid.[94] The section on operating support focuses on electricity as relevant schemes are used primarily to support the growth of renewable electricity production.

5.5.3   Investment Support

Financial backing for research and development for early stage renewable energy technologies and for the trialling of pre-commercial prototypes is often hard to obtain because of the high risk that monies invested will not be recovered. Alternatively, investors may only be prepared to provide investment at rates of return that would make it difficult to recover development costs through energy sales without substantial operating support (see section 5.5.4). Member States of the European Union use measures collectively referred to as 'investment support' to make it easier for innovators to develop new renewable energy technologies and to encourage consumers to use them through the provision of public financial support.[95]

Grant schemes provide funding to developers for the development and testing of eligible renewable energy technologies. The provision of long-term public loans enables them to access investment at much lower rates of return than would be available to them through private finance. Alternatively, guarantees of repayment from public bodies in the event that a borrower defaults may enable developers to access private finance more cheaply. Long-term grants are also used to encourage the purchase of renewable technologies by domestic and business consumers. For example, most Member States offer investment grants to promote the take-up of renewable energy heating systems such as biomass boilers.[96]

In addition to making public funds available, governments provide tax exemptions and reductions to support investment by reducing the financial burden on developers. More

---

[93]   Commission, European Commission guidance for the design of renewable support schemes – Accompanying the document 'Communication from the Commission: Delivering the internal market in electricity and making the most of public intervention', SWD (2013) 439 final.

[94]   Treaty on the Functioning of the European Union [2008] OJ C115/47.

[95]   Commission, SEC (2011) 131 final (n 76), 4–6; Commission, SWD (2013) 439 final (n 93), 11–12.

[96]   Commission, SEC (2011) 131 final (n 76), 6, 9–10; current information on Member State support schemes can be found at www.res-legal.eu/ accessed 17 June 2021.

importantly, reduced tax rates can be used to encourage the consumption of renewable energy rather than fossil fuels. The use of biofuels is commonly promoted by applying lower rates of fuel tax to them as compared to petrol, or by allowing tax offsets for their consumption.[97]

Investment support is generally viewed as playing a supporting role to operating support. Its most important function to date has been to enable technologies to reach the point where they are perceived as a lower risk by investors because they are capable of commercial-scale operation. However, in its most recent guidance on financial support for renewable energy, the Commission encourages the wider use of investment support because it does not distort the operation of energy markets.[98]

5.5.4   Operating Support

Feed-in tariffs

Feed-in tariffs (FITs) provide renewable energy operators with a specified price for each unit of electricity fed in to the electricity grid to which they are attached over a specified duration (typically 1–20 years). FIT schemes may be funded by the State concerned. Alternatively, some schemes oblige operators of electricity networks or suppliers to pay the tariff, with the costs of this being recovered through consumers' electricity bills. Tariff rates may vary according to the level of financial support that a technology needs to become established or according to policy goals.

FITs have proved to be a successful means of increasing renewable electricity consumption.[99] The stability provided by receipt of a guaranteed return for a guaranteed period attracts lower-cost investment due to the lower level of market risk. However, the Commission does not favour the use of FITs because it considers that the provision of a definite sales price distorts markets for electricity in the EU. It is also concerned that a lack of market exposure could be to the detriment of the renewable electricity sector by removing an incentive for generators to improve the efficiency of energy production through technological and operational innovation.[100]

The Commission's guidance document of November 2013 on the design of support schemes recommends that FITs should be used only to support small-scale renewable energy development, and that their use should otherwise be phased out. Its guidance on State aid in the field of energy and environment (see section 5.5.5) entrenches this position by making it clear that new FIT schemes should only be introduced for developments of below 3MW or 3 units for wind energy or 500kW for other technologies (for example, domestic-/community-scale

---

[97]   Commission, SEC (2011) 131 final, (n 76), 6.

[98]   Commission, SWD (2013) 439 final (n 93), 11–12.

[99]   Commission, SEC (2011) 131 final (n 76), 6; V. Lauber, 'The European Experience with Renewable Energy Support Schemes and Their Adoption: Potential Lessons for Other Countries' (2011) 2 *Renewable Energy Law and Policy* 120.

[100]   Commission, SWD (2013) 439 final (n 93), 11–12.

developments).[101] FIT schemes that were approved under preceding State aid regimes may continue to accept new entrants, and recipients of tariffs under them will not be affected. However, existing schemes should be 'brought into line' (for example, replaced with less market-distorting alternatives) with the new State aid regime at points where their approval under the preceding regime expires or if they are adapted. The 2018 RES Directive confirms the guidance by only allowing Member States to use feed-in tariffs for the small-scale installations identified in the State aid guidelines (Art. 4(3) and recital 17).

Obligation/certificate schemes

Some Member States have enacted laws that oblige electricity sector actors (usually suppliers, but sometimes also producers and consumers) to include a specified proportion of renewable electricity in their overall production, supply or consumption of energy.[102] Compliance with the obligation is demonstrated by the provision of certificates. These are issued to renewable electricity producers who may sell them to obligated actors either together with or separately from the related electricity. The idea behind such schemes is that the receipt of two separate revenue streams should enable developers of renewable energy installations to recover monies invested in them. This type of scheme is also widely used in connection with energy for transportation, with suppliers of fuel being obliged to include a proportion of energy from renewable sources in their supplies. Meeting these obligations is largely achieved by the blending of petrol and diesel with biofuels as permitted under the Fuel Quality Directive.[103] The 2018 RES Directive legally entrenches this approach at the European level by requiring Member States to set an obligation on fuel suppliers to ensure that the share of renewable energy within the final consumption of energy in the transport sector is at least 14 per cent by 2030 (minimum share) with the growth of the share following an indicative trajectory to be set by the Member State and to be calculated as provided for under the Directive (Art. 25(1)).

Obligation schemes can be technology neutral or can give differing levels of support for different technologies (a practice known as banding). This is generally done by providing that well-established technologies such as onshore wind will receive fewer certificates for each unit of electricity produced than will newer technologies such as wave and tidal energy, which require stronger initial support to become established.

The Commission has long preferred obligation schemes over FIT schemes because it believes that the market exposure under them will incentivise producers to reduce costs resulting in lower energy prices. However, in its 2013 guidance document on the design of support schemes it recognises that the cost of capital tends to be higher for projects funded by obligation schemes because of the risk that uncertain revenue streams from electricity and cer-

---

[101]   Commission, 'Guidelines on state aid for environmental protection and energy 2014–2020', 2014/C 200/01, 25 para. 125.

[102]   Commission, SWD (2013) 439 final (n 93), 10–11, 25.

[103]   Directive 98/70/EC of 13 October 1998 relating to the quality of petrol and diesel fuels and amending Council Directive 93/12/EEC [1998] OJ L350/58.

tificate sales will not cover development costs.[104] As a result, this may limit renewable energy development to large-scale incumbents that can afford to finance projects themselves or that are able to obtain investment on more favourable terms than new market entrants. Cheaper, well-established technologies are also preferred over newer technologies whose electricity is more costly under such schemes because of the greater chance of securing a profit through the combined revenue from electricity and certificate sales. In view of this, the Commission endorses the use of banding mentioned above, and that of measures such as a floor price for tradable certificates to enhance confidence that cost recovery will be possible for technologies at whatever stage of development they may have reached.

Premium schemes

Premium schemes are increasingly seen by Member States and the Commission alike as a potential alternative to FIT and obligation schemes for supporting renewable energy technologies that are able to be developed at commercial scales.[105] Electricity from generating plant supported under such schemes is sold into markets, but the risks associated with exposure to price volatility are tempered by the payment of a premium (typically by the operators of transmission systems or suppliers under a legal obligation) for each unit of sold electricity. The payment may be fixed at a specified level or may be a 'floating' amount that falls as electricity and carbon prices increase.

The Commission prefers the use of premium schemes over other options for renewable energy technologies that are capable of commercial deployment because it considers that they strike an appropriate balance between the market exposure that may drive renewable energy producers to improve efficiency and the higher level of risk under obligation schemes that may discourage investment in newer technologies.[106] Its guidelines on State aid in the field of energy and the environment advise that aid should be granted as a premium in addition to the market price from 1 January 2016, save for low-capacity/small-scale developments that may continue to receive FITs.[107] The 2018 RES Directive confirms the guidance by requiring Member States to support electricity from renewable sources other than small-scale and demonstration installations in the form of a market premium (Art. 4(3)). In view of this, premium schemes are likely to become the dominant means of providing operating support for renewable energy during the next decade, although they will continue to work alongside obligation schemes, which the Commission's State aid guidelines and the 2018 RES Directive also endorse.

Competitive allocation

One of the Commission's aims in calling for a shift to subsidy schemes which give market exposure to renewable generators is to promote technological development and increasing efficiency in renewable electricity production, with a view to reducing the cost of a low-carbon energy transition. It further seeks to advance this aim in the 2013 State aid guidelines by

[104] Ibid., 8–9.

[105] Ibid., 8–9.

[106] Commission, 2014/C 200/01 (n 101), 25 para. 126.

[107] Ibid., para 76.

requiring that from 1 January 2017 operating support for renewable electricity generation should be granted 'in a competitive bidding process on the basis of clear, transparent and non-discriminatory criteria' as a default position.[108] States are exempted from this requirement on grounds including the small scale of the supported development programme or that it would lead to higher support levels or to low project realisation. Bidding processes open to all renewable electricity generators could disadvantage new technologies that have not yet been able to follow the same cost reduction and increasing efficiency pathways that established technologies such as onshore wind and solar PV have taken. In view of this, the guidelines also allow tendering to be limited to specific technologies where to do otherwise 'would lead to a suboptimal result', and with a view to advancing other desirable goals for an energy transition such as the longer-term potential of new technologies and diversification in energy supplies.[109] In addition, aid for wind farms with an installed capacity of 6MW or six generating units and for all other installations of less than 1MW and demonstration projects may be granted without a competitive bidding process.[110] The 2018 RES Directive entrenches the legal status of competitive allocation by advising that 'Member States shall ensure that support for electricity from renewable sources is granted in an open, transparent, competitive, non-discriminatory and cost-effective manner' (Art. 4(4)). This is subject to opt-outs for small-scale and demonstration projects and to permission for States to limit tenders to certain technologies on the same grounds as those set out in the guidelines (Art. 4(5)).

5.5.5   Support Schemes and the TFEU

Permission under the RES Directives for Member States to use support schemes does not mean that they have free rein to provide financial support for renewable energy as they see fit. Measures employed by Member States must be compatible with provisions of EU Treaties that enshrine principles of the internal market and that seek to prevent anti-competitive behaviour by States. Other sources present much fuller analyses of the complex interaction between these provisions and support schemes and of the case law of the European Court of Justice (ECJ) that interprets the relevant laws than can be provided in this chapter.[111] The following section concentrates on the two main respects in which EU Treaty law has influenced and continues to shape the design of support schemes by Member States.

Free movement of goods

Article 34 TFEU prohibits measures by Member States that have an equivalent effect to a quantitative restriction on imports. Schemes under which access to financial support is restricted to indigenous energy producers may contravene this provision because they limit scope for electricity to be imported. However, they may still be found to be lawful, either because they

---

[108]   Ibid.

[109]   Ibid.

[110]   Ibid., para. 77.

[111]   K. Talus, *Introduction to EU Energy Law* (Oxford: Oxford University Press, 2016); T.M. Rusche, *EU Renewable Energy Law and Policy* (Cambridge: Cambridge University Press, 2015).

fall under a derogation to Article 34 or because the ECJ finds some other basis for justifying a departure from the Article.

The *PreussenElektra* judgment of the ECJ is the leading authority on the effect of Article 34 on support schemes.[112] This case concerned a requirement under German law that operators of grids should purchase electricity generated by plants attached to them at a fixed tariff. The ECJ found that this constituted a clear breach of Article 28 of the EC Treaty (the corresponding provision to Article 34 TFEU in its predecessor) as the preference for producers attached to the national grid restricted market access for imported electricity. The legal constraint was not permitted under the EC Treaty itself as derogations from Article 28 did not include measures taken for environmental protection. Surprisingly, this remains the case under TFEU, although environmental protection has a much higher profile in the EU now than it had in 2001.[113] However, the Court found that the measure could be justified because of the important public interest it served of securing the decarbonisation of energy supplies. Its reasoning was consistent with other cases in which it legitimised measures whose principal purpose is to address environmental problems that would otherwise have fallen foul of laws preventing constraints on trade.[114]

The ECJ was willing to endorse the restrictive law in *PreussenElektra* because of the perceived difficulty, as at 2001, with distinguishing between electricity generated from renewable and from non-renewable sources. This meant that the renewable status of imported electricity could not be verified. Its decision of July 2014 in the *Ålands Vindkraft* case considers whether the subsequent introduction of arrangements for issuing documents, which confirm that electricity was produced from a renewable source (guarantees of origin[115]), removes this justification for measures that would otherwise contravene Article 34.[116] The ECJ concluded that the position has not changed since its *PreussenElektra* judgment because it remains difficult to determine the origin of electricity at the point of consumption. The production of guarantees of origin does not prove that imported electricity has contributed physically to national support schemes' typical objective of promoting the growth of renewable energy production within the State concerned.

---

[112]   Case C-379/98 *PreussenElektra v Schleswag* [2001] ECR I-2099. See A.C. Johnston and G. Block, *EU Energy Law* (Oxford: Oxford University Press, 2012), pp. 342–53 for a detailed account of the case.

[113]   Derogations are listed under Article 36 TFEU. These include measures that can be justified on grounds of 'the protection of health and life of humans, animals or plants', but this falls short of a general derogation for measures taken to protect the environment.

[114]   Johnston and Block (n 112), pp.343–50.

[115]   See Section 5.4.2 of this chapter.

[116]   [2014] EUECJ C-573/12, *Ålands Vindkraft AB v Energimyndigheten*, ECLI: EU: C: 2014: 2037.

In addition, its decision emphasises the exclusive authority given under Article 3(3) of the 2009 RES Directive to Member States to decide if their support schemes should be accessible to energy produced in a different State, in terms of the following:

- the allocation of legally binding targets to each Member State including by reference to their ability to finance renewable energy development;[117]
- the corresponding importance of Member State control of the effects and costs of their support schemes according to their differing potentials;[118] and
- the possibility for States to agree to open up their support schemes to electricity produced in other Member States under the Joint Support Scheme cooperation mechanism.[119]

The ECJ gave a further decision in September 2014 concerning the compatibility of an obligation/certificate scheme that restricts access to electricity produced nationally with the prohibition on measures preventing the free movement of goods.[120] In the *Essent* case, only certificates issued for renewable electricity generated in the territory covered by the scheme could be used by suppliers to discharge their obligations. The operator fined the complainant when it attempted to meet its obligation by submitting guarantees of origin for electricity produced outside the territory within which the scheme operated. The court used the same line of reasoning employed in *Ålands Vindkraft* in concluding that while the scheme constituted a prima facie breach of the prohibition, it could be justified because it provided a proportionate means of advancing an overriding requirement (protecting the environment by mitigating climate change).[121] The proportionality analysis approach was endorsed and applied by the ECJ in a subsequent judgment of 2016, but with the opposite conclusion that a law discriminating in favour of domestically produced renewable electricity fed directly into distribution networks by waiving distribution charges was not proportionate.[122]

State aid

Article 107(1) TFEU declares that the grant of aid by Member States either directly or through State resources is incompatible with the common market where this would: (a) distort or threaten to distort competition by favouring certain undertakings or the production of certain goods; and (b) affect trade between Member States (for example, measures that have only an internal impact are not unlawful under this Article). Compliance with this provision is promoted by a requirement that States should provide advance notification to the Commission of their intention to provide aid so that the compatibility of proposed measures with Article

---

[117] See Section 5.3.2 of this chapter.

[118] 2009 RES Directive, Recital 25.

[119] See Section 5.4.6 of this chapter.

[120] [2014] EUECJ Case C-204/12 *EssentBelgium NV v Vlaams Reguleringsinstanties voor de Elektriciteits en Gasmarkt*, ECLI: EU: C: 2014: 2192.

[121] Ibid., paras 77–116.

[122] [2016] EU ECJ Case C-492/4 *EssentBelgium NV v Vlaams Geweste*, EU:C: 2016: 732.

107(1) can be assessed.[123] The Commission simplifies this process by waiving the notification requirement for measures that fall within specified categories of aid. These are identified in a law known as the 'General Block Exemption Regulation' that is drafted, adopted and revised periodically by the Commission. The most recent regulation came into effect on 1 July 2014.[124]

Other sources offer a much fuller analysis of this complex area of EU law than can be provided in this chapter.[125] This section confines itself to consideration of the two key respects in which the articles of EU Treaties on State aid have influenced Member States' provision of financial support for renewable energy. First, the ECJ's decision in *PreussenElektra* considered whether a legal requirement by a State for transmission and distribution system operators to purchase electricity produced by renewable generating plants attached to their networks could be regarded as a measure to which Article 87 of the EC Treaty (the corresponding provision to Article 107(1) TFEU in its predecessor) would apply.[126] It concluded that this could not be regarded as State aid because the costs of the scheme were borne ultimately by consumers through the inclusion of costs for the electricity in their bills rather than being financed by the State. This decision legitimised the use of feed-in tariffs as a means of supporting renewable energy from a State aid perspective. It has not been overturned by subsequent decisions, although the ECJ's decision has been criticised for excluding the use of legislative powers to require private actors to finance support schemes from the scope of Article 107(1).

Second, the Commission assists Member States with designing measures that do not contravene constraints on State aid. It does this by producing guidelines that indicate types of measures that are likely to be regarded as compatible with the internal market. The guidelines are not legally binding, but they inevitably have a strong influence on the form of Member State support schemes in view of the Commission's role in approving their compatibility with Article 107(1). New guidelines on State aid in the field of energy and the environment took effect on 1 July 2014.[127]

The already-mentioned requirements for premium or obligation schemes to be used rather than FITs for all measures notified after 1 January 2016, and for operating support to be made available through competitive allocation, will bring radical change to the way in which Member States support renewable energy.[128] The Commission's aim in proposing these changes is to further reduce the market-distorting effect of support schemes that prevent completion of the internal market.[129] It also hopes that increasing the exposure of renewable energy technologies to market conditions will require them to become more efficient, with

---

[123] TFEU, Article 108.

[124] Commission Regulation (EU) No 651/2014 of 17 June 2014 declaring certain categories compatible with the Internal Market in application of Articles 107 and 108 of the Treaty [2014] OJ L187/1.

[125] J.J. Piernas Lopez, *The Concept of State Aid under EU Law: From Internal Market to Competition and Beyond* (Oxford: Oxford University Press, 2015).

[126] Case C-379/98 (n 112); Johnston and Block (n 112), pp.351–3.

[127] Commission, 2014/C/01 (n 101).

[128] Section 5.4.4.

[129] The Commission's reasons for making these changes are explained in SWD (2013) 439 final (n 93).

resulting reductions in the cost of energy. However, related concerns arise that greater market integration and cost reductions may be achieved at the expense of efforts to achieve a rapid decarbonisation of energy supplies.

5.6 CONCLUSION

The 2018 RES Directive, coupled with the 2018 Governance Regulation and the 2019 Electricity Directive and Regulation, creates a legal framework for driving further growth in renewable energy consumption in the EU to at least 32 per cent. Member States and sectoral actors are required to create more favourable conditions for attracting investment in renewable energy and for its integration into energy systems and markets.

Early experience with working under the new framework has been positive in spite of the absence of legally binding national targets. Member States' proposed contributions to renewable energy growth by 2030 were assessed by the Commission to exceed the 32 per cent goal by up to 1.7 per cent.[130] Member States that have fallen behind the trajectory for renewable energy growth set out in their climate and energy plans have an obligation under the Governance Regulation to implement additional measures within one year to increase the proportion of renewable energy in national consumption to the planned level. The obligation is triggered by notification from the Commission that cumulative progress has fallen below the trajectory for reaching the EU's 32 per cent goal.[131] Now that cumulative planned contributions exceed this level, the EU's 2030 renewable energy goal is likely to be met without Member States having taken on legally binding national targets.

However, the legal framework for the 2030 goal will face a stiffer test if the EU's ambition for renewable energy growth increases. In this regard, the European Council agreed in December 2019 to reach net zero greenhouse gas emissions by 2050.[132] This requires a 55 per cent reduction in greenhouse gas emissions by 2030 and a minimum rise in renewable energy by 2030 to between 38 per cent and 40 per cent.[133] Proposals for revising Clean Energy for All Europeans laws will be produced once the European Council has agreed on the level of new 2030 goals.[134] The 2020s are therefore very likely to see further change in the EU's legal framework for

---

[130] Commission, 'An EU-wide Assessment of National Energy and Climate Plans: Driving Forward the Green Transition and Promoting Economic Recovery through Integrated Energy and Climate Planning', COM (2020) 564 final, 2–3.

[131] (EU) Regulation 2018/1999 (n 43), Article 32(3)(c).

[132] European Council, 'European Council meeting (12 December 2019) – Conclusions'. EUCO 29/19, 12 December 2019, available at www.consilium.europa.eu/media/41768/12-euco-final-conclusions-en.pdf accessed 17 June 2021.

[133] Commission, 'Stepping Up Europe's 2030 Climate Ambition – Investing in a Climate-neutral Future for the Benefit of Our People', COM (2020) 562 final, 9.

[134] Ibid., 25.

renewable energy as the EU needs to transit to 'an integrated energy system largely based on renewables already by 2030' to get the Union on track for net zero by 2050.[135]

CLASSROOM QUESTIONS

1. How may the lack of legally binding national targets for increasing renewable energy consumption at the EU level affect achievement of the EU renewable energy 2030 target set in the 2018 RES Directive?
2. How does the 2018 RES Directive seek to: (a) promote renewable energy consumption by requiring Member States to reform administrative procedures and introduce guarantee of origin schemes; and (b) deter unsustainable production of biofuels, bioliquids and biomass fuels?
3. Why does the European Commission view feed-in tariff schemes with concern? Why does it promote premium schemes, certificate/obligation schemes and competitive allocation as preferable alternatives for supporting renewable energy?

SUGGESTED READING

Books

Crossley P, *Renewable Energy Law: An International Assessment (Cambridge University Press 2019)*.
Rusche TM, *EU Renewable Electricity Law and Policy: From National Targets to a Common Market (Cambridge University Press 2015)*.

Articles and chapters

Caldes N, del Rio P, Lechon Y and Gerbeti A, 'Renewable Energy Cooperation in Europe: What Next? Drivers and Barriers to the Use of Cooperation Mechanisms' (2019) 12 *Energies* 22.
Martini A and Romera BM, 'Fifty Shades of Binding: Appraising the Enforcement Toolkit for the EU's 2030 Renewable Energy Targets' (2020) 29 *Review of European, Comparative and International Environmental Law* 221.
Webster E, 'Transnational Legal Processes, the EU and the REDII: Strengthening the Global Governance of Bioenergy' (2020) 29 *Review of European Comparative and International Environmental Law* 86.

Policy documents

Commission, 'A policy framework for climate and energy in the period from 2020 to 2030' COM (2014) 15 final.
Commission, *Clean Energy for All Europeans* (Luxembourg: Publications Office of the European Union, 2019).
Commission, 'Stepping up Europe's 2030 climate ambition: Investing in a climate neutral future for the benefit of our people' COM (2020) 562 final.

---

[135]   Ibid., 18–19.

6
Energy efficiency

ABSTRACT

- EU Energy Efficiency Law consists of the general Energy Efficiency Directive 2012/27/EU and the other measures concerning specific energy efficiency standards set for certain products and goods, such as buildings, electrical appliances and tyres;
- EU energy efficiency policy is formulated with a dual climate change and security of supply objective in mind;
- The Energy Efficiency Directive provides for an EU-wide energy efficiency target, but most importantly lays down non-binding minimum energy efficiency contributions for the Member States as well as a methodology for calculating such contributions;
- The Energy Efficiency Directive and the other EU acts concerning energy efficiency standards envisage the introduction of market-based mechanisms to ensure cost-effective implementation of energy efficiency measures;
- The Directives contain a methodology for determining cost-effective investments in energy efficiency;
- The Energy Efficiency Directive and the other measures take into account the higher costs of increasing energy efficiency for those Member States where the level of energy efficiency is already relatively high, thus taking into account the need to ensure solidarity between the Member States;
- Solidarity at the level of individual energy consumers is also taken into account where network tariffs and their impact on energy efficiency is concerned;
- As far as concrete measures are concerned, the focus of energy efficiency legislation is on buildings, since these account for a large percentage of energy use;
- The effectiveness of the energy efficiency measures can be doubted as a result of:
    - The non-binding nature of the national energy efficiency contributions;
    - The general reluctance by EU Member States to submit to strict energy efficiency targets;
    - The EU's poor track record on increasing energy efficiency.

6.1   INTRODUCTION

EU energy efficiency policy goes back quite some time, with the 1991 SAVE programme the earliest example of EU involvement.[1] Such action finds its legal basis in what is currently Article 191 TFEU, insofar as it refers to the 'prudent and rational utilisation of natural resources' and the source principle, requiring environmental problems to be dealt with at the source. In addition to the environmental legal basis, Article 194(1)(c) TFEU now explicitly enables the EU to adopt measures to 'promote energy efficiency and energy saving'. Both the environmental and the energy legal basis thus enable the EU to adopt measures in the field of energy efficiency. In addition to these legal bases, energy efficiency-related acts have been adopted on the basis of the Common Commercial Policy (external trade).[2] This underlines the dual nature of energy efficiency measures as measures at the crossroads of product standards, environmental protection and security of supply. EU energy efficiency legislation can be separated into a general Directive as well as the Directives and Regulations applicable to the different products or goods for which specific rules have been formulated:

- Directive 2012/27/EU (Energy Efficiency Directive)[3]
- Directive 2010/31/EU (Energy Performance of Buildings Directive)[4]
- Regulation 2017/1369 (Labelling Regulation)[5]
- Directive 2009/125 (Eco-design Directive)[6]
- Measures concerning various products for which energy efficiency standards have been set pursuant to the Eco-design Directive or the Labelling Regulation, such as electronic displays.[7]

We will explain that the main problems arising in relation to energy efficiency regulation tie in with the differing levels of energy efficiency between the Member States and the resulting

---

[1]   SAVE is the wonderfully far-fetched acronym for Specific Actions for Vigorous Energy Efficiency and was adopted by means of Decision 91/565, OJ 1991 L 307/34. Note that as early as 1985 the Council called upon the Member States to promote energy saving policies.

[2]   For example, the Energy Star Agreement: see Case C-281/01 *Commission v Council* (Energy Star) ECLI:EU:C: 2002:761.

[3]   As amended by Directive 2013/12, OJ 2013 L 141/28; Directive 2018/844, OJ 2018 L 156/75; and Directive 2018/2002, OJ 2018 L 328/210. A consolidated version is available from: https://eur-lex.europa.eu/legal-content/EN/TXT/?uri=CELEX%3A02012L0027-20201026&qid=1607943292093 last accessed 22 June 2021.

[4]   As amended by Directive 2018/844, OJ 2018 L 156/75. A consolidated version is available from: https://eur-lex.europa.eu/legal-content/EN/TXT/?uri=CELEX%3A02010L0031-20181224&qid=1607943350319 last accessed 22 June 2021.

[5]   OJ 2018 L 198/1. This repeals Directive 2010/30, OJ 2010 L 153/1, the Labelling Directive.

[6]   OJ 2009 L 285/10, as amended by Directive 2012/27/EU on energy efficiency, amending Directives 2009/125/EC and 2010/30/EU and repealing Directives 2004/8/EC and 2006/32/EC, OJ 2012 L 315/1.

[7]   Annex II A of Commission Regulation (EU) 2019/2021 laying down ecodesign requirements for electronic displays, OJ 2019 L 315/241, contains the maximum energy efficiency index for such screens.

differences in costs to increase energy efficiency. In some Member States buildings are on average better insulated, or heat and electricity generation and distribution are more efficiently organised, than in others. Also, the differing climatic situations in the Member States impact the demand for heat, for example. This results in cost-effectiveness and solidarity playing a major role, not only because increasing the energy efficiency of an already highly efficient system involves relatively high costs, but also because the investments required for extra energy efficiency must be affordable in relation to the GDP of that Member State. Below we will first examine the basics of energy efficiency (section 6.2) and then take a closer look at the Energy Efficiency Directive (section 6.3). After that we will study the Energy Efficiency of Buildings Directive (section 6.4) and the Labelling Regulation (section 6.5). Finally, the other energy efficiency-related measures will be scrutinised in section 6.6. Section 6.7 concludes.

6.2   BASICS OF ENERGY EFFICIENCY

Energy efficiency is the relation between the amount of energy put into a certain process and the result of that process. Generating electricity, for example, often involves the combustion of primary energy sources with the resultant thermal energy being converted to kinetic energy that drives turbines. These in turn drive generators, with the production of electrical energy as the final result. This is then transported to the end-user, but resistance losses mean that the amount of electrical energy delivered to the end-user will be lower than the amount fed into the grid. The end-user, finally, will want to convert that electrical energy into light or motion and thus personal utility, but this will also involve losses. Incandescent light bulbs, for example, produce not just visible light, but also an overwhelming amount of thermal energy (heat). The losses provided in this example all come with negative environmental effects arising from the combustion of fossil fuels and thus carbon dioxide emissions, without resulting in any utility for the end-user. If we require extra utility at the same level of or with a reduced energy efficiency, that means more waste and emissions. Increased energy efficiency, on the other hand, replaces the need to use other energy resources, resulting in some authors calling energy efficiency an (or first) energy source in its own right.[8]

Energy saving through increased energy efficiency is often said to be the best way to deal with the climate change impact of energy use, as well as with security of supply and energy poverty.[9] It therefore features prominently in the EU's 20-20-20 goals.[10] The recently adopted Clean Energy for All Europeans Package also devotes significant attention to energy efficiency, setting a 32.5 per cent target for 2030.[11] In addition, increased energy efficiency plays an

---

[8]   M. Yan and X. Yu, *Energy Efficiency*, Springer, 2015, pp.11 et seq.

[9]   Commission proposal for the Energy Efficiency Directive, COM (2011) 370 final, p.1. See more recently, the Commission's European Energy Security Strategy, COM (2014) 330 final, chapter 3. Note that energy efficiency was also relevant during the oil crisis of the 1970s.

[10]   See Chapter 2.

[11]   This package was proposed by means of COM (2016) 860 final and acquired political approval on 30 November 2018: see Commission press release IP/18/6870. In relation to energy efficiency, it has resulted in Directive 2018/2002,

important geopolitical role for the EU, as reduced energy consumption allows for a reduction of imported (fossil) fuels and thus a higher level of supply security. Increased energy efficiency may well be the way in which we can maintain as much as possible of our existing way of life at a lower carbon cost, but it is very much a technical issue with a different impact for each Member State. Increasing energy efficiency means, among others, investing in insulation of buildings and industrial installations; changing the way buildings are heated – for example by using residual heat; and changing the way in which electricity networks are operated. All of these measures require considerable investments, for example in the creation of a heat network. This translates into fairly technical rules that entail a prominent role for cost–benefit analyses, as we will see below. The EU's energy efficiency target is not part of the 2009 climate and energy package and is otherwise set apart from the instruments in the 2009 climate and energy package.[12] The reason for this is that it deals with energy efficiency at a relatively high level of abstraction that leaves considerable room for the widely different backgrounds of the Member States in terms of climate conditions and current levels of energy efficiency. Moreover, it envisages an important role for cost-effectiveness considerations.[13] The importance of cost-effectiveness results from the observation that increasing energy efficiency involves significant investments that will pay off only in the long term. This is compounded by the fact that such investments will be more significant when the level of energy efficiency is already higher. When the level of energy efficiency is relatively low, simple and cheap investments can lead to significant increases in energy efficiency. However, transforming an already highly energy-efficient building into a zero-energy building may very well be impossible and in fact require the complete reconstruction of the building. It is not difficult to see that the costs of the latter are far greater than those involved in putting in double glazing and adding insulation.

Energy efficiency can be increased in a number of ways, but by and large they boil down to technical, network, financial and behavioural solutions. On the technical side, energy efficiency can be increased by changing to more energy-efficient equipment, such as switching from incandescent light bulbs to LEDs or installing more efficient turbines in power plants, or simply adding insulation. On the network side, losses in transmission and distribution networks can be reduced by dispatching power from generation sources that are closer to the demand centres. Another network-based solution would be to use the thermal energy (waste heat) from, for example, power plants to heat buildings or as an input in another industrial process. This requires a network to transport not only the electrical energy, but also the heat from that plant. As far as financial incentives are concerned, investments in increased energy

---

OJ 2018 L 328/210, which has raised the EU objective to a 32.5per cent increase in energy efficiency for 2030 compared to the 2007 baseline.

[12]   The 2009 package was published in [2009] OJ L 140 and does not address energy efficiency as such. Note that the Commission's proposal did include energy efficiency as part of a package that addressed greenhouse gas abatement and renewable energy production: COM (2008) 30 final.

[13]   Cf. recital 21 of the preamble to Directive 2012/27 and Article 1(1) of the Energy Performance of Buildings Directive.

efficiency are almost always profitable as they will result in lower energy costs. However, the capital costs may outweigh the reduction in energy costs in the short term, and thus make it financially unattractive to invest in energy efficiency. Similarly, in relation to rented buildings, there is a financial impediment resulting from the so-called split incentive. The split incentive results from the fact that in buildings that are rented, the investment in increased energy efficiency is financed by the owner whereas the tenant enjoys the advantages. On the behavioural front, an obvious solution would be to get end-users to switch off the lights when nobody is in a certain room. This requires an increased level of awareness on the part of that end-user. This would also address the energy efficiency paradox or rebound effect, according to which increases in energy efficiency actually trigger more energy consumption.[14] The bottom line is that energy efficiency is notoriously difficult to implement in practice. This is evidenced, for one, by the progress reports drawn up by the Commission. Each year since 2015, these reports have concluded that progress was insufficient to attain the energy efficiency targets set by the Member States or at the EU level.[15]

6.3   THE ENERGY EFFICIENCY DIRECTIVE

Directive 2012/27/EU (the Energy Efficiency Directive) replaces Directive 2004/8/EC on Cogeneration and Directive 2006/32/EC (the 2006 Energy Efficiency Directive).[16] It essentially lays down rules to ensure that the EU achieves its 32.5 per cent energy efficiency target[17] and rules intended to overcome the market failures[18] impeding energy efficiency increases. The Directive is adopted on the basis of Article 194(2) TFEU, the energy legal basis, and explicitly allows for stricter measures, that is, higher energy efficiency standards, insofar as these meas-

---

[14]   This is also referred to as Jevons' Paradox after the British economist who noticed that coal consumption did not decrease after coal-fired steam engines had become more efficient. For a survey see: B. Alcott, 'Jevons' Paradox', *Ecological Economics*, 2005, vol. 54, pp.9–21.

[15]   Cf. 2015 assessment of the progress made by Member States towards the national energy efficiency targets for 2020, (COM(2015) 574); 2016 assessment of the progress made by Member States towards the national energy efficiency targets for 2020, (COM(2017/056); 2017 assessment of the progress made by Member States towards the national energy efficiency targets for 2020, (COM(2017)687); 2018 assessment of the progress made by Member States towards the national energy efficiency targets for 2020, (COM(2019) 224); 2019 assessment of the progress made by Member States towards the national energy efficiency targets for 2020 (COM(2020) 326).

[16]   OJ 2012 L 315/1, see Arts 27 and 28 on repeals and transposition. Note that this Directive was amended by Directive 2013/12/EU, OJ 2013 L 141/28 to reflect the accession of Croatia and most recently by means of Directive 2018/844, OJ 2018 L 156/75, to implement the Clean Energy for All Europeans Package. This last amendment essentially entailed the moving of the provisions on long-term renovation strategies for buildings, the old Article 4, to the Energy Performance of Buildings Directive: see recital 8 of the preamble to Directive 2018/844.

[17]   This 32.5 per cent target is provided for in the Climate Governance Regulation, Regulation 2018/1999, OJ 2018 L 328/1, Recital 6 to the preamble, Article 2(11) and 4(b)(1), see Chapter 11.

[18]   Such as the split incentive in buildings that are rented out: see section 6.4 below and in energy efficiency increases in the transmission, distribution and transformation of energy, see section 6.3.3 below.

ures are compatible with EU law. It contains rules on the indicative national energy efficiency targets (discussed below in 6.3.1), efficiency in energy use (discussed below in 6.3.2), efficiency in energy supply (analysed in 6.3.3 below), efficiency provisions that apply generally (discussed in 6.3.4 below) and some final provisions that will not be dealt with separately.

6.3.1 EU Energy Efficiency Targets and National Contributions Thereto

Defining energy efficiency

As energy efficiency is essentially the relation between the amount of energy put into a certain process and the useful energy obtained from that process, the Energy Efficiency Directive first defines a number of common standards for determining energy efficiency. This efficiency can be expressed in a number of terms, which relate to the production or consumption of energy and the difference between these. In this regard the Directive refers to, respectively, *primary* energy consumption, *final* energy consumption and *primary* or *final* energy saving or energy intensity.[19] 'Energy' is given a very broad definition that basically includes all forms of energy.[20] Primary energy consumption is defined as gross inland consumption, excluding non-energy uses. This concept encompasses all the oil, gas and other primary energy sources as well as electricity produced in or imported into a country. Final energy consumption refers to all energy supplied to final consumers, excluding supply to the energy sector and the energy transformation sector.[21] These sectors are excluded as they do not use energy as final consumers, but rather as intermediary parties that will ultimately supply final consumers. Energy saving consists of the difference between the (estimated or measured) amount of energy used before and after efficiency improvement measures have been implemented.[22] This means that – at the household level, for example – energy savings can be calculated as the difference between the amount of electrical energy consumed before and after an old inefficient washing machine is replaced with a new energy-efficient one. Another way of measuring energy efficiency is by means of energy intensity. This is the amount of energy used in a country divided by the gross domestic product of that country. This directly relates energy use to the economic productivity achieved with that input and follows from the fact that increased economic growth by and large connects to increased energy consumption. A decrease in energy intensity thus points to a decoupling of economic growth and energy consumption that equates to increased energy efficiency.[23]

---

[19] Article 3(1) Energy Efficiency Directive.

[20] Article 2(1) Energy Efficiency Directive.

[21] The energy transformation sector is a concept that is undefined in the directive. It refers to the industry that transforms one form of energy (e.g. kinetic energy) into another form (e.g. electrical energy).

[22] Article 2(5) Energy Efficiency Directive.

[23] It could, for example, mean that more goods are produced and transported with the same or a reduced energy consumption.

Energy efficiency targets

Although the EU has set itself a 32.5 per cent energy efficiency target, it requires the Member States to actually achieve this target.[24] Despite the various buildings operated by the EU institutions,[25] the majority of energy consumption obviously takes place in the Member States. It is therefore at Member State level that the most significant energy savings can be achieved. This is also where we can see obvious differences between the Member States, both in terms of energy intensity and of the reductions in energy intensity that have been achieved.[26] Moreover, we also see obvious differences in the means used to attain energy efficiency increases, with the two smaller island states of Malta and Cyprus being mentioned specifically and receiving special targets.[27]

To this end, Article 3(1) requires the Member States to set indicative energy efficiency contributions (referred to as an 'indicative national energy efficiency target' in the text) in their national energy and climate plans, in such a way that the EU does not exceed a certain amount of energy consumption defined as million tonnes of oil equivalent in 2020. As a result, the Member States, irrespective of how they define their energy efficiency target, also have to express their targets in an absolute level of primary or final energy consumption. However, apart from the hard target set at EU level, the Member States shall also take into account various other factors such as 'other measures to promote energy efficiency'.[28] With regard to the 2030 32.5 per cent target, the Member States shall set indicative national energy efficiency targets in a similar way and report their progress to the Commission.[29] We see the cost–benefit analysis prominently where Article 3(6) of the Directive provides for the possibility of an upward revisioning of the target, inter alia, 'in the event of substantial cost reductions resulting from economic or technological developments'.

Furthermore, they may also take into account factors affecting energy consumption, such as economic growth forecasts, changes in energy imports and exports and the development of renewables, nuclear energy and CCS. The targets were to be set for the first time in 2014 in order to enable the Commission to assess whether the EU is likely to achieve the 20 per cent target.[30] It may be noted that the Commission is only allowed to issue recommendations and that the text of the Directive does not in any way guarantee that the EU will attain its objec-

---

[24]   Directive 2018/2002, recital 6 and Article 1(1).

[25]   The buildings in Brussels are in fact rented from the Brussels capital region and thus fall under the Belgian rules implementing the Energy Efficiency of Buildings Directive. See further the answer by Energy Commissioner Oettinger to Parliamentary question E-007805-13, available at: www.europarl.europa.eu/sides/getDoc.do?type=WQ&reference =E-2013-007805&language=EN last accessed 22 June 2021.

[26]   For example S.R. Schubert, J. Pollak and M. Kreutler, *Energy Policy of the European Union*, Palgrave, 2016, p.181.

[27]   Directive 2018/2002, recital 13.

[28]   Article 3(1)(d) Energy Efficiency Directive.

[29]   Article 3(5) Energy Efficiency Directive.

[30]   Article 3(2) Energy Efficiency Directive.

tive.[31] In this respect the Directive confines itself to laying down a reporting obligation for the Member States as well as a duty to include their contributions to the EU target in the integrated national energy and climate plans they need to submit on the basis of the Governance Regulation.[32]

6.3.2 Efficiency in Energy Use

The directive recognises the importance of energy efficiency in energy use but also the wide range of energy uses, ranging from building to industries and appliances. Therefore, other EU laws (directives and regulations) are also relevant when discussing efficiency in energy use. One example is energy efficiency in buildings, which is governed by the Energy Efficiency Directive and the Energy Performance of Buildings Directives (see further section 6.4).

The Energy Efficiency Directive provides some guidance as to how Member States can achieve efficiency contribution in energy use. For example, it recognises that central governments can play an important role in their purchasing decisions. Article 6 requires the Member States to ensure that central government public procurement results in the purchase of products, services and buildings with high energy efficiency performance insofar as this is consistent with other considerations laid down in Annex III. This is closely connected to the public procurement procedures in general, as is evidenced by the fact that the thresholds laid down in the Public Procurement Directive determine whether the obligation applies. Green public procurement is fairly well established and widely accepted by the European institutions.[33] The interesting trade-off surfaces when we see that such purchasing is to be consistent with, *inter alia*, sufficient competition. This refers to the fact that Annex III requires the purchase to concern only those products that comply with the highest EU standards, but this may well exclude several competitors (that do not meet these standards), thus potentially harming the public procurement process itself. Again, lower levels of government are not directly concerned, and the Member States shall only 'encourage' them to purchase highly energy-efficient equipment, services and buildings.

Of more relevance to non-state actors is Article 7, which prescribes the actual energy savings obligation that the Member States are to achieve. The article contains detailed rules on how these reductions are to be calculated and assigned to the various reporting periods. This is where we also find the derogation for Cyprus and Malta.[34] Article 7(10) requires the Member States to achieve these energy savings by means of an energy efficiency obligation or an alternative measure; in doing so, they are to take into account the need to alleviate energy poverty.[35]

---

[31] Article 24(3) Energy Efficiency Directive.

[32] Article 3(5) Energy Efficiency Directive. Confusingly, the Directive also refers to the National Energy Efficiency Action Plans as another means of communicating the progress on meeting the EU target.

[33] More information, including the criteria that have been developed for green public procurement, can be accessed from http://ec.europa.eu/environment/gpp/index_en.htm last accessed 22 June 2021. See further Case C-513/99 *Concordia Bus* ECLI:EU:C:2002:495 and Case C-448/01 *Wienstrom* ECLI:EU:C:2003:651.

[34] Article 7(1)(b) Energy Efficiency Directive.

[35] Article 7(11) Energy Efficiency Directive.

Pursuant to Article 7a(2), an energy efficiency obligation may be imposed upon energy distributors and/or retail energy supply companies. The term energy distributor encompasses all distribution system operators (DSOs)[36] as well as other companies responsible for selling and transporting energy to consumers outside the context of the regulated electricity and gas networks. The retail energy sales company is normally referred to as a supply company in electricity and gas law, but again the wider scope of the Directive necessitates a broader term.[37] Under these obligation schemes the obligated companies are responsible for achieving a cumulative energy savings target by the end of 2020.[38] It may be noted that Article 7(6) allows for some flexibility in meeting these targets.[39] This flexibility allows them to take into account energy savings by third parties as well as energy savings in earlier or later years (banking).

The last sentence of Article 7(6) provides for further flexibility since it allows the Member States to take measures to reduce the costs of direct energy savings in energy-intensive industries exposed to international competition. This aligns with the powers to grant state aid to such industries.

Whereas the scheme of obligated companies already offers considerable flexibility, member states are not bound to this instrument and instead can opt for alternative instruments pursuant to Article 7b. Compared to the older version of the Energy Efficiency Directive, the framework for such alternative instruments is even less well-defined. This is problematic as research shows that the majority of Member States have opted for a fund as an alternative instrument.[40] The main proviso that such measures must be equally effective and comply with the conditions laid down in the Directive, for example, has been dropped after the amendments brought by Directive 2018/2002. The only remaining condition in the current version is that the alternative instruments must ensure that the energy savings are achieved among final customers.[41] The old version of the Directive suggested the use of energy taxes and voluntary agreements as alternative instruments. These come with different conditions for compliance with EU law as an energy tax will have to comply with Article 110 TFEU whereas the agreement must be compatible with Article 101 TFEU. For the Member States, this offers considerable leeway that will be difficult to challenge for a private party on the basis of the Directive, as the *Saras* case shows.[42] Saras Energía sought to challenge the Spanish implementation of this provision. This implementation essentially forced Saras to pay into the Energy Efficiency National Fund

---

[36]   Article 2(20) Energy Efficiency Directive.

[37]   See Article 2(19) of Directive 2009/72, OJ 2009 L 211/55.

[38]   Companies become obligated following a designation on the basis of objective and non-discriminatory criteria: Article 7(2) Energy Efficiency Directive.

[39]   This is a mechanism comparable to that limiting the use of Kyoto flexible units as part of the EU ETS cf. Article 5 of the Effort Sharing Decision.

[40]   J. Rosenow and F. Kern, 'EU Energy Innovation Policy: The Curious Case of Energy Efficiency', in R. Leal-Arcas and J. Wouters (eds), *Research Handbook on EU Energy Law and Policy*, Edward Elgar Publishing, 2017, p.506.

[41]   Article 7b(1) Energy Efficiency Directive.

[42]   Case C-561/16, *Saras Energía*, ECLI:EU:C:2018:633.

without offering it any possibility to achieve energy efficiency targets by actually saving energy. This was made even more unacceptable to Saras because only retail energy sales companies had to contribute, whereas energy distributors would not have to pay.[43] The Court ruled essentially that the member states have a broad margin of discretion that is, however, constrained by the Directive's bottom line.[44] This bottom line is that companies should save energy and just paying into a fund does not save energy as such. This means that the Court then tests whether the contribution requirement qualifies as an alternative measure within the meaning of the Directive.[45] All in all, the Directive provides the Member States with very considerable discretion and imposes only a minimal burden on Member States to show compliance. This is arguably a rather low threshold, supporting the conclusion that the effectiveness of this Directive can be doubted.[46] This considerable margin for discretion is maintained, if not expanded,[47] in the current version of the Energy Efficiency Directive.

Concerning the imposition of the duty to contribute on only some energy companies, the Court finds that a Member State's discretion in this regard is limited by the duty to designate the obligated companies on the basis of explicitly stated objective and non-discriminatory criteria.[48] This, in the view of the Court, means that the specifics of the national energy market must be taken into account in deciding which companies are included and excluded from the duty to contribute. As a result, considerations regarding the level playing field of the national energy markets are closely linked to the Member State's decision regarding the actual imposition of the duties that are required to attain the energy efficiency targets.[49]

Awareness of energy use

Much energy efficiency results from awareness and behavioural change. To this end the Directive requires the Member States to ensure that high-quality, cost-effective energy audits are available to all final customers. The Directive lays down rules on the qualification and impartiality of the experts as well as the quality assurance for the audits in Article 8. Whereas such audits are voluntary for households as well as small and medium-sized enterprises, they are required for larger companies from 2015 onwards (Article 8(4)), unless such companies

---

[43] Case C-561/16, *Saras Energía*, ECLI:EU:C:2018:633, para. 14.

[44] Case C-561/16, *Saras Energía*, ECLI:EU:C:2018:633, paras 24 and 27–29.

[45] Case C-561/16, *Saras Energía*, ECLI:EU:C:2018:633, paras 29–34. Note that this refers to the old version of the Energy Efficiency Directive. The abandonment of these conditions following the entry into force of Directive 2018/2002 means that the Member States have even more leeway.

[46] These fears were expressed during the negotiations of the Directive: S.R. Schubert, J. Pollak and M. Kreutler, *Energy Policy of the European Union*, Palgrave, 2016, pp.184, 185.

[47] See the comments above in relation to Article 7b of the Energy Efficiency Directive.

[48] Case C-561/16, *Saras Energía*, ECLI:EU:C:2018:633, para. 53.

[49] Notably the reference to recital 20 of the preamble supports this conclusion: Case C-561/16, *Saras Energía*, ECLI:EU:C:2018:633, para. 54.

have a certified environmental management system.[50] In all cases, the audits must comply with the requirements in Annex VI. Further awareness-raising takes place on the basis of Article 12 on consumer information and empowering.

Another awareness-raising provision can be seen in Article 9 on metering. This provision requires the Member States to ensure, insofar as this is technically possible and economically sensible, that every energy user gets what is often referred to as a smart meter. This is a meter that provides real-time information on actual energy consumption. Strictly speaking, the Directive does not require the installation of a smart meter, but rather a 'competitively priced individual meter' that only needs to provide real-time information on energy usage. Smart meters add a two-way communication functionality to this real-time information functionality.[51] The Directive contains rules on the roll-out of such meters that essentially imply that their installation is mandatory for new buildings and buildings that undergo major renovation. For all other cases, individual meters must be installed whenever an old meter is replaced unless this is technically impossible or not cost-effective in the light of the estimated long-term savings. This element of cost-effectiveness is central to the Energy Efficiency Directive and will be addressed in some more detail below.

From Article 9(2) we can furthermore gather that the rules in the Electricity Directive determine whether or not smart meters will be rolled out. The Directive contains some rules on the functionality of such smart meters, including requirements concerning data and privacy protection that have nothing to do with energy efficiency as such. Whereas individual meters are most easily envisaged in relation to electricity and gas, individual metering can play an important role in relation to district heating and/or cooling and domestic hot water consumption, as is envisaged by Article 9a.

The provisions on metering are reflected in those on billing. Here, again, the information needs to reflect individual energy use as much as possible, including information on historical energy use, where this is technically possible and economically justified. Such billing information shall be disseminated to the final customers free of charge so that the consumer gets insights into the actual energy use and is thus enabled and incentivised to save energy. These rules largely concern energy consumption and the ability and incentive to reduce energy consumption. The potential for energy efficiency increases is not confined to the final consumer, but also includes energy suppliers.

6.3.3   Efficiency in Energy Generation, Transmission, Distribution and Supply

Of course, the final energy user is only one of the links in the energy chain where energy efficiency matters. From production and generation at the upstream level to transmission and distribution to end-consumers, energy is lost and thus energy efficiency gains can be achieved.

---

[50]   Rules for such EMASs are laid down in Regulation 1221/2009, OJ 2009 L 342/1; see further Commission Decision 2013/131, OJ 2013 L 76/1.

[51]   See also Chapter 9. Note that the roll-out of smart meters is also regulated in Directive 2009/72, the Electricity Directive, Article 3(1)) and Annex I, point 2.

This starts with the production of heat and cooling as part of the generation of electricity (residual heat), but also with the production of heat and cooling as such. Member States are obliged to assess the potential for high-efficiency cogeneration and efficient district heating and cooling, including a cost–benefit analysis.[52] If the benefits outweigh the costs, the Member States are under an obligation to take adequate measures for the development of efficient district heating and or cooling and high-efficiency cogeneration. The Directive also contains provisions on the connection to the grid of high-efficiency cogeneration.[53] By and large, Article 14 of the Energy Efficiency Directive is characterised most prominently by the central role played by cost–benefit analyses, and not so much by the actual impact of these measures in terms of energy efficiency.

Apart from the production and generation of energy, the transformation, transmission and distribution sectors also involve potential for energy efficiency. Article 15(1) requires the Member States to ensure that national energy regulators pay due regard to energy efficiency and that there are incentives for grid operators to make system services available to network users in the context of smart grids. These system services concern the activities that are needed to ensure the reliable functioning of the grid, such as balancing input and consumption and voltage control.

Currently, most of these services are provided at the upstream level, which entails potential reductions in energy efficiency. A smart grid would also allow for such services to be provided at the downstream level. Smart grids are energy (primarily electricity) grids that enable bi-directional flows from decentralised (renewable) production and demand response. Smart grids allow a wider uptake of intermittent renewable energy and allow for a more efficient operation of the grid, for example by reducing the need for spinning reserves.[54] Instead of using centralised generation and large-scale demand response connected to medium to high-voltage grids, local low-voltage demand response is used to balance peaks in renewable energy that is fed into the grid at lower voltages. This, for one, reduces the need to use transformers, and the losses inherent in these.

Including demand response, which concerns the flexibility in energy demand, can improve the efficiency of these system services and thus system operation. Tariffs should be amended so that there is no incentive for inefficiency in the system operation and to ensure that demand response is encouraged.[55] In relation to solidarity between network users, we may refer to Article 15(3) that allows the Member States to maintain network tariffs that serve a social purpose. This could relate to lower tariffs for people with a lower income. In this regard, the Member State is under a duty to investigate the effects on energy efficiency. Relatively low

---

[52]   Article 14(1) and (3) of the Energy Efficiency Directive.

[53]   These provisions are elaborated in Annex XII to the Directive.

[54]   This refers to installations where reserve generating capacity is spinning, thus using primary energy sources, without being coupled to the grid and thus not producing usable energy. It essentially means that there is a power plant with a continuously running generator that is not delivering electricity to the grid unless it is asked to do so by the system operator.

[55]   This is elaborated in Annex XI to the Directive.

tariffs reduce the incentive to invest in energy efficiency. However, such investments may not be available to lower incomes in the first place, which raises the question whether subsidisation of such investments may not be more (energy-) efficient than maintaining socially motivated tariff structures.

Finally, the Directive creates a framework to encourage demand-side management and aggregators.[56] Most importantly, system security remains the prime objective and the energy efficiency measures may never jeopardise the safe and reliable functioning of transmission and distribution systems.

6.3.4   Markets and Cost-effectiveness in Energy Efficiency

The provisions mentioned above often refer to cost-effectiveness as a guiding principle. However, the problem is that there is an asymmetric relation between costs, benefits and thus incentives to invest in energy efficiency. In a nutshell, the costs for increased energy efficiency often fall upon the energy end-user, who then reaps the benefits of reduced energy consumption and thus has an incentive to invest in such energy saving measures. Smart meters or individual meters, however, are most likely to be rolled out by the supply side of the market: distribution system operators, supply companies or other market participants. These parties essentially have no incentive to invest in such meters if the result is that their energy sales (or distribution) decrease. In other words, if the roll-out of smart meters works as it is supposed to, these companies will actually see a reduction of their income as they will sell or transport lower amounts of energy. This is exactly why cost-effectiveness plays such a central role in the Energy Efficiency Directive. It is operationalised by means of the framework and criteria laid down in Annex IX to the Directive. However, rather than mentioning cost-effectiveness every time, the introduction of market mechanisms is also generally assumed to deliver efficiencies. This follows from the fact that a competitive market will generally ensure optimal efficiency of the energy services provided whereas the provision of these services themselves may generate the income that compensates the reduced income following from the lower energy consumption.

Such mechanisms were explicitly envisaged by the current Directive's predecessor in the form of a 'white certificate scheme', whereby certified energy savings would be tradable throughout the EU.[57] This Union-level scheme is rejected in the current Directive on the basis of the excessive administrative costs involved as well as the risk that energy savings would be concentrated in a few Member States only.[58] Still, Member States are allowed to have their own white certificate trading schemes. This nicely highlights the tension between the two themes central to this book: cost-effectiveness assumed in market-based mechanisms and solidarity between the Member States. We see that the EU legislator envisages that using a market-based

---

[56]   Article 15(8) of the Energy Efficiency Directive.

[57]   Article 6(2)(b) of Directive 2006/32 envisaged a market-oriented scheme, such as one based on white certificates, that would enable increased energy efficiency at the production, supply and distribution levels. For an analysis of this and similar schemes see L.-G-Giraudet and D. Finon, 'The Costs and Benefits of White Certificates Schemes' [2012] *Energy Efficiency*, 179–99.

[58]   Recital 20 of the preamble.

mechanism, widely considered to be an efficient mechanism, may result in disproportionate administrative costs. This shows at least the limits to market-based mechanisms. Furthermore, the risk that energy efficiency gains would be concentrated appears very realistic in such a market-based scheme. Member States where levels of energy efficiency are comparatively low would be able to attain considerable efficiency gains at relatively low costs. This would mean that the market mechanism directs investments in energy efficiency to those Member States, to the detriment of investments in energy savings in those Member States that are already relatively energy efficient and thus face higher marginal costs for extra energy efficiency. Here we see a clear limit to the solidarity that Member States are willing to accept, particularly in view of the fact that the brunt of such investments would be borne by the energy consumer.

However, this does not mean that the market-based approach is fully rejected. The energy efficiency obligation schemes still allow for the use of energy savings attained by energy service providers to be counted toward the obligation resting on obligated parties.[59] The idea underlying this is that energy services would be offered by energy utilities (a term that includes, but is not limited to, supply companies). This would offer them an alternative source of income to compensate for the loss in income resulting from energy supply as the latter would decrease with increased end-user energy efficiency. Another effect would be increased competition in the energy market, as the energy services would enable the supply companies to differentiate their products.[60]

6.4 ENERGY EFFICIENCY OF BUILDINGS

Buildings account for a significant amount of the EU's energy usage and offer the biggest potential for energy savings.[61] Also, in view of the fact that buildings account for approximately 40 per cent of the EU's energy use, the potential for energy efficiency increases is significant. This explains why buildings feature prominently in the Energy Efficiency Directive and are the subject of a specific directive: Directive 2010/31/EC (the Energy Performance of Buildings Directive), which aims to increase the energy efficiency of buildings.[62]

The directive does so by providing, *inter alia*, a framework for the calculation of the energy efficiency and by enabling minimum energy efficiency requirements. The minimum energy efficiency requirements are to be set by the Member States and may differentiate between

---

[59] Article 7a(6)(a) of the Energy Efficiency Directive. See section 6.3.2 for more information on energy efficiency obligation schemes.

[60] Recital 20 of the preamble. Supply companies can essentially differentiate their products on the basis of the price and the extent to which the energy they supply is renewable. Adding energy services would thus provide them with an extra quality aspect that allows them to differentiate their products.

[61] Commission Energy Efficiency Plan 2011, COM (2011) 109, p.3. See also the Commission Impact Assessment for this plan, SEC (2011) 277, p.9.

[62] OJ [2010] L 153/13, as amended by Directive 2018/844, OJ 2018 L 156/75. It recasts Directive 2002/91 into a consolidated text in the interest of clarity since this Directive had been amended in the past and new amendments were needed.

various categories of buildings (Article 4(1), third paragraph). This already highlights the considerable latitude that the EU leaves the Member States in this matter, resulting in significant Member State discretion. Buildings are central, together with transport,[63] to the Union's energy efficiency policy.[64] This is one of the explanations for why the Energy Performance of Buildings Directive has been much more comprehensively updated by means of the Clean Energy for All Europeans Package than the Energy Efficiency Directive.[65]

This situation, with continuing national discretion even following a significant overhaul, may be contrasted with the observation by the Commission, in the proposal for the Directive, that the energy-saving potential of Directive 2002/91/EC had not been fully achieved.[66] As a result, the proposal for the recast went beyond mere consolidation and included proposals to increase the level of energy efficiency, for example by an increase of the scope of the various measures.[67] It also deserves mentioning that the Commission explicitly noted the low level of ambition regarding the implementation of Directive 2002/91 by the Member States.[68] On a similar note, complexity and lack of clarity concerning the wording of Directive 2002/91 was noted.[69] Following two readings, the Energy Performance of Buildings Directive was adopted by the European Parliament on 18 May 2010. On the basis of the Directive, the Commission adopted Delegated Regulation 244/2010 establishing a comparative methodology framework for calculating cost-optimal levels of minimum energy performance requirements.[70]

Energy performance of buildings can be improved by renovating existing buildings. Article 2a(8) of the Energy Performance of Buildings Directive thus requires the Member States to establish a long-term strategy for building renovation as part of their integrated national energy and climate plans. As part of a leading-by-example policy, public bodies' buildings are subject to special and stricter rules in terms of renovation with a view to increasing energy efficiency. Article 5 requires the Member States to ensure that 3 per cent of the total floor area of buildings operated by the central government is renovated to meet the energy efficiency standards laid down in the Energy Performance of Buildings Directive.[71] In any case, the rules only

---

[63]   Cf. M. Ntovantzi et al., 'Do We Have Effective Energy Efficiency Policies for the Transport Sector? Results and Recommendations from an Analysis of the National and Sustainable Energy Action Plans', ECEEE Summer Study Proceedings, p.895.

[64]   M. Economidou, V. Todeschi, P. Bertoldi, D. D'Agostino, P. Zangheri and L. Castellazzi, 'Review of 50 Years of EU Energy Efficiency Policies for Buildings', *Energy and Buildings*, 2020, vol. 225.

[65]   Directive 2018/844, OJ 2018 L 156/75, contains more than ten amendments to the Energy Performance of Buildings Directive and only one amendment to the Energy Efficiency Directive.

[66]   OJ 2002 L 1/65.

[67]   COM (2008) 780, at pp.3 and 5.

[68]   COM (2008) 780, at p.3. See further SEC (2008) 2864, at p.4.

[69]   Commission Impact Assessment to COM (2008) 780, SEC (2008) 2865, at p.27.

[70]   OJ 2012 L 81/18. This regulation was accompanied by Commission Guidelines published in OJ 2012 C 115/1.

[71]   This directive is discussed below in section 6.3. Note that the application of this directive to buildings by lower governments is optional.

apply to larger buildings, that is, those with a floor space exceeding 500 m². [72] Furthermore, historical or otherwise architecturally special buildings and buildings used by the armed forces or as places of worship may be excluded. Finally, the Directive creates considerable flexibility as a result of the possibility to use excessive (that is, more than 3 per cent) renovations in any of the three previous or following years. [73] To add to the already considerable flexibility, alternatives to renovation are also enabled. [74]

Below we will study in more detail the Energy Performance of Buildings Directive (section 6.4.1), and then briefly look at how cost-effectiveness is operationalised (section 6.4.2).

6.4.1   The Energy Performance of Buildings Directive

Objective

The main objective of the Directive consists of the promotion of energy performance of buildings, taking into account local conditions, indoor climate conditions and cost-effectiveness. [75] This objective is to be attained through a common framework for the methodology to calculate the integrated energy performance and the cost-optimal level of energy performance. [76] Member States are thus required to take measures that will ensure that minimum energy performance requirements are set with a view to achieving cost-optimal levels, that is, the minimum energy efficiency required of all buildings that is efficient in relation to the investments necessary. [77] To this end, the methodology adopted on the basis of Article 3 is to be applied. Furthermore, it follows from the above that this common methodology framework respects the subsidiarity principle and allows local conditions, which will vary widely between the 27 Member States, to influence the methodology. [78] At the same time it ensures that the methodologies are sufficiently comparable in order not to jeopardise the attainment of the energy efficiency objectives of the Directive and the internal market. [79]

It also shows, just as we have seen in relation to the Energy Efficiency Directive, that the solidarity that underpins much of EU climate law has its limits, not only because of the climatic differences between Member States, but also because of the differences in energy performance

---

[72]   This threshold was lowered to 250 m² from 9 July 2015.

[73]   Articles 5(2) and (3) Energy Efficiency Directive.

[74]   Article 5(6) Energy Efficiency Directive.

[75]   Article 1.

[76]   Article 3 Energy Performance of Buildings Directive. This framework is elaborated to some extent in Annex I to the Directive.

[77]   Article 4.

[78]   In a nutshell, the subsidiarity principle ensures that the EU only acts where this is more effective than national or regional action by the Member States.

[79]   The internal market consists of an area with free movement of goods, services, people and capital with undistorted conditions of competition. The latter may be affected by having widely different methodologies for calculating cost-effective levels of energy efficiency.

at this moment.[80] Whereas there is a degree of solidarity inherent in the EU-wide approach, the latter's limitations in defining a methodology means that the Member States are largely free to determine their policies on a national basis. Finally, we may note that this methodology can be set at the national or the regional level.[81] Use of the latter option may complicate the implementation process, but it is standing case law that the division of powers at the national level does not affect a Member State's responsibility to ensure correct implementation.[82]

Article 5 of the Directive constitutes a major innovation in relation to the previous Directive. It contains a framework for the calculation of cost-optimal levels of minimum energy performance requirements. It envisages the adoption by the Commission of a comparative methodology and this has resulted in Delegated Regulation 244/2012.[83] Interestingly, this provision has been significantly expanded in comparison to the Commission proposal. Particularly noteworthy is the requirement for the Member States to compare the minimum energy performance standards in force with the cost-optimal levels calculated[84] and to justify such differences where they are 'significant'. This vague concept is clarified in the preamble at the level of 15 per cent.[85] However, in view of the limited legal effects of preambles to EU legislation according to the Court's standard case law, there still remains considerable uncertainty.[86]

Different types of buildings

The directive applies to buildings but also to those parts that belong to buildings, such as systems for heating, cooling and water supply (generally referred to as technical building systems).[87] In addition, the Directive differentiates between new and existing buildings, but also 'nearly zero-energy buildings'.

For new buildings, Member States, pursuant to Article 6(1), are under an obligation to take the necessary measures to ensure that new buildings meet the minimum efficiency requirements set on the basis of Article 4. Only some specific buildings are exempted from this obligation. These include buildings with special architectural or historical merit and places of worship (Article 4 para 2). In addition, the Member States are obliged to ensure that the feasibility of alternative high-efficiency systems is taken into account. Given that this feasibility involves technical, environmental and economic aspects, cost-effectiveness is again clearly pivotal to the Directive.

---

[80]   For an overview see Commission Working Document on Financial Support for Energy Efficiency in Buildings, SWD (2013) 143 final, pp.6, 7.

[81]   An analysis of the possible heterogeneity of transposition can be found in E. Annunziata, 'Towards Nearly Zero-Energy buildings: The State-of-Art of National Regulations in Europe' [2013] *Energy*, 125–33.

[82]   For example Case C-225/96 *Commission v Italy* ECLI:EU:C:1997:584.

[83]   See further below section 6.4.2.

[84]   Article 5(2) Energy Performance of Buildings Directive.

[85]   Recital 14 of the preamble.

[86]   For example Case C-162/97 *Nilsson and Others* ECLI:EU:C:1998:554, para. 54 and Case C-136/04 *Deutsches Milch-Kontor* ECLI:EU:C:2005:716, para. 32.

[87]   See Article 2 and Article 9.

The regime for existing buildings is laid down in Article 7. The Directive provides the Member States with a large degree of flexibility. This follows, among others, from the fact that the Member States shall only *encourage the consideration* of high-efficiency alternative systems, again insofar as this is technically, environmentally and economically feasible.

A major part of the energy use of buildings comes from what the Directive calls 'technical building systems'. This essentially means the heating, cooling and lighting equipment, as defined in Article 2(3) of the Directive, in the building. Article 8 has been amended quite extensively to include rules on the requirements of charging points for electrical vehicles. These reflect the obvious costs involved in these measures and the fact that not every owner or occupant will be able to fund these costs.[88] These provisions, again, clearly indicate the importance of cost-effectiveness in the scheme of the Directive.

A special category of buildings is regulated under the heading of 'nearly zero-energy buildings'. The Directive sets deadlines before which new buildings have to meet the standards for nearly zero-energy buildings. Also in line with the Energy Efficiency Directive we see that the government should lead by setting the example. This means that from 2019 onwards the new buildings owned and occupied by public authorities have to be nearly zero-energy, whereas for all other new buildings this deadline is 2021.[89] To attain this objective, plans that are part of the integrated national energy and climate plans have to be drawn up and will include, inter alia, intermediate energy efficiency targets for new buildings in 2015. Again, we see that a cost–benefit analysis is integrated into the provision and may result in the disapplication of the central obligation.[90] Interestingly, the Directive does not address the situation where a new building is partly occupied by a public authority and a commercial party before 2021. Would such a building then also have to be nearly zero-energy, or does this obligation apply only when a certain threshold is exceeded?

A recurring problem with energy efficiency is that the initial investment may be significant and recouped only over (a long) time. This especially holds true for the investments in buildings. It is further exacerbated in relation to rented housing, where the investment is undertaken by the landlord whereas the tenant reaps the benefits of a lower energy bill. To overcome these problems, Article 10 envisages a duty for the Member States to take the appropriate steps, which may include financial instruments, such as subsidy schemes.

Energy labels for buildings

Again, awareness is a major issue and to this end the Directive contains a number of provisions concerning energy performance certificates. According to the Directive, all buildings need to be awarded an energy performance certificate, often referred to as an energy label. The Directive defines the minimal amount of information that needs to be in such a certificate and the issuing of such certificates. Similar to the rules on nearly zero-energy building,

---

[88] Cf. the possibility for the Member States to exclude buildings owned and occupied by small and medium-sized enterprises from the obligation to include such charging infrastructure.

[89] Article 9(1)(a) and (b) Energy Performance of Buildings Directive.

[90] Article 9(6) Energy Performance of Buildings Directive.

stricter rules apply to public buildings.[91] In all cases, the Member States are to ensure that such certificates are handed out or made available to new tenants or buyers of a building. These certificates are to be issued by independent and accredited experts in accordance with Articles 17 and 18. In this regard it is interesting to know that the Member States are to ensure effective, proportionate and dissuasive sanctions for violations of the rights and obligations laid down in the Directive, including the obligation to present a certificate to the new owner or tenant of a building.[92]

The Directive contains a specialised inspection regime for heating and air-conditioning systems (Articles 14–16) that exceed the household scale. In this regard we note that certain heating systems may also be subject to the Directive on Industrial Emissions, which sets further and specific inspection and reporting obligations.[93]

6.4.2   Defining Cost-effectiveness

The fact that cost-effectiveness is central to the Directive, in combination with the leeway provided to the Member States in defining what exactly is cost-effective, necessitates more guidance. This follows from the fact that determining what is cost-effective may also impact general cost structures and thus undistorted conditions of competition. This is addressed by Article 5 of the Directive, which essentially enables the Commission to establish a comparative methodology framework on the basis of the steps set out in Annex III.

On the basis of this the Commission has adopted Delegated Regulation 244/2012/EU,[94] supplemented by Guidelines.[95] Both the Regulation and the Guidelines are highly technical and detailed documents that set out a methodology to be applied by the Member States that will enable them to compare, for different categories of buildings, the relative costs and benefits of various energy efficiency measures.

Beyond the energy system as a whole and buildings as major sources of energy consumption, the myriad devices and appliances that we use also consume energy and thus offer potential for energy savings.

6.5   ENERGY EFFICIENCY OF APPLIANCES

With regard to appliances, there are essentially two categories of energy efficiency-related rules. First, there are rules that regulate the actual energy efficiency of certain appliances. Second, we find rules that basically prescribe an energy labelling obligation for certain prod-

---

[91]   Article 12(1)(b) and 13(1) Energy Performance of Buildings Directive.

[92]   Article 27 Energy Performance of Buildings Directive.

[93]   Directive 2010/75 on industrial emissions replacing the Directive on Integrated Pollution Prevention and Control, OJ 2010 L 334/17. This Directive applies to industrial combustion installations with a rated thermal input exceeding 50 MW.

[94]   OJ [2012] L 81/8.

[95]   OJ [2012] C115/1.

ucts, in line with the energy performance certification for buildings. The latter category ties in with the observation, made above, that consumer awareness has a major role to play in bringing about increased energy efficiency. This awareness may result from clearer energy bills, real-time information on energy consumption and energy performance certificates for buildings. However, there are many other products and appliances that use energy, to varying degrees of efficiency.

6.5.1   Labelling Regulation

Regulation 2017/1369 (the Labelling Regulation) intends to create a clear and uniform framework to enable the potential purchaser of these products to make an informed choice taking into account energy efficiency.[96]

The Labelling Regulation replaces the Labelling Directive from 2010, which in turn replaced Directive 92/75/EC on the energy labelling of household products for the sake of clarity, as this Directive had been amended several times.[97] The Labelling Regulation not only updates the Labelling Directive to take account of technological advances, but also changes the legal instrument to a regulation in order to ensure a more level playing field in the EU.[98] The Labelling Regulation is a framework measure that confines itself to stating the basic rules and enabling the Commission to adopt delegated acts and take various implementing measures that then, for example, set the rules applicable to the labelling of specific categories of products or the rescaling of these labels.[99]

The framework character of the Labelling Regulation means that the actual rules are laid down in delegated acts. These are acts adopted by the Commission and that are binding for the entire European Union.[100] Such delegated acts have been adopted for numerous products, such as heaters, vacuum cleaners[101] and tumble driers.[102] These delegated acts will set out a basic obligation for the suppliers of the products concerned to ensure that there is an energy efficiency label. Furthermore, the layout, colours and text of the labels concerned are regulated. Most importantly, the methodology for calculating the applicable energy efficiency class is set out, as well as the boundaries for the different classes. According to Article 16 of the

---

[96]   OJ 2017 L 198/1. This replaces the Labelling Directive, Directive 2010/30, OJ [2010] L 153/1.

[97]   Recital 1 of the preamble to the Labelling Directive. Directive 92/75/EC on the indication by labelling and standard product information of the consumption of energy and other resources by household appliances was published in OJ [1992] L 297/16.

[98]   Recital 6 of the preamble to the Labelling Regulation.

[99]   Articles 10–18 Labelling Regulation.

[100]   See, in general, Article 290 TFEU. For a discussion see F. Amtenbrink and H. Vedder, *EU Law: A Textbook*, Eleven Publishing, 2021, p.189.

[101]   An indication of the level of technical detail that is involved can be seen in the appeal lodged by vacuum cleaner manufacturing company Dyson in relation to the testing of such cleaners with empty bags only: see Case T-544/13, OJ [2013] C 344/68.

[102]   An overview is available at: http://ec.europa.eu/energy/efficiency/labelling/doc/overview_legislation_energy _labelling_household_appliances.pdf last accessed 22 June 2021.

Regulation, such acts shall be drawn up for all products for which there is a significant potential for energy saving, where the products available have a wide range of energy efficiency and
where the energy efficiency does not result in negative consequences in terms of functionality
or affordability.[103] In other words: labelling shall only be required where significant energy
savings are possible and where there is a choice available in terms of energy efficiency.

Energy efficiency labels

The Regulation provides for a system of energy efficiency labelling whereby the energy performance of products is categorised as being in a scale that ranges from A (most efficient)
to G (least efficient). The Labelling Regulation has led to a rescaling of the existing labels.
Experience with the earlier Directive has taught the legislature that the top levels of the labels
tend to be overcrowded and – at times – even outperformed, leading to multiple products in
the A+++ category. To avoid this, the Regulation envisages a regular rescaling of the labels to
the effect that at the time of the rescaling no products qualify for the A label.[104] The idea behind
this is that the regular rescaling encourages competition on innovation to increase energy efficiency and avoids the need for regular rescaling.[105] The Labelling Regulation first and foremost
requires the suppliers to ensure that energy efficiency labels are attached to all products for
which labelling requirements have been set on the basis of the Regulation or the delegated acts
adopted on the basis of it.[106] This labelling requirement also applies to retailers, who are under
the obligation to display such labels in a proper and clearly visible manner.[107] The information
in the label must be correct and drawn up in accordance with the methodology laid down
in the Directive and the delegated acts. The responsibility for gathering the information and
drawing up the energy efficiency label rests with the supplier, who is to provide these labels
free of charge to dealers.[108] Most probably in response to what is called 'Dieselgate', suppliers
are now explicitly prohibited from placing on the market products for which the performance
has been automatically altered to 'cheat' the energy efficiency tests.[109]

Given that such labels must be attached to products and that these products may not be
marketed without the relevant energy efficiency labels, it should not come as a surprise that the
Labelling Regulation contains a free movement clause. According to Article 7, the marketing
of products falling under the Regulation shall not be restricted or impeded. This means that

---

[103] Article 16(2) Labelling Regulation.

[104] Article 11(8) Labelling Regulation.

[105] Recital 18 of the preamble to the Labelling Regulation.

[106] Article 3 Labelling Regulation.

[107] Article 5 Labelling Regulation. On the temporal scope of this obligation see Case C-319/13, *Udo Rätzke v S+K
Handels*, ECLI:EU:C:2014:210, paras 27–41, where the Court held that the duty to display the label applies only to
products dispatched into the sales chain from the date of applicability of the delegated act that sets the label for that
product.

[108] Article 3(1) and (2) Labelling Regulation.

[109] Article 3(5) Labelling Regulation.

any product that complies with the Regulation and the relevant delegated act benefits from the free movement of goods.

Interestingly, the Labelling Directive required the Commission to take account of voluntary, industry-based energy efficiency measures, to avoid adopting EU-wide measures. Such industry-based initiatives are no longer relevant under the Regulation. This, together with the ban on products that automatically alter their performance during testing, appears to indicate a lower degree of trust in the industry.

EU case law

Concerning energy efficiency labelling, quite a few cases have been brought by Dyson in relation to the energy efficiency labelling of vacuum cleaners. These cases all turn on the unique bagless construction of Dyson vacuum cleaners and the fact that other vacuum cleaners will lose suction power when the bag fills up. This obviously reduces energy efficiency, raising the question to what extent the testing method laid down in the Commission delegated regulation actually corresponds to reality. After several instances, Dyson obtained a ruling from the Court to the effect that the Commission's method, which entailed measuring energy consumption with an empty bag, did not correspond to the 'actual conditions of use'.[110] As a result, the Court found that the Commission had overstepped the boundaries of the powers delegated to it by the Labelling Directive.[111] On a similar note, the Court had to rule whether the sale of a bagged vacuum cleaner with an A label that was awarded in accordance with the Labelling Directive and the delegated regulation could mislead consumers and thus constitute an unfair commercial practice. This case boils down to the question whether the EU labelling rules themselves result in misleading advertising. The Court comes to the conclusion that this is not the case.[112] It further finds that the information on the label may not be supplemented by extra information on the energy and environmental performance of the products in question if that extra information is likely to mislead customers.[113] These cases show above all the increasing importance of energy efficiency, both as a design requirement and an element of competition between manufacturers.

6.6 ENERGY EFFICIENCY OF VARIOUS OTHER PRODUCTS

The EU allows Member States to make use of a wide range of instruments to enhance the energy efficiency of products and sectors. These include the use of voluntary agreements. Additionally, pursuant to the Labelling Regulation, standards have been set for all sorts of

---

[110] Case C-44/16P *Dyson v Commission* ECLI:EU:C:2017:357, paras 60–68.

[111] The General Court, hearing the case when it had been referred back to it by the Court for final adjudication, came to the conclusion that the Commission had indeed overstepped the limits of the powers that had been delegated to it: Case T-544/13 RENV *Dyson v Commission* ECLI:EU:T:2018:761.

[112] Case C-632/16 *Dyson v BSH Home Appliances* ECLI:EU:C:2018:599, paras 43–46.

[113] Case C-632/16 *Dyson v BSH Home Appliances* ECLI:EU:C:2018:599, paras 56–58.

different products. In this regard we may refer to the Energy Star package concerning ICT equipment.[114] This is a voluntary scheme that enables manufacturers to place the Energy Star label on their office equipment provided that the criteria are complied with. The Energy Star label rules go beyond a mere labelling scheme as the label does not indicate the actual energy efficiency of the product, but rather imposes a singular threshold that is either complied with or not. Given that the Energy Star scheme is of US origin, its application in the EU required the signing of an international agreement that would govern cooperation on setting the standards and other practicalities involved.[115]

The signature of this agreement gave rise to a case on the correct legal basis for doing so.[116] This case centred essentially on the issue of whether such energy efficiency requirements are part of the common commercial policy or rather fall under the environmental protection heading. Applying its standard centre of gravity test, the European Court of Justice came to the conclusion that the measure was of a predominant common commercial policy nature, with the environmental effects arising only indirectly as a result of changes in purchasing and use of the equipment concerned. This case shows the essence of all energy efficiency labelling regulations in that they pursue an environmental benefit indirectly, through influencing manufacturing and purchasing, and thus market-based decisions. The Energy Star Regulation does not contain any energy efficiency criteria itself but instead provides for a framework that should enable such standards to be developed. This involves the EU Energy Star Board, the Commission and the US Environmental Protection Agency. As participation in the scheme is voluntary, there is no free movement clause as there is in the Labelling Directive.

Similar measures have been adopted for other categories of products as well, such as car tyres.[117] Of a slightly different nature is the Regulation that sets minimum emission performance standards for cars and other vehicles.[118] This works on the basis of the CO_2 efficiency of the fleet of cars sold by a given manufacturer, rather than setting specific targets for the individual cars that are marketed. This means that a manufacturer, or a group of manufacturers, must ensure that on average the cars sold by it will become more energy efficient (emit less CO_2 per kilometre) over the years. This measure has resulted in the appearance of many energy-efficient cars on the market, as manufacturers need to ensure that the total fleet of cars marketed by them meets the increased energy efficiency standard. Of course, marketing such more efficient cars is relatively easy for large manufacturers with a wide array of models, whereas the producers of more exotic, high-powered vehicles may find marketing

---

[114] Currently Regulation 106/2008 on a Community energy-efficiency labelling programme for office equipment, OJ [2008] L 39/1, as amended by Regulation 174/2013, OJ [2013] L 63/1, hereafter: Energy Star Regulation.

[115] The signature of this agreement was approved by means of Decision 2006/1005/EC, OJ [2006] L 381/24.

[116] Case C-281/01 *Commission v Council* ('Energy Star') ECLI:EU:C:2002:761.

[117] Regulation 1222/2009 on the labelling of tyres with respect to fuel efficiency and other essential parameters, OJ [2009] L 342/46.

[118] Regulation 443/2009 setting emission performance standards for new passenger cars as part of the Community's integrated approach to reduce CO_2 emissions from light-duty vehicles, OJ [2009] L 140/1, as amended by Commission Delegated Regulation 2018/649, OJ 2018 L 108/14.

more economic cars more challenging. This has resulted in some exceptions to the scope and the possibility of joining forces to meet the tightening efficiency objectives. Here we see that, rather than relying on market forces, the EU legislator has thus adopted – albeit with a large degree of flexibility – a more traditional regulatory approach that simply imposes a binding environmental performance requirement upon the industry.

In addition to the special rules for the automobile sector, there are energy efficiency requirements set for energy using products pursuant to the Eco-design Directive.[119] This Directive is, just like the energy-labelling rules, primarily aimed at completing the internal market by avoiding distortions of competition or restrictions on the free movement of goods as a result of diverging national rules on the eco-design of products.[120] Such eco-design rules cover a wider range of concerns than just energy use, such as the ease with which a product can be recycled; also the consumption of, inter alia, energy throughout the life cycle is an aspect of the eco-design.[121] The idea behind this Directive is that producers will design their products in a way that minimises the environmental impact, for example by saving energy.[122] This is implemented by means of Commission regulations that prescribe specific eco-design criteria for categories of products that must be complied with. For dishwashers, for example, the relevant regulation prescribes the presence of an 'eco-mode' from 1 March 2021 onwards. Moreover, manufacturers will need to meet certain (energy-)efficiency conditions with the eco-programmes on the dishwashers they market from that date onwards.[123]

Finally, we refer to the various voluntary agreements on energy efficiency that have been drawn up by the industry involved. There was such an agreement between the European car manufacturers (ACEA) and the Japan (JAMA) and Korean (KAMA) car manufacturers associations. However, the failure of those agreements to deliver results triggered the adoption of the Regulation on energy efficiency standards for cars.[124] This shows that increases in energy efficiency require a strict and binding framework for the industry to deliver. This follows from the observation already made above that in general more energy-efficient products are more advanced and thus command a higher purchasing price that will only be recouped in the longer term.[125] This in turn means that there is a market for cheap, inefficient products, the demand for which is met by industry.

The stringency of the measures required can be seen in the *CECED* case, which involved the decision by the European Washing Machines Industry Association to phase out energy-inefficient washing machines. The conditions attached to this agreement were so

---

[119]   Directive 2009/125, OJ 2009 L 285/10, as amended by Directive 2012/27/EU, OJ 2012 L 315/1.

[120]   Article 6 of the Eco-design Directive thus contains a free movement clause for products that comply with the rules laid down in the Directive.

[121]   Eco-design Directive, Annex I, part 1.3(c).

[122]   Recital 4 of the preamble to the Eco-design Directive.

[123]   Commission Regulation (EU) 2019/2022 laying down eco-design requirements for household dishwashers, OJ 2019 L 315/267; these conditions are included in Annex II to that Regulation.

[124]   Cf. the comments above in section 6.5 on the generally reluctant stance towards industry-based initiatives.

[125]   Section 6.3.

stringent that they were considered to restrict competition within the meaning of Article 101 TFEU. However, the environmental benefits resulting from increased efficiency meant that the agreement qualified for an exemption on the basis of Article 101(3) TFEU.[126]

6.7 CONCLUSION

Despite the rationality and increasing (political) importance of investing in energy efficiency, actual increases in energy efficiency are currently insufficient. In addition, the effectiveness of most of the measures involved can be doubted. The initially higher purchasing costs of, for example, an eco-friendly washing machine will be recouped only by means of reduced energy costs in the longer term, which apparently insufficiently informs purchasing decisions. Increasing energy efficiency ultimately requires an 'enlightened consumer' who makes a rational choice for longer-term savings. It may be doubted whether regulation will change purchasing behaviour to subsequently change production decisions, triggering competition for energy efficiency. However, the *CECED* case and the Regulation on emissions performance standards for cars shows that changes in production are realistic, provided there is a strict legal framework.

These findings translate into an ultimate question on the degree of solidarity and cost-effectiveness involved in defining energy efficiency standards that may be at odds with the subsidiarity principle. An increasingly uniform definition of cost-effectiveness necessitates a Europe-wide decision on inter-state solidarity, much in line with what has been seen regarding the Effort Sharing legislation. At the same time, such a decision will impact the myriad energy-related decisions that are taken throughout the EU on a daily basis, greatly reducing national sovereignty on these matters. Perhaps in response to these national concerns, the EU has provided a more prominent place to funding schemes to enhance energy efficiency. Such schemes are less intrusive and influence the decentralised decision-making that drives much of the energy efficiency policies, and notably those relying on incentivising energy-efficient consumption.

We conclude that the effectiveness of energy efficiency measures can be doubted as a result of: (a) the non-binding nature of the national energy efficiency targets; (b) the general reluctance on the part of the Member States to submit to strict and binding energy efficiency targets with concomitant measures; and (c) the poor EU track record for increasing energy efficiency.

---

[126] Commission Decision 2000/475 *CECED*, OJ [2000] L 187/47. For a fuller discussion see H.H.B. Vedder, *Competition Law and Environmental Protection in Europe; Towards Sustainability?* Europa Law Publishing, 2003, pp.125 and further.

Classroom Questions

1. What are the major EU instruments to increase energy efficiency and how do these relate to each other?
2. How and why may stricter EU energy efficiency law meet with opposition from the Member States?
3. Design a more effective EU energy efficiency policy with a specific measure. How would you ensure higher degrees of energy efficiency using legal instruments?

SUGGESTED READING

Books

Schubert VSR, Pollak J and Kreutler M, *Energy Policy of the European Union (Palgrave 2016).*

Articles and chapters

Huhta K, 'Prioritising Energy Efficiency and Demand Side Measures over Capacity Mechanisms under EU Energy Law' (2017) 35(1) Journal of Energy & Natural Resources Law 7–24.

Roggenkamp MM, 'Regulating Energy Efficiency in the European Union' in MM Roggenkamp, KJ de Graaf and R Fleming (eds), *Energy Law and the Environment (Edward Elgar Publishing 2021).*

Schomerus T, 'Energy Efficiency and Energy Saving: The "First Fuel"' in M Peeters and M Eliantonio (eds), *Research Handbook on EU Environmental Law (Edward Elgar Publishing 2020).*

Policy documents

D-G Energy website: http://ec.europa.eu/energy/efficiency/index_en.htm.

Economidou M, Todeschi V, Bertoldi P, D'Agostino D, Zangheri P and Castellazzi L, 'Review of 50 Years of EU Energy Efficiency Policies for Buildings (2020) 225 Energy and Buildings.

Ntovantzi M et al. 'Do We Have Effective Energy Efficiency Policies for the Transport Sector? Results and Recommendations from an Analysis of the National and Sustainable Energy Action Plans (2015) ECEEE Summer Study Proceedings 895–906.

7
Carbon capture and storage

ABSTRACT

- Directive 2009/31/EC (CCS Directive) is at the heart of the EU regulatory framework for carbon capture and storage (CCS);
- The CCS Directive leaves a number of important regulatory choices to the Member States, for instance on CO_2 stream purity and financial securities for CO_2 storage;
- From a cost-effectiveness point of view, this has both positive effects in the form of tailor-made regulation and negative effects as it increases transaction costs;
- From a solidarity point of view, the lack of a more detailed and common EU approach for regulating CCS makes it more difficult for Member States with little own CO_2 storage capacity to deploy CCS technologies;
- Offshore deployment of CCS is the latest trend. Not only is the selection of offshore CO_2 storage sites challenging, but there are also several regulatory hindrances that could delay offshore CCS deployment in the EU:
  - The 1996 London Protocol prohibits contracting parties from exporting CO_2 to other countries for offshore storage, although a temporary legal solution to this problem seems to have been found;
  - The EU Greenhouse Gas Emissions Trading System (EU ETS) does not recognize CO_2 transport per ship to a storage location;
  - The selection of offshore CO_2 storage sites is challenging; and
  - It has to be decided how to bridge the temporal gap between the end of hydrocarbon mining and the start of CO_2 storage.

7.1 INTRODUCTION

This chapter explores EU regulation of carbon capture and storage (CCS) technologies. CCS consists of: (a) the capture of CO_2 from industrial installations; (b) its transport to a storage site; and (c) injection into a suitable underground geological formation for the purpose of permanent storage.[1] CO_2 transport and storage technologies have been applied for several decades in the US oil industry for mining purposes. Since the late 1990s, CCS has increasingly

---

[1] Rec. 4 of the preamble to European Parliament and Council Directive 2009/31/EC of 23 April 2009 on the geological storage of carbon dioxide and amending Council Directive 85/337/EC, European Parliament and Council Directives 2000/60/EC, 2001/80/EC, 2004/35/EC, 2006/12/EC, 2008/1/EC and Regulation (EC) No 1013/2006 [2009] OJ L140/114.

come to the fore as an instrument to reduce CO_2 emissions, both in the EU and in the rest of the world. In 2007 the European Commission published a CCS action plan, requiring the development of an EU regulatory framework for CCS.[2] In 2009 this regulatory framework came into being with the entry into force of Directive 2009/31/EC on the geological storage of CO_2 (CCS Directive).

In the following, we examine the EU regulatory framework for CCS in the light of the overarching EU climate policy principles of cost-effectiveness and solidarity. Section 7.2 gives an overview of the concept and technologies of CCS. It briefly explores the different capture, transport and storage technologies to provide a technological background to the regulatory issues addressed in later sections. Section 7.3 explores the legislative approach underlying the CCS Directive, the Directive's scope and EU instruments for stimulating CCS deployment. Section 7.4 examines the Directive's regulation of CO_2 capture, transport and storage. Section 7.5 discusses the 2015 review of the CCS Directive. Section 7.6 deals with several regulatory hindrances to the offshore deployment of CCS technologies. Finally, section 7.7 concludes by answering the question to what extent the overarching principles of EU climate policy are honoured in the CCS Directive and what balance has been struck between them.

7.2 BASICS OF CARBON CAPTURE AND STORAGE

The main application of CCS is likely to be at large sources of CO_2 emissions such as fossil fuel power plants, fuel processing plants and plants for the manufacture of iron, steel, cement and bulk chemicals.[3] The CCS chain is roughly made up of three parts: capture, transport and storage of CO_2. In each part of the CCS chain different entities are active (capturers, transporters and storage operators)[4] and for each part, different technologies exist.

7.2.1 Capture

The *capture* of CO_2 requires separating CO_2 from industrial or energy production-related emissions into relatively pure streams, which can then be pressurized for transport (compression). There are four basic methods for capturing CO_2 from use of fossil fuels: capture from industrial process streams, post-combustion capture, oxy-fuel combustion capture and pre-combustion capture.[5]

---

[2]   Recital 7 Directive 2009/31/EC.

[3]   Intergovernmental Panel on Climate Change (IPCC), *IPCC Special Report on Carbon Dioxide Capture and Storage. Prepared by Working Group III of the Intergovernmental Panel on Climate Change* (Cambridge University Press 2005), pp.111–13.

[4]   Hypothetically, it is possible that one undertaking performs all activities in the CCS chain.

[5]   R. Surampalli, S. Brar and B. Gurjar, *Carbon Capture and Storage: Physical, Chemical, and Biological Methods* (ASCE 2015), p.9. The below four paragraphs are based on this source.

An example of CO_2 *capture from industrial process streams* is the removal of CO_2 from raw natural gas (natural gas purification). Most of the techniques for CO_2 capture from industrial process streams are similar to those used in pre-combustion capture.

In *post-combustion capture technology*, the flue gas, instead of being discharged directly into the atmosphere, is passed through equipment which separates most of the CO_2. This is usually done by means of a chemical solvent that reacts with CO_2.

The *oxy-fuel combustion process* eliminates nitrogen from the flue gas by combusting a hydrocarbon fuel in pure oxygen (instead of air). This produces a nitrogen-free flue gas with water vapour and a high concentration of CO_2 as its main components. The resulting flue gas is then further concentrated into an almost pure stream of CO_2, which can be transported for storage.

Pre-combustion capture involves the removal of CO_2 from the primary fuel before combustion takes place. The fuel is reacted with air or oxygen to produce a fuel that contains carbon monoxide and hydrogen. The resulting gas is then reacted with steam to produce a mixture of CO_2 and hydrogen, after which the CO_2 is separated and transported for storage.

The differences between the various capture techniques lie in factors such as: (a) the installation they are being applied to (CO_2 capture from industrial process streams versus CO_2 capture in electricity production facilities); (b) the phase in which the CO_2 sequestration takes place (before or after fuel combustion); and (c) the substances used (pure oxygen instead of air).

In addition to these four basic capturing methods, a promising new capture technology is being developed: *chemical looping combustion* involves the use of a metal oxide or other compound to transfer oxygen from combustion air to fuel.[6] Since direct contact between fuel and combustion air is avoided, the products from combustion (CO_2 and H_2O) will be kept separate from the rest of the flue gases. The advantage of this technology is that CO_2 removal by means of a chemical solvent is no longer required.

It is important to underline that CO_2 capture and compression require energy and thus reduce the efficiency of a power plant in converting fuel into electricity (thermal efficiency).[7] CO_2 capture and compression, in other words, increase the amount of fossil fuel used to achieve a given power generation output. This is the so-called energy penalty of CCS.

7.2.2   Transport

There are many ways to *transport* captured CO_2, such as by pipeline, marine tanker, train, truck or compressed gas cylinder. Pipelines and tankers are generally considered to be the most economically viable modes for transporting the large quantities of CO_2 that are associated with centralized collection hubs or CO_2 emitters such as large power plants.[8]

---

    6   T. Mendiara et al, 'Chemical Looping Combustion of Biomass: An Approach to BECCS', *Energy Procedia*, Volume 114, July 2017, p.6022.

    7   Kurt Zenz House, 'The Energy Penalty of Post-combustion CO_2 Capture & Storage and Its Implications for Retrofitting the U.S. Installed Base', *Energy & Environmental Science*, Issue 2, 2009, p.193.

    8   www.globalccsinstitute.com/about/what-is-ccs/transport/ accessed 6 January 2021.

7.2.3   Storage

CO_2 *storage* can also take place by various means. Captured CO_2 can, for instance, be used for the production of biomass or even food and beverages. Nevertheless, geological storage is generally seen as the most promising form of CO_2 storage in terms of storage safety, capacity and duration. In geological storage, CO_2 is injected under high pressure into deep, stable rocks in which there are countless tiny pores that trap the CO_2. Oil fields, depleted gas fields, deep coal seams and saline formations (sediment or rock body containing brackish water or brine) are all possible storage formations.[9] There are however significant differences between depleted oil/gas fields and saline formations. There are four primary components that determine a storage site's appropriateness for CO_2 storage: containment, injectivity, connectivity and capacity.[10] During the storage site selection process, these criteria have to be taken into account.

A combination of physical and geochemical trapping mechanisms determines the effectiveness of geological CO_2 storage. *Physical* trapping mechanisms range from the CO_2 being trapped under impermeable rocks to processes whereby CO_2 dissolves in the water present in the storage formation and then migrates with the groundwater or is trapped in the tiny pores of rocks. Eventually, the injected CO_2 will be confined by *geochemical* trapping mechanisms, consisting of a sequence of geochemical interactions between CO_2 on the one hand and rock and formation water on the other.

The first engineered injection of CO_2 into subsurface geological formations took place in Texas (US) in the early 1970s, as part of a project that used CO_2 to recover remaining oil reserves. This is known as 'enhanced oil recovery' (EOR). Ever since, geological storage of CO_2 has taken place at many locations worldwide. The world's first large-scale CO_2 storage project was initiated by Statoil (and its partners) at the Sleipner gas field in the North Sea in 1996. This gas purification project stores about 1 million tonnes of CO_2 per year. In 2017, the Sleipner project had stored over 20 million tonnes of CO_2.[11] The Global CCS Institute identified 19 large-scale CCS projects world-wide in 2019.[12] Due to public opposition to onshore CCS, project developers have become more interested in storing CO_2 offshore. In particular, the future for CCS in Europe seems to lie in the North Sea, with its many suitable storage locations.

EU policy on CCS originally centred on two political objectives.[13] The first goal was that by 2020 all new fossil fuel power plants should capture their CO_2 emissions. Second, by 2015 up to 12 CCS demonstration projects in commercial power generation should have been realized. These projects were intended to test the integration of CCS in coal- and gas-fired power gener-

---

[9]   Intergovernmental Panel on Climate Change (IPCC), *IPCC Special Report on Carbon Dioxide Capture and Storage. Prepared by Working Group III of the Intergovernmental Panel on Climate Change* (Cambridge University Press 2005), p.199.

[10]   IEAGHG, 'Re-use of Oil & Gas Facilities for CO_2 Transport and Storage', 2018/06, July 2018, p.28.

[11]   Global CCS Institute, 'The Global Status of CCS: 2018', p.75.

[12]   Global CCS Institute, 'The Global Status of CCS: 2019', p.71.

[13]   European Council, 'Brussels European Council 8/9 March 2007', 7224/1/07, p.22.

ation.[14] In the next sections, we discuss the regulation that applies to these and other EU CCS projects.

7.3   THE CARBON CAPTURE AND STORAGE DIRECTIVE

At the heart of the EU regulatory framework for CCS is Directive 2009/31/EC, commonly referred to as the CCS Directive. The CCS Directive is one of the first dedicated legal frameworks for the deployment of CCS worldwide. It provides a legal framework for the safe geological storage of CO_2 in the EU (Art. 1(1)). In the following, we explore the CCS Directive by examining, respectively, the Commission's approach in drafting its original proposal for the Directive, the Directive's scope, instruments for stimulating CCS deployment and the way in which the three parts of the CCS chain are regulated.

7.3.1   The Commission's Legislative Approach

As its name suggests, the Directive on the geological *storage* of CO_2 is primarily a Directive on the storage of CO_2. In drafting its proposal for the CCS Directive, the Commission took a conservative approach.[15] The default option for regulating each of the components of the CCS chain was the existing EU legal framework regulating activities of a similar risk.[16]

Since CO_2 capture, in its view, presented risks similar to those of the chemical and power generation sector, the Commission considered then Directive 96/61/EC (IPPC Directive)[17] a suitable regulatory framework for the activity. Likewise, since it deemed CO_2 transport to present risks similar to those of natural gas transport, the Commission chose to regulate it in the same manner.

By contrast, in relation to CO_2 storage, the Commission believed the existing legal framework (IPPC Directive and EU waste legislation)[18] was not well adapted to regulating the risks of this particular activity. Therefore, it decided to develop a free-standing legal framework for CO_2 storage – the CCS Directive – and to remove CCS deployment under the CCS Directive from the scope of EU waste legislation.

---

[14]   Commission, 'Communication from the Commission to the Council and the European Parliament – Sustainable power generation from fossil fuels: aiming for near-zero emissions from coal after 2020' COM (2006)843 final, p.6.

[15]   Commission, 'Proposal for a Directive of the European Parliament and of the Council on the geological storage of carbon dioxide and amending Council Directives 85/337/EEC, 96/61/EC, Directives 2000/60/ EC, 2001/80/EC, 2004/35/EC, 2006/12/EC and Regulation (EC) No 1013/2006' COM (2008)18 final.

[16]   Ibid, pp.2–3.

[17]   Council Directive 96/61/EC of 24 September 1996 concerning integrated pollution prevention and control [1996] OJ L257.

[18]   The latter refers to European Parliament and Council Directive 2006/12/EC of 5 April 2006 on waste [2006] L114/9 (Waste Directive) and European Parliament and Council Regulation (EC) 1013/2006 of 14 June 2006 on shipments of waste [2006] L190 (Shipments of Waste Regulation).

In addition, the Commission chose to give itself a large role in the application of the Directive. The Commission, among other things, reviews draft permits for CO_2 storage as well as draft decisions to transfer responsibility for the storage site to the competent authorities (post-closure),[19] while it has the possibility to issue guidelines on various matters.[20] Given the limited experience with the large-scale deployment of CCS technologies, the Commission wanted to ensure a consistent application of the CCS Directive in the various Member States and help enhance public confidence in the safety of CCS as an instrument to reduce CO_2 emissions.[21]

7.3.2   Scope of the CCS Directive

Article 2 provides that the CCS Directive applies to the geological storage of CO_2 in the territory of the Member States, their exclusive economic zones and their continental shelves within the meaning of the United Nations Convention on the Law of the Sea (UNCLOS). Storage in a storage site extending beyond these areas is not allowed. Following the OSPAR Convention, the CCS Directive further expressly prohibits CO_2 storage in the water column.[22] By virtue of the definition of the term 'water column' in Article 3(2), this prohibition includes CO_2 storage in the deep water of or on the bottom of lakes. Also, small-scale R&D CO_2 storage projects (<100 kilotonnes) are excluded from the scope of the Directive.

Recital 20 of the preamble to the CCS Directive states that the injection of captured CO_2 to recover hydrocarbons such as oil and gas (enhanced hydrocarbon recovery) is not included in the Directive. Yet, it also provides that where enhanced hydrocarbon recovery is combined with geological storage of CO_2, the provisions of the CCS Directive should apply. Article 3(1) of the Directive defines the 'geological storage of CO_2' as 'injection accompanied by storage of CO_2 streams in underground geological formations'. Following the definition in Article 3(1), the question is whether the distinction made in recital 20 between enhanced hydrocarbon recovery and geological CO_2 storage is tenable.

In practice, both CO_2 storage projects and enhanced hydrocarbon recovery projects will likely lead to a significant amount of the injected CO_2 being geologically stored. Enhanced hydrocarbon recovery has a long history,[23] and historically most CO_2 has been stored in the context of enhanced hydrocarbon recovery.[24] On the basis of the definition in Article 3(1), we

---

[19]   Articles 10 and 18 of the Directive 2009/31/EC. We further discuss this in section 7.4.

[20]   See Articles 12(2) and 18(2) Directive 2009/31/EC.

[21]   See recital 33 of the preamble to the CCS Directive and Commission, 'Questions and Answers on the Proposal for a Directive on the Geological Storage of Carbon Dioxide', MEMO/08/36 https://ec.europa.eu/commission/presscorner/detail/en/MEMO_08_36 accessed 6 January 2021.

[22]   The OSPAR Convention is a regional treaty on the protection of the marine environment in the North-East Atlantic. By virtue of Decision 2007/1 of the OSPAR Commission, the placement of CO_2 streams in the water column or on the seabed is, with a few exceptions, prohibited. The EU and 13 EU Member States are contracting parties to the OSPAR Convention, named OSPAR from the original Oslo ('OS') and Paris ('PAR') Conventions.

[23]   IEA, '20 Years of Carbon Capture and Storage', 2016, p.20.

[24]   Global CCS Institute, 'The Global Status of CCS: 2018', p.18.

argue that enhanced hydrocarbon recovery activities, contrary to that alleged in recital 20,[25] do fall within the scope of the CCS Directive. This would mean that the storage permitting requirements outlined in Section 7.4 would also be applicable to these activities.

7.3.3   Incentivizing CCS Deployment

The main incentive for CO_2 emitters to deploy CCS technologies is provided by the EU ETS.[26] The idea is that every tonne of geologically stored CO_2 will count as not having been emitted under the EU ETS. Accordingly, it does not have to be covered by an emission allowance. By storing their CO_2 emissions, holders of an EU ETS emissions permit, such as power plants and refineries, can either sell the emission allowances they keep or not buy any allowances in the first place.

If the price of an emission allowance under the EU ETS is higher than the costs of capturing, transporting and geologically storing a tonne of CO_2, holders of an EU ETS emissions permit will have a financial incentive to deploy CCS technologies. Once CCS becomes a cost-effective option to reduce CO_2 emissions, the participants in the EU ETS could start deploying these technologies.

In phase II of the EU ETS (2008–12), individual CCS projects could already be recognized under the emissions trading scheme through prior approval by the Commission.[27] To provide a long-term incentive for the further development of the different technologies, CCS has been fully recognized as an instrument to reduce CO_2 emissions in the amended (post-2012) EU ETS.[28] Article 12(3)(a) of the EU ETS Directive provides that no emission allowances have to be surrendered for CO_2 emissions that are permanently stored in accordance with the CCS Directive.

In addition to the allowance price incentive, the EU has arranged for various forms of subsidies for CCS deployment. These subsidizing schemes were initially related to phase III of the EU ETS (2013–20). First, Article 10(3) of the EU ETS Directive states that Member States should use at least 50 per cent of the revenues from the auctioning of emission allowances for a number of climate policy instruments, among which is CCS.[29] Second, by virtue of Article 10(a)(8) of the EU ETS Directive up to 300 million allowances in the new entrants' reserve (NER) were available for the construction and operation of the up to 12 planned EU CCS demonstration projects, as well as for innovative renewable energy demonstration projects.

---

[25]   In this regard, it needs to be stressed that the preamble to an EU Directive, contrary to its provisions, has no binding legal force. See, for instance, Case C-162/97 *Nilsson* [1998] ECR I-07477.

[26]   On the (functioning of the) EU ETS, see Chapter 3 of this book.

[27]   See Article 24 of European Parliament and Council Directive 2003/87/EC of 13 October 2003 establishing a scheme for greenhouse gas emission allowance trading within the Community and amending Council Directive 96/61/EC [2003] OJ L275 (EU ETS Directive).

[28]   To this end, capture, transport and geological storage of greenhouse gases (among which CO_2) have been added to Annex I to the EU ETS Directive, which contains the list of activities for which an EU ETS permit is required.

[29]   On the basis of the wording of Article 10(3) Directive 2003/87/EC, we would argue that the Member States, however, are not obliged to spend any auctioning revenues on the development of CCS technologies.

Finally, under the European Economic Recovery Plan, €1 billion has been granted to six EU CCS demonstration projects.

However, in 2018 the European Court of Auditors concluded that progress had so far been insufficient.[30] From the six projects granted subsidies under the European Economic Recovery Programme, four projects ended due to termination of the grant agreement, one project ended without completion and the only project that was completed was not a commercial project but a pilot project. Progress under the NER300 program has likewise been poor. One project in the UK was awarded a €300 million grant, but due to a change in the UK support scheme in 2015 the project was cancelled.

In the future, CCS activities will be further incentivized through an amendment of the EU ETS and its auxiliary subsidizing regime. At the time of writing this chapter, the goal of the fourth phase (2021–30) of the EU ETS is to reduce emissions by 43 per cent compared to 2005 levels (based on the EU's overall reduction target of 40 per cent by 2030, which will probably be increased to 55 per cent). The emissions cap will be lowered to meet these strengthened emission targets, which is likely to make emissions more expensive and CCS more attractive. In addition, two auxiliary financial support mechanisms have been put in place: the ETS Innovation Fund and the Modernisation Fund. The ETS Innovation Fund is to replace the NER300 programme. The amount of funding available under the ETS Innovation Fund will correspond to the market value of at least 450 million emission allowances. The ETS Innovation Fund will support renewable energy and CCS development for the period 2021–30. The Modernisation Fund will support investment in modernizing the power sector and wider energy systems, boosting energy efficiency and facilitating the transition in carbon-dependent regions of ten lower-income Member States.[31]

These subsidies are meant to bridge the gap between the costs of CCS deployment (per tonne of CO_2 emissions avoided) and the emission allowance prices under the EU ETS. While the different subsidies contribute to making CCS a financially more attractive instrument for reducing CO_2 emissions, they distort the EU emissions trading market. The effect of subsidizing specific climate technologies under the EU ETS is greater deployment of relatively expensive technologies, eventually leading to lower emission allowance prices and raising overall CO_2 emissions reduction costs.[32] EU subsidization of CCS technologies, in other words, comes at the expense of the cost-effectiveness of emissions reductions realized under the EU ETS. At the time of writing, the combination of subsidies and the EU ETS together prove to be insufficient for incentivizing the large-scale deployment of CCS in Europe.[33] Project initiators are,

---

[30]   European Court of Auditors, 'Demonstrating Carbon Capture and Storage and Innovative Renewables at Commercial Scale in the EU: Intended Progress Not Achieved in the Past Decade', 2018, p.50.

[31]   These ten lower income Member States are: Bulgaria, Czechia, Croatia, Estonia, Hungary, Latvia, Lithuania, Poland, Romania and Slovakia.

[32]   See: Heleen Groenenberg and Heleen de Coninck, 'Effective EU and Member State Policies for Stimulating CCS', *International Journal of Greenhouse Gas Control* (2008) Volume 2, Issue 4, pp.653–4.

[33]   European Court of Auditors, 'Demonstrating Carbon Capture and Storage and Innovative Renewables at Commercial Scale in the EU: Intended Progress Not Achieved in the Past Decade', 2018, pp.31–2.

therefore, looking into alternative methods of bridging the gap, such as enhanced hydrocarbon recovery or a combination with re-use of infrastructure for CO_2 storage.

7.4 REGULATION OF CO_2 CAPTURE, TRANSPORT AND STORAGE

In the following, we explore the way in which the CCS Directive regulates all three parts of the CCS chain. Most substantive rules in the CCS Directive concern CO_2 storage. Yet, as a consequence of a number of amendments to existing EU (environmental) regulation, the CCS Directive indirectly also regulates the capture and transport phases. In the sections that follow we first deal with the Directive's rules affecting CO_2 capture and transport, after which we examine in more detail the way in which CO_2 storage is regulated.

7.4.1 CO_2 Capture

To enable CO_2 capture, the CCS Directive amends several EU environmental Directives, such as Directives 85/337/EEC (Environmental Impact Assessment Directive),[34] 2008/1/EC (Integrated Pollution Prevention and Control Directive)[35] and 2001/80/EC (Large Combustion Plants Directive).[36] These Directives contain, among other things, emission standards for large industrial installations.

In the following, we examine two issues that are central to EU regulation of CO_2 capture. First, we explore the CCS Directive's requirements on the composition of the CO_2 stream for storage.[37] Second, we address the possibility to oblige large CO_2 emitters to capture their CO_2 emissions. In examining these two issues, we discuss the CCS Directive's most important provisions on CO_2 capture.

---

[34] European Parliament and Council Directive 85/337/EEC of 27 June 1985 on the assessment of the effects of certain public and private projects on the environment [1985] OJ L175 (EIA Directive). In 2011, this Directive was replaced by European Parliament and Council Directive 2011/92/EC of 13 December 2011 on the assessment of the effects of certain public and private projects on the environment [2012] OJ 26/1.

[35] European Parliament and Council Directive 2008/1/EC of 15 January 2008 concerning integrated pollution prevention and control [2008] OJ L24/8 (IPPC Directive). In 2011, the IPPC Directive was replaced by European Parliament and Council Directive 2010/75/EU of 24 November 2010 on industrial emissions (integrated pollution prevention and control) (Recast) [2010] OJ L334/17 (IE Directive), with effect from January 2011.

[36] European Parliament and Council Directive 2001/80 of 23 October 2001 on the limitation of emissions of certain pollutants into the air from large combustion plants [2001] OJ L309 (LCP Directive). This Directive sets limits on sulphur dioxide, nitrogen oxides and dust emissions. Like the IPPC Directive, the LCP Directive was replaced by the IE Directive (Directive 2010/75/EU) in 2011.

[37] Article 3(13) Directive 2009/31/EC defines a CO_2 stream as '[...] a flow of substances that results from CO_2 capture processes'.

CO$_2$ stream purity

The required purity of the CO$_2$ stream for storage is of great importance to CO$_2$ capture. CO$_2$ purity refers to the substances other than CO$_2$ – also known as impurities – that are allowed to be in the captured CO$_2$ stream. The composition of the captured CO$_2$ stream can vary from project to project as a consequence of differences in, for instance, fuel type, capture process and post-capture processing.

Article 12(1) of the CCS Directive provides that a CO$_2$ stream for storage has to consist 'overwhelmingly' of CO$_2$ and may not be used for the disposal of waste. This term was first used in the London Protocol, the first international marine environment protection instrument[38] to allow for the offshore geological storage of CO$_2$. It was deliberately chosen by the London Protocol's Scientific Group to prevent authorities setting arbitrary levels for CO$_2$ stream components.[39]

Some storage reservoirs may, for instance, be rich in hydrogen sulphide at levels far greater than are present in the CO$_2$ stream. Purity criteria that would require the amount of hydrogen sulphide in the CO$_2$ stream to stay below these levels would increase the energy penalty for CO$_2$ capture[40] as well as emissions from the capture process (both leading to higher capturing costs), while not necessarily increasing the safety of CCS deployment.

By virtue of Article 12(1), the CO$_2$ stream may contain substances deriving from the source, capture or injection process (incidental substances) and substances added to keep track of the CO$_2$ stream in the underground (trace substances). Article 12 does not further define the term 'overwhelmingly', other than that the concentrations of these incidental and added substances shall stay below levels that would either:

 a. adversely affect the integrity of transport infrastructure or storage site;
 b. pose a significant risk to the environment or human health; or
 c. breach the requirements of applicable EU legislation.

The criteria in (a) and (b) are not very clear. The Directive does not specify when incidental substances or tracers adversely affect the integrity of transport and storage infrastructure. Furthermore, the Directive's definition of the crucial term in criterion (b), 'significant risk', is vague: 'a combination of a probability of occurrence of damage and a magnitude of damage that cannot be disregarded without calling into question the purpose of this Directive for the storage site concerned'.[41]

---

[38] Shortly after the London Protocol, the OSPAR Convention was amended to allow for the offshore geological storage of CO$_2$. The term 'overwhelmingly' is also used in the OSPAR Convention. See Articles 3(2)(f)(ii) and 3(3)(b) of, respectively, Annexes II and III to the OSPAR Convention.

[39] Tim Dixon and others, 'International Marine Regulation of CO$_2$ Geological Storage. Developments and Implications of London and OSPAR', *Energy Procedia* (2009) Volume 1, 4503, 4506.

[40] See section 7.2.1 above.

[41] Article 3(18) Directive 2009/31/EC.

Yet, when does disregarding a foreseeable risk run contrary to the purpose of the Directive for the storage site concerned? Article 1 provides that the CCS Directive establishes a legal framework for the environmentally safe geological storage of CO_2, the purpose of which is the permanent containment of CO_2. This seems to suggest that incidental and trace substances in the CO_2 stream should stay below levels that raise doubt as to whether the injected CO_2 will be permanently contained in the storage reservoir.

The criterion in (c) refers to the requirements imposed on the operators of CO_2 capture installations by several EU environmental Directives. First, Article 31 of the CCS Directive amends the EIA Directive so as to include in the latter Directive capture installations, as well as CO_2 pipelines (including booster stations) and storage sites. This means that an environmental impact assessment, an assessment of the likely environmental effects of a capture project, has to be carried out in the capture permit process.

In addition, Article 37 of the CCS Directive amends the IPPC Directive so as to bring the capture of CO_2 by IPPC installations under the scope of this Directive. This ensures that best available techniques (BATs)[42] to improve the composition of the CO_2 stream have to be established and applied.[43] Also, combustion plants with a so-called rated thermal input of at least 50 MW are required to meet the emission limits set by the LCP Directive. The effect of the emissions limits in the environmental permits issued under the IPPC and LCP Directives is that CO_2 capturers will generally already have to reduce flue gas concentrations of substances such as sulphur dioxide and nitrogen oxides. These environmental standards thus directly influence the composition of the CO_2 stream for storage.

Article 12(2) of the CCS Directive allows the Commission to issue guidelines to help the Member States apply the criteria in Article 12(1). In 2011 the Commission published a set of guidance documents, one of which partly deals with CO_2 stream composition.[44] The legally non-binding guidance document gives concrete form to the purity requirements in Article 12(1). It, for instance, lists indicative compositions of CO_2 streams from the main capture technologies. Also, the guidance document proposes a certain procedure for the Member States to follow in determining the concentration limits of incidental substances. This procedure starts with a proposal for CO_2 stream composition by the operator of the storage site in question, which is followed by a review of key parameters related to, for instance, the integrity of the CO_2 storage site as well as human and environmental safety.

This procedure probably represents the way in which the criteria in Article 12(1) are normally put into practice. The CO_2 storage operator proposes a certain concrete CO_2 stream

---

[42]   Article 2(10) Directive 2010/75/EU defines the term 'best available technique'. In essence, it boils down to the requirement that the installations concerned use the best – from an environmental protection point of view – technology for reducing emissions, as long as it is economically and technically viable to do so.

[43]   Recital 27 Directive 2009/31/EC.

[44]   European Commission, 'Implementation of Directive 2009/31/EC on the Geological Storage of Carbon Dioxide – Guidance Document 2: Characterisation of the Storage Complex, CO_2 Stream Composition, Monitoring and Corrective Measures' (2011), p.61. Accessible at: https://ec.europa.eu/clima/sites/clima/files/lowcarbon/ccs/implementation/docs/gd2_en.pdf accessed 6 January 2021.

composition to the competent authority of the Member State concerned, which is then tested for safety and the general criteria in Article 12(1). The proposed CO_2 stream composition is approved only if it complies with applicable EU environmental standards and does not pose a risk to the integrity of transport and storage infrastructure or the environment and public health. If the latter is the case, the proposed CO_2 stream composition can be included in the CO_2 storage permit,[45] possibly through the introduction of a minimum CO_2 content and maximum values for, or bans on, substances other than CO_2. Importantly, Article 12(3) provides that the storage operator may inject a CO_2 stream only if its composition has been analysed and a risk assessment has shown that the criteria in Article 12(1) are met.

The above-described procedure for determining the specific CO_2 stream purity requirements would be in line with the case-by-case approach[46] underlying Article 12. The term 'overwhelmingly' is vague and does not set an absolute standard for CO_2 content. Nor do the other requirements in Article 12(1) set maximum values for, or bans on, impurities. Yet, the general and vague character of the purity requirements in Article 12 provides the Member States and industry with the flexibility to jointly set the concrete standards in accordance with the characteristics of each specific CCS project. This should avoid the setting of general standards that might not necessarily improve safety, but do raise project costs. On the other hand, the uncertainty caused by the general and vague character of the purity requirements in Article 12 could raise transaction costs[47] for CCS market players, as the discussion on the contents of the stream lengthens negotiations and forces parties to take the stream criteria into account in their risk assessments.

There is still a lot of uncertainty about the long-term effects of impurities on the safety of CO_2 storage. Research suggests that the presence of impurities in the CO_2 stream could both increase and decrease the risks of CO_2 storage.[48] In that light, it seems preferable to set the specific purity requirements at the lowest possible level in accordance with the characteristics of each specific project.

In sum, the purity requirements in Article 12 appear to have different effects. On the one hand, they contribute to a more cost-effective CO_2 capture process, and thus to CO_2 emission

---

[45] We further discuss the (storage) permitting process in section 7.4.3 below. In this regard, see also Article 9(4) Directive 2009/31/EC, which provides that the storage permit shall contain the requirements for CO_2 stream composition.

[46] Dixon and others argue that the formula that a CO_2 stream for storage consists overwhelmingly of CO_2 allows for a case-by-case assessment of the levels of impurity, recognizing the natural variation in storage-site characteristics and different transport constructions. See: Tim Dixon and others, 'International Marine Regulation of CO_2 Geological Storage: Developments and Implications of London and OSPAR', *Energy Procedia* (2009) Volume 1, 4503, 4506.

[47] Transaction costs can be described as the costs of transferring resources between individuals. See: William Jack Tavneet Suri, 'Risk Sharing and Transactions Costs: Evidence from Kenya's Mobile Money Revolution', *American Economic Review* (2014) Volume 104 Issue 1, p.183.

[48] See: Zaman Ziabakhsh-Ganji and Henk Kooi, 'Sensitivity of the CO_2 Storage Capacity of Underground Geological Structures to the Presence of SO2 and Other Impurities' *Applied Energy* (15 December 2014) Volume 135, pp.43–52; IEA GHG, 'CCS Site Characterization Criteria', Report No. 2009/10, p.6.

reductions through the deployment of CCS technologies being realized more cost-effectively. On the other, they could increase the transaction costs of CCS market players.

Obliging CO_2 capture

During the negotiations on the CCS Directive, one of the much debated issues was incentivizing CCS by introducing an emissions performance standard, which would de facto oblige new, large coal-fired power plants to capture their CO_2 emissions. The Commission had originally proposed to insert in the LCP Directive a provision that obliged each new large combustion plant to have suitable space on its site for capture and compression equipment and required the availability of suitable transport facilities and storage sites and assessment of the technical feasibility of retrofitting the installation with capture equipment to have been assessed (capture readiness).[49] The intention was to make sure that large industrial installations initially constructed without CO_2 capture equipment could still decide to start capturing their CO_2 emissions at a later moment.

The European Parliament (hereinafter: Parliament) had sought to amend the proposed Article 32 of the CCS Directive by replacing the capture readiness requirement with an emissions performance standard obliging all new large power plants not to emit more than 500g CO_2/kWh, as from January 2015. Parliament's amendment, however, was not adopted in the final text of the CCS Directive. Instead, the capture-readiness requirement proposed by the Commission was weakened.

The resulting Article 33 requires the operators of new large combustion plants to reserve space on the installation site for CO_2 capture and compression equipment under three conditions only: (1) a suitable storage site is available; and it is (2) technically and (3) economically feasible to both retrofit the installation with capture equipment and develop a CO_2 transport facility. Considering the high costs associated with CO_2 capture, we would argue that the requirement of economic feasibility seriously weakens the capture-readiness obligation in Article 33.

Besides proposing an EU emissions performance standard for new large power plants, Parliament also suggested deleting Article 9(3)(iii) of the IPPC Directive.[50] This provision required permits issued under the IPPC Directive not to contain an emission limit value[51] for emissions of greenhouse gases covered under the EU ETS, unless necessary to prevent significant local pollution. By deleting Article 9(3)(iii), Parliament sought to clear the way for Member States to introduce CO_2 emissions performance standards through permits granted under the IPPC Directive. Emissions performance standards can be seen as a subcategory of the prohibited emission limit values.

---

[49]   Article 32 of the Commission's original proposal for Directive 2009/31/EC, COM(2008) 18 final.

[50]   European Parliament, Report A6-0414/2008 (16-10-2008), proposed amendments 124 and 126. Article 9(3)(iii) has been replaced by Article 9(1) Directive 2010/75/EU.

[51]   Article 2(8) Directive 2010/75/EU defines an emission limit value as '[...] the mass, [...] concentration and/or level of an emission, which may not be exceeded during one or more periods of time'.

Yet, like the proposal for an EU emissions performance standard, this amendment did not make it into the final text of the CCS Directive. Article 9(1) of the IE Directive still prohibits the inclusion of an emission limit value for EU ETS greenhouse gases[52] in the permit granted under the Directive.[53]

As an instrument to boost CCS deployment, a CO_2 emissions performance standard for installations participating in the EU ETS would have cost-effectiveness effects similar to those of CCS subsidies. The effect of any instrument tilting the mix of options to reduce greenhouse gas emissions on an emissions trading market towards a specific technology will be a greater deployment of relatively expensive technology. A CO_2 emissions performance standard for EU ETS installations would affect the cost-effectiveness of the EU ETS as a policy instrument for reducing greenhouse gas emissions.

7.4.2 CO_2 Transport

EU regulation of CO_2 transport for storage is essentially based on the idea that captured CO_2 is transported by pipeline and not, for instance, by ship. Several provisions in the CCS Directive demonstrate this point.

First, Article 3(22) defines the transport network as 'the network of pipelines, including associated booster stations, for the transport of CO_2 to the storage site'. Even though shipping routes, including associated CO_2 hubs or shipping terminals used to load captured CO_2 onto tankers, could just as well be part of a CO_2 transport network, there is no mention of them in Article 3(22).

Second, by virtue of Directive 2009/29/EC (amending the EU ETS Directive), only the transport of greenhouse gases (for storage) *by pipeline* has been added to the list of activities in Annex I of the EU ETS Directive. Accordingly, no EU ETS emissions permit is required for the transport of greenhouse gases (for storage) by ship.

Third, nowhere does the CCS Directive mention the option of CO_2 transport per ship. This is remarkable, considering the explicit reference to shipping as one of the two main kinds of modes of CO_2 transport in the Commission's impact assessment to the CCS Directive.[54] Since

---

[52] On 1 January 2013, a number of new activities responsible for emissions of nitrous oxide and perfluorocarbons were included in the EU ETS. The permit issued under Directive 2010/75/EU may also not contain an emission limit value for these greenhouse gases.

[53] Nonetheless, a lively (practical) debate has developed on the legality of national CO_2 emissions performance standards under EU law. On this discussion, see, for instance, Lorenzo Squintani, Marijn Holwerda and Kars de Graaf, 'Regulating Greenhouse Gas Emissions from EU ETS Installations: What Room Is Left for the Member States?' in: Marjan Peeters, Mark Stallworthy and Javier de Cendra de Larragán (eds), *Climate Law in EU Member States: Towards National Legislation for Climate Protection* (Edward Elgar Publishing 2012), p.67.

[54] European Commission, 'Commission Staff Working Document Accompanying Document to the Proposal for a Directive of the European Parliament and of the Council on the geological storage of carbon dioxide', COM(2008)18 final, p.23.

the future for CCS in Europe is most likely offshore, more attention to this type of transport is needed.[55]

Later in this chapter we will discuss the consequences of CO_2 transport by ship not having been added to the EU ETS Directive, but we will first explore EU requirements on third-party access to CO_2 transport infrastructure.

Third-party access[56]

Third-party access refers to access to the relevant infrastructure by parties who do not own or control that infrastructure. In relation to CO_2 transport, issues of third-party access can occur, for instance, if spare capacity exists in a CO_2 pipeline or network of pipelines and the relevant authority has an interest in optimizing the use of that capacity.[57] An example is a new CO_2 capture plant connecting to an existing nearby CO_2 pipeline passing the facility. Another example is an increase in the capacity of a pipeline prior to construction in order to accommodate another project in the same area.

Article 21 of the CCS Directive addresses third-party access to CO_2 transport networks and storage sites. Brockett and Roggenkamp have argued that Article 21 intends to apply the principles of negotiated rather than regulated third-party access, as applied to the upstream gas sector.[58] Under a negotiated access regime, the majority of binding access terms and conditions are set by the parties themselves through negotiation.[59] Under a regime of regulated third-party access, access terms and conditions are set externally by, for instance, independent regulatory bodies, state ministries or competition authorities.

However, Article 21 explicitly states that it is up to each Member State to determine the way in which its regime for third-party access to CO_2 transport and storage infrastructure is designed, as long as such access is provided in a fair, open and non-discriminatory manner. As a consequence, Article 21 does not rule out the option of regulated third-party access to CO_2 transport and storage infrastructure. Ultimately, national law determines whether the relevant operator has the duty to supply the transport service or capacity (regulated third-party access)

---

[55]   For an overview of the regulatory barriers see: Element Energy, 'Shipping CO_2–UK Cost Estimation Study', November 2018, pp.51–5. Available at: https://assets.publishing.service.gov.uk/government/uploads/system/uploads/attachment_data/file/761762/BEIS_Shipping_CO_2.pdf accessed 6 January 2021.

[56]   Even though this section falls under the heading of CO_2 transport, most of its findings are also applicable to CO_2 storage (infrastructure). The reason is that the third-party access provisions in the CCS Directive concern access to CO_2 transport as well as storage infrastructure.

[57]   IEA, 'Model Regulatory Framework', November 2010, p.46.

[58]   See: Scott Brockett, 'The EU Enabling Legal Framework for Carbon Capture and Geological Storage' *Energy Procedia*, Volume 1, Issue 1, February 2009 p.4439 and Martha M. Roggenkamp, 'The Concept of Third Party Access Applied to CCS' in: Martha M. Roggenkamp and Edwin Woerdman (eds), *Legal Design of Carbon Capture and Storage* (Intersentia 2009), pp.297–9.

[59]   Thomas W. Waelde and Andreas J. Gunst, 'International Energy Trade and Access to Energy Networks', *Journal of World Trade* (2002) Volume 36 Issue 2, 191, 197–8.

or to – at least – enter into negotiations over the service or capacity requested (negotiated third-party access).

When drafting its proposal for the CCS Directive, the Commission did not consider it necessary to prescribe a detailed third-party access regime.[60] According to the Commission, the market for CCS was at an early stage and indications were that there would, in practice, be separate operators for the capture phase on the one hand and for transport and storage on the other. The relevance of this is that a CO_2 capturer controlling a CO_2 transport pipeline might have an incentive to refuse access to another capturer with whom he competes in the market for electricity production. If this competitor needs access to the relevant pipeline to transport his captured CO_2 to a storage site and it is more costly for him to buy an EU ETS emissions allowance than to capture, transport and store a tonne of CO_2, the refusal to grant access will raise his production costs and thus damage his competitive position in the market for electricity production ('raising rivals' costs'). As the Commission did not expect CO_2 capturers to be active in the other parts of the CCS chain, it was not concerned about operators of CO_2 transport infrastructure abusing their position by refusing access to capturers with which they competed in related markets, such as the electricity production market.

Importantly, Article 21(2) requires each Member State to take into account a number of elements when designing their third-party access regime:

a) the (potentially) available storage capacity;
b) the proportion of its international and EU CO_2 emission reduction obligations the Member States intends to meet through CCS;
c) the need to refuse access in case of a technical incompatibility that cannot reasonably be overcome; and
d) the requirement to respect the needs of the transport or storage operator and the interests of all other users of the infrastructure.

Subparagraphs (b) and (c) appear to provide explicit grounds for refusal of access to CO_2 transport and storage infrastructure. The wording of Article 21(2)(b) is not very precise, but the provision seems to require Member States to take into account their own (anticipated) need of national transport and storage capacity when developing a third-party access regime. Accordingly, it appears to give them the possibility to require the operators of CO_2 transport and storage infrastructure to refuse to grant access to the relevant infrastructure when the capacity concerned is needed to meet part of the Member State's international and EU obligations to reduce greenhouse gas emissions. Think, for instance, of national legislation dedicating certain (or perhaps even all) CO_2 transport and storage infrastructure to national use only.

Likewise, Article 21(2)(c) seems to provide the operators of CO_2 transport and storage infrastructure with the possibility to refuse to grant access to the relevant infrastructure when the technical specifications of the specific CO_2 stream are incompatible with the required

---

[60] European Commission, 'Commission Staff Working Document Accompanying Document to the Proposal for a Directive of the European Parliament and of the Council on the geological storage of carbon dioxide', COM(2008)18 final, pp.44–5.

technical standard and the incompatibility 'cannot reasonably be overcome'. An example of a technical incompatibility would be a CO_2 stream that contains more water than is required by the technical specifications of the relevant CO_2 transport pipeline (corrosion prevention).

The question, of course, is when an incompatibility of technical specifications can no longer be 'reasonably' overcome and who determines if this is the case. When negotiating the CCS Directive, Parliament was concerned that this ground for refusal of access could be used by CO_2 transport operators to discriminate between CO_2 capturers. Therefore Parliament, unsuccessfully, proposed to have new CO_2 transport pipelines designed in such a manner that they would be suited to take any CO_2 stream of a given minimum CO_2 quality.[61]

It is not difficult to imagine both grounds for access refusal creating (significant) hindrances to the cross-border transport of captured CO_2 in the EU. Vedder has argued that Article 21(2)(b) could effectively foreclose an entire Member State's carbon storage market.[62] For certain industries, such as the cement industry, CCS might well be the only viable means of achieving deep emission cuts. At the same time, potential CO_2 storage capacity is unevenly spread over Europe and certain Member States have significantly less potential storage capacity than others.[63] The necessity to deploy CCS in order to achieve deep emission cuts in certain sectors, in combination with the uneven spread of potential CO_2 storage capacity across Europe, means that for some Member States access to CO_2 transport and storage infrastructure in other Member States might be crucial for achieving deep national emissions reductions.

As for the ground for refusal of access in Article 21(2)(c), the vagueness of the requirement that the technical incompatibility cannot be reasonably overcome creates a large margin of discretion for national authorities and possibly also for transport operators to determine under what circumstances access to CO_2 transport and storage infrastructure can be refused on technical grounds. What exactly is reasonable (in terms of costs and effort) in overcoming technical incompatibilities is something that will have to be determined through case law.

In addition, subparagraph (a) suggests that a lack of transport or storage capacity can also be a ground for refusing access to CO_2 transport and storage infrastructure. This is confirmed by Article 21(3), which explicitly provides that CO_2 transport and storage operators may refuse access on the grounds of a lack of capacity. Yet, by virtue of Article 21(4), an operator refusing access due to a lack of capacity has to increase capacity if it is economic to do so. However, the article does not specify the time by which the operator has to increase capacity.

Finally, Article 21(2)(d) refers to the requirement to respect the needs of the transport or storage operator and of all users of the relevant infrastructure. Again, Article 21 is not particularly clear. The provision seems to primarily refer to a transport or storage operator's demand for capacity on his own infrastructure. A transport or storage operator who is also a capturer

---

[61]   Parliament's proposed amendment was somewhat changed and can now be found in recital 38 Directive 2009/31/EC.

[62]   Hans Vedder, 'An Assessment of Carbon Capture and Storage under EC Competition Law', *International Energy Law Review* (2008) Volume 8, pp.316–17.

[63]   See, for instance, T. Vangkilde-Pedersen, 'Assessing European Capacity for Geological Storage of Carbon Dioxide – the EU GeoCapacity Project', *Energy Procedia* (2009) Volume 1, 2665.

would normally want to have the guarantee of having sufficient transport or storage capacity to fulfil his own needs. If he does not have that guarantee due to (stringent) third-party access requirements, he will likely be less inclined to invest in the development of infrastructure.

Similarly, stringent access requirements, for instance in the form of a (low) regulated tariff for transport or storage services, will create disincentives for infrastructure investments. One of the incentives for transport and storage operators to invest in the development of infrastructure is that they can recover the costs from their investments quickly.[64] In that light, it is not beneficial to force these operators to open networks to new connections from other players at regulated prices which are lower than they would normally have charged. Article 21(2)(d) appears to require Member States to take such considerations into account when designing their third-party access regime.

In sum: by choosing not to prescribe a particular regime, the Commission has avoided imposing the regulated third-party access regime applied to EU gas and electricity transport infrastructure. This is likely to be beneficial from a cost-effectiveness point of view. The more stringent regulated third-party access requirements would increase costs for CO_2 transport and storage operators, thus raising the costs of CCS deployment. Yet, at the same time, the absence of more detailed third-party access rules could raise transaction costs for CCS market players. In addition, Article 21 does not prevent the Member States from adopting this administratively quite burdensome type of access regime. If the Commission indeed had an implicit preference for the negotiated third-party access regime, it would perhaps have done better to prescribe that regime.

As for the solidarity between Member States, Article 21(2)(b) appears to be problematic. As we have argued above, it could lead to the foreclosure of a Member State's entire CO_2 storage market, possibly making it more difficult and expensive for other Member States to achieve deep national greenhouse gas emission reductions. Similarly, the vague wording of Article 21(2)(c) provides opportunities for discriminatory refusal of access to (cross-border) CO_2 transport and storage infrastructure. From a solidarity point of view, these provisions are not particularly helpful.

7.4.3 CO_2 Storage

The point of departure for the geological storage of CO_2 in the EU is the full discretion of Member States to select CO_2 storage sites, including the right not to allow for any storage in their territory.[65] The provisions in the CCS Directive cover the three basic phases of any CO_2 storage project: (1) pre-storage; (2) operation; and (3) closure and post-closure. In the following, we explore the legal requirements in all three phases of CO_2 storage.

---

[64]   IEA, 'Model Regulatory Framework', November 2010, p.47.

[65]   Article 4(1) Directive 2009/31/EC. As witness recital 19 of the preamble to the Directive, Member States might want to put potential CO_2 storage reservoirs to other uses, such as natural gas storage or the production of geothermal energy. These considerations were not included in the text of recital 19 as originally proposed by the European Commission.

Pre-storage

Before any CCS project can start injecting CO_2 into the underground, two steps have to be taken. First, a proper storage site has to be selected. Second, all permitting procedures need to have been completed. From a safety point of view, the selection of a storage site with the appropriate geological characteristics is crucial.[66] Article 4(4) of the CCS Directive determines that a storage site may only be selected if there are no significant risks of leakage and no significant environmental or health risks.

Whether no such risks exist is to be determined through a geological characterization of the site by means of three-dimensional computer modelling, as described in Annex 1 to the Directive. This modelling is used to predict the underground behaviour of injected CO_2. Site characterization, arguably, is the most time-consuming and costly part of the CO_2 storage site selection process.[67]

If it is necessary to obtain more information about the potential storage site by, for instance, performing drilling activities or injection tests, such activities may only be undertaken on the basis of an exploration permit.[68] Likewise, no CO_2 storage site may be operated without a storage permit, the Directive's central instrument for guaranteeing the environmentally safe geological storage of CO_2 in the EU.[69] Both permitting procedures appear to be based on the regime applied in Directive 94/22/EC (Hydrocarbon Licensing Directive).[70] Permits are exclusive, granting the holder the sole right to perform the exploration or storage activities, and are awarded through competitive procedures based on objective, published and transparent criteria. Applicants have to show that they are financially sound and technically competent, as well as reliable, to operate and control the site. To encourage investments in exploration, exploration permit holders have priority in the storage permitting procedure. As part of the storage permit procedure, applicants have to submit to the authorities, among other things, the results of the compulsory environmental impact assessment of the potential storage site.

Importantly, Article 19 requires potential storage operators to provide financial security. This is to ensure that the operator can meet all obligations arising from the permit during operation, closure and post-closure, until responsibility for the storage site has been transferred to the competent authority. Should a storage operator default on his obligations under the Directive, the competent authority can draw on the financial security. The financial security requirements in Article 19 are controversial and have been heavily debated by industry. Below, we will further examine Article 19.

---

[66]   See recital 19 Directive 2009/31/EC.

[67]   IEA, 'Model Regulatory Framework', November 2010, pp.95–100.

[68]   Article 5(1) Directive 2009/31/EC.

[69]   Article 6(1) and recital 24 Directive 2009/31/EC.

[70]   Martha M. Roggenkamp, Kars J. de Graaf and J. Marijn Holwerda, 'Afvang, Transport en Opslag van CO_2: De Implementatie van Richtlijn 2009/31/EG in Nederland', *Milieu en Recht*, Volume 37, 2010, 548, 553. This is European Parliament and Council Directive 94/22/EC of 30 May 1994 on the conditions for granting and using authorizations for the prospection, exploration, and production of hydrocarbons [1994] OJ L164.

As indicated earlier, the Commission reviews all draft storage permits.[71] It may issue a non-binding opinion, from which the Member States may not depart without stating their reasons. The Commission's opinion is based on a review by a scientific panel, which assesses whether the information on the basis of which the permit decision is to be taken is comprehensive and reliable.[72] In this way, the Commission hopes to exchange information and best practices as well as to enhance public confidence in early CCS deployment.

Operation

As soon as a storage permit has been awarded, CO_2 injection into the storage reservoir may start. In the operating phase, all efforts are directed at ensuring that the injected CO_2 remains safely underground. Crucial in this regard are the monitoring requirements in Article 13 of the CCS Directive. This provision requires storage operators to monitor the injection facilities, the storage complex[73] (including the injected CO_2) and the surrounding environment. This is to verify whether the underground behaviour of the injected CO_2 matches the modelling done in the pre-storage phase and to keep a close eye on possible adverse effects on the environment. The storage operators need to report the monitoring results to the competent authority at least once a year.[74]

In addition, Member States have to organize a system of routine and non-routine inspections of all storage complexes covered by the CCS Directive.[75] Surface installations, such as the injections facilities, are visited and all relevant records kept by the storage operator are checked. Routine inspections are conducted at least once a year, until three years after closure of the site or five years after the transfer of responsibility for the storage site to the competent authority. Non-routine inspections are carried out in case of problems such as leakage and serious complaints related to the environment or human health. After each inspection, the competent authority has to prepare a report on the results of the inspection, evaluating whether the operator (still) complies with the requirements of the storage permit. Under Article 11(3), it has the power to withdraw the permit should the storage operator not meet his obligations.

Article 16 of the CCS Directive deals with the measures that are to be taken in case of leakage[76] or other 'significant irregularities'.[77] In the event of a risk of leakage or damage to

---

[71] Article 10 Directive 2009/31/EC.

[72] European Commission, 'Commission Staff Working Document Accompanying Document to the Proposal for a Directive of the European Parliament and of the Council On the geological storage of carbon dioxide', COM(2008)18 final, p.36.

[73] Article 3(6) Directive 2009/31/EC defines the storage complex as '[…] the storage site and surrounding geological domain which can have an effect on overall storage integrity and security […]'.

[74] See Article 14(1) Directive 2009/31/EC.

[75] See Article 15 Directive 2009/31/EC. This, for instance, excludes the small-scale R&D storage projects mentioned in section 7.3.2.

[76] By virtue of Article 3(5) Directive 2009/31/EC, leakage is '[…] any release of CO_2 from the storage complex'.

[77] Article 3(17) of the CCS Directive defines a significant irregularity as '[…] any irregularity in the injection or storage operations or in the condition of the storage complex itself, which implies the risk of a leakage or risk to the environment or human health'.

the environment or human health or actual leakage, the operator has to immediately notify the competent authority and take corrective measures. The latter are basically measures that prevent or stop injected CO_2 from leaking from the reservoir. If the competent authority decides to take any corrective measures, it will draw on the financial security required under Article 19 in order to recover any costs made.

Besides the obligation to take corrective measures, storage operators bear three types of liability in relation to the geological storage of CO_2 pursuant to the CCS Directive:

1. environmental liability;
2. climate liability; and
3. civil liability.

Environmental liability refers to liability for damage to protected species and natural habitats, water and land. By virtue of Article 34 of the CCS Directive, liability for environmental damage is regulated by Directive 2004/35/EC (Environmental Liability Directive).[78] As a consequence, the Environmental Liability Directive's strict liability regime applies to CO_2 storage activities. This means that the operator of a storage site may be held financially liable for environmental damage caused by CO_2 storage activities, even if he is not at fault. In case of (an imminent threat of) environmental damage occurring, he is required to take (preventive and) remedial measures pursuant to Articles 5(1) and) 6(1) of the Environmental Liability Directive.

Climate liability arises when stored CO_2 leaks into the atmosphere. The 2009 amendment of the EU ETS Directive added the geological storage (as well as the capture and transport) of greenhouse gases to Annex I to the EU ETS Directive. As a consequence, storage operators need to surrender emissions allowances for any leaked greenhouse gas emissions. Industry has in the past perceived climate liability as a significant financial obstacle to deploying CCS technologies.[79] This is mainly due to the uncertainties surrounding the size of the required compensation, which are in part caused by the unpredictability of EU ETS allowance prices. We explore this issue further later in the chapter.

Finally, civil liability arises when CO_2 storage causes damage to individuals or property. Underground storage of CO_2 could, for instance, cause earthquakes and ground movement, which could damage property. Recital 34 explicitly states that liabilities other than those covered by the CCS Directive, EU ETS Directive and Environmental Liability Directive should be dealt with at national level. Civil liability is therefore solely regulated under national law.

Closure and post-closure

Under Article 17 of the CCS Directive, a storage site shall be closed when:

a) the conditions in the storage permit have been met;
b) an operator's request for closure is authorized by the competent authority; or
c) the competent authority decides so after withdrawal of a storage permit.

---

[78] This is European Parliament and Council Directive 2004/35/EC of 21 April 2004 on environmental liability with regard to the prevention and remedying of environmental damage [2004] OJ L143.

[79] ClimateWise, 'Managing Liabilities of Carbon Capture and Storage', 2012, p.47.

Importantly, the operator of a storage site remains responsible for that site until responsibility is transferred to the competent authority. After the closure of a site, the storage operator would normally not want to retain responsibility for the site indefinitely, since this would be a considerable business risk.[80] Therefore, responsibility for the storage site is to be transferred to the competent authority when:[81]

(a) the stored CO_2 is likely to be completely and permanently contained;
(b) a minimum period of at least 20 years, or less if the competent authority so decides, has elapsed;
(c) the operator has made available a financial contribution covering the costs of the competent authority after transfer of responsibility;[82] and
(d) the site has been sealed and injection facilities have been removed.

The transfer of responsibility to the competent authority includes environmental and climate liabilities, but excludes civil liability. The latter is, as indicated above, not regulated by the CCS Directive.

It is important to underline that the transfer of responsibility, including environmental and climate liabilities, does not mean that the storage operator will never face post-closure costs. Where there has been fault on the part of the operator, including cases of deficient data, concealment of relevant information, negligence, wilful deceit or a failure to exercise due diligence, the competent authority has to recover incurred costs from the former storage operator.[83]

In order to demonstrate the likelihood of stored CO_2 staying completely and permanently contained (first criterion in Article 18(1)), the storage operator has to demonstrate that:

a) the injected CO_2 behaves as modelled;
b) there are no detectable leakages; and
c) the store site is evolving towards a situation of long-term stability.[84]

Consistent application of these criteria is guaranteed in two ways. First, Article 18(2) provides the Commission with the opportunity to issue guidelines on the assessment of these matters, which it did in 2011.[85] As shown by the guidelines, permanent containment of stored CO_2 is demonstrated through modelling, based on the modelling done in the pre-storage phase.

---

[80] What is more, the lifespan of a company is probably shorter than the (post-closure) period during which a storage site needs to be monitored and watched.

[81] Article 18(1) Directive 2009/31/EC.

[82] See Article 20 Directive 2009/31/EC.

[83] Article 18(7) Directive 2009/31/EC. See also recital 35 Directive 2009/31/EC.

[84] Article 18(1) Directive 2009/31/EC.

[85] European Commission, 'Implementation of Directive 2009/31/EC on the Geological Storage of Carbon Dioxide – Guidance Document 3: Criteria for Transfer of Responsibility to the Competent Authority' (2011) https://ec.europa.eu/clima/sites/clima/files/lowcarbon/ccs/implementation/docs/gd3_en.pdf accessed 6 January 2021.

Second, by virtue of Article 18(4), the Commission reviews all draft decisions to transfer responsibility in a way similar to its review of draft storage permits.

Finally, just as there is a provision ensuring that the storage operator can meet all financial obligations as long as he is responsible for a storage site, there is a provision that guarantees that the competent authority is able to do the same after it has become responsible for the site. Under Article 20, the storage operator has to make a financial contribution (financial mechanism) to the competent authority before responsibility for the storage site is transferred. The contribution may be used to cover the costs incurred by the competent authority in ensuring that the CO_2 is completely and permanently contained.[86] This financial mechanism should cover at least the anticipated cost of monitoring for a period of 30 years. Below we further examine Article 20 and its consequences for large-scale CCS deployment in the EU.

When exploring EU regulation of the geological storage of CO_2, the general character and lack of detail of a number of provisions in the CCS Directive are both quite visible. In relation to several issues, exact standards must be set by the Member States, which do have the Commission's legally non-binding guidelines at their disposal.[87] Examples are the specifics of the storage permitting procedure (Article 6), the design of the routine and non-routine inspections of storage complexes (Article 15) and the arrangements for implementing the financial security and financial mechanism requirements (Articles 19 and 20). From a cost-effectiveness point of view, we would argue that this has different effects. On the one hand, the discretion left to the Member States makes it possible to set the exact standards in a way that better takes account of the specifics of each individual storage project. This probably diminishes the chances of unnecessary (high) administrative burdens being imposed. On the other hand, the uncertainty caused by the general and imprecise character of the CCS Directive is likely to raise transaction costs for CCS market players.

From a solidarity point of view, the Directive's provisions on CO_2 storage do not appear to make a particularly good impression. First, Article 4(1) demonstrates the individual approach that is taken to storage site selection and development. As we have seen, it underlines the right of Member States to determine the areas from which storage sites may be selected, including the right not to allow for CO_2 storage at all. There is no common EU approach for storage site selection and development, as a result of which Member States with only little storage capacity might encounter difficulties in having their captured CO_2 stored. The possibility for CO_2 storage operators to refuse third-party access to their infrastructure when the capacity is needed to meet a Member State's EU or international CO_2 emissions reduction obligations (Article 21(2)(b)) is not beneficial.

---

[86]   To help the Member States implement Articles 19 and 20 the Commission also adopted guidelines for these provisions in 2011. See Commission, 'Implementation of Directive 2009/31/EC on the Geological Storage of Carbon Dioxide – Guidance Document 4: Article 19 Financial Security and Article 20 Financial Mechanism' (2011) https://ec.europa.eu/clima/sites/clima/files/lowcarbon/ccs/implementation/docs/gd4_en.pdf accessed 6 January 2021.

[87]   In Member States with a strong tradition in oil and gas mining, such as the Netherlands and the UK, there is obviously significant experience with and knowledge of the extraction and storing of hydrocarbons. This experience and knowledge can be of great value when considering the storage of CO_2 in depleted oil or gas fields.

This is also evidenced by fact that the CCS Directive contains only two provisions dealing, in a very rudimentary way, with the cross-border deployment of CCS in the EU. By virtue of Article 22(2), Member States are to consult each other in case of a cross-border dispute over access to cross-border CO_2 transport or storage infrastructure. Article 24 merely requires Member States to cooperate in case of cross-border CO_2 transport or cross-border storage complexes. The CCS Directive enables those Member States that seek to deploy CCS technologies to do so in an environmentally safe manner. Yet, there seems to be no overall (centralized) EU approach to storage site selection and development. This lack of inter-Member State solidarity in the CCS Directive might come at the expense of Member States with little own storage capacity.

Long-term liability

Whereas the CCS Directive addresses climate and environmental liabilities in relation to the geological storage of CO_2, it does not deal with civil liability. The former two types of liability can be transferred to the competent authority after, among other things, the storage site has been closed and a minimum period of 20 years – or less if so determined by the competent authority – has elapsed.[88]

The European legislator is clear on this matter: each of the parties in the CCS chain bears climate liability, including the storage operator who is responsible for the largest quantity of CO_2 for the longest period of time.[89] Yet, in practice, storage operators involved in CCS demonstration projects in the EU do not always seem to be inclined to take on climate liability. They appear to only want to enter into storage contracts when the capturer bears the financial consequences of the storage operator's climate liability.

The problem with long-term climate liability for CO_2 storage is the difficulty of quantifying the potential size of the liability.[90] If CO_2 leaks from a storage reservoir, the storage operator will have to surrender the equivalent number of EU ETS emission allowances to the national emissions authority. Should the operator not have these allowances at his disposal, he will have to buy them in the market at prevailing market prices.

What is more, storage operators will have to wait for at least 20 years before they can transfer responsibility for the site – and thus climate liability – to the competent authority. Making an accurate estimation of the potential liability for possible leakages in the far away future with varying emission prices is impossible. Both of these factors make it difficult to make a reliable estimate of the potential size of long-term climate liability for CO_2 storage.

---

[88]   See Article 18 Directive 2009/31/EC.

[89]   All three activities have been added to Annex I to the EU ETS Directive.

[90]   In this regard, see also CO_2 Capture Project, 'Regulatory Challenges and Key Lessons Learned from Real World Development of CCS Projects' (2012) 21.

Financial security

Article 19(1) of the CCS Directive requires storage permit applicants to provide financial security to show that they can bear the costs resulting from the following obligations under the CCS Directive:

1) Operating obligations (monitoring, corrective measures);
2) Closure and post-closure obligations ([premature] decommissioning,[91] monitoring, corrective measures);
3) EU ETS obligations (monitoring and remedying of leakage).

In essence, the CCS Directive requires a storage permit to be awarded only if the applicant proves able to finance the operation of the storage site, as well as closure and post-closure obligations and possible corrective measures.

Likewise, we saw that Article 20 of the CCS Directive requires the storage operator to make a financial contribution (financial mechanism) to the competent authority before responsibility for the storage site is transferred. This contribution may be used to cover the costs incurred by the competent authority once responsibility for the site has been transferred and should cover at least the anticipated cost of monitoring for a period of 30 years.

The security requirements under Articles 19 and 20 raise two main issues. First, there is the general regulatory uncertainty about the size, form and timing of the financial security, the financial mechanism and the transfer of responsibility for the storage site to the competent authority. Second, there is the uncertainty caused by the possibility that the size of the financial security required under Article 19 is based on a worst-case leakage scenario, in which leakage occurs and large numbers of EU ETS emission allowances have to be surrendered, with no regard given to the actual probability of hypothesized incidents occurring.

7.5 REVIEW OF THE CCS DIRECTIVE

In November 2015, the Commission transmitted its report on the implementation of the CCS Directive to Parliament and the Council.[92] The report evaluates the CCS Directive for its effectiveness, efficiency, coherence, relevance and EU added value under the Commission's Regulatory Fitness and Performance Programme. Unsurprisingly, the report is very brief indeed.

---

[91] When the injection has ended and the storage site has been closed, CO_2 injection facilities have to be removed. This process is referred to as decommissioning. During the operational phase, the operator reserves money for later decommissioning. When the site is closed prematurely, decommissioning will take place earlier than initially anticipated.

[92] Commission, 'Report on the review of Directive 2009/31/EC on the geological storage of carbon dioxide – Accompanying the document Report from the Commission to the European Parliament and the Council' COM(2015)576 final. This review results follows from Article 38 of the CCS Directive.

In its review, the Commission concludes that the lack of practical experience of projects going through the regulatory process precludes a robust judgement of the performance of the CCS Directive. It admits that the number of CCS installations achieved to date is much lower than had been expected when the CCS Directive was passed. In 2015, only one project – the Rotterdam Capture and Storage Demonstration Project in the Netherlands – had practical experience with the CCS Directive, other than with exploration permits and the feasibility of retrofitting large combustion plants with CCS.

Nevertheless, the Commission concludes that the CCS Directive:

- creates the necessary legal certainty for investors to construct large-scale CCS installations;
- is internally coherent; and
- provides sufficient minimum requirements and guidance to ensure a common approach while leaving Member States sufficient freedom to adapt them to their national circumstances.

Considering the very limited deployment of CCS in the EU to date, the first conclusion seems questionable.

Importantly, the Commission states in its review that there is clear stakeholder concern that reopening the CCS Directive now could be counterproductive as it would bring a period of uncertainty for CCS, which would not be helpful in a sector where investor confidence is already low. Accordingly, significant amendments to the CCS Directive are not to be expected any time soon.

Finally, the Commission makes clear that the next review of the CCS Directive will be carried out when more experience with CCS is available in the EU.

7.6 OFFSHORE CCS DEPLOYMENT

At the time of writing, CCS technologies are not yet being deployed on a large scale in the EU. In recent years, several planned onshore EU CO_2 storage projects were cancelled.[93] The main reasons for the belated deployment of CCS technologies are public opposition and high technology costs. Plans for development of CO_2 storage sites now predominantly focus on offshore CO_2 storage. However, this brings its own set of challenges for project initiators. Although facing less public opposition, these offshore projects have to deal with high technology costs, sometimes even higher than the costs associated with onshore projects.

This section explores several hindrances to the offshore deployment of CCS.[94] We will first discuss two obstacles to the offshore *transportation* of captured CO_2: the 1996 London

---

[93]   See for example the CCS project in Barendrecht, the Netherlands, where public opposition was so strong that the government had to withdraw the project. *Kamerstukken II* 2010/11, 28 982, nr. 113, p.2.

[94]   These are not the only possible regulatory hindrances to offshore CCS deployment, but we have chosen these matters following discussions with (part of) the European CCS stakeholder community.

Protocol and the EU ETS. The London Protocol currently prohibits the export of CO_2 from contracting parties to other countries for the purpose of offshore geological storage, whereas the EU ETS does not regulate CO_2 transport by ship. Second, we explore two obstacles to offshore *storage* of CO_2: the selection of an appropriate storage reservoir and the temporal gap between the end of hydrocarbon mining and the start of CO_2 storage.

7.6.1   The 1996 London Protocol

The 1996 London Protocol (London Protocol) is a protocol to the 1972 Convention on the Prevention of Marine Pollution by Dumping of Wastes and other Matter (London Convention). The London Protocol essentially prohibits dumping of waste at sea, except for an approved list of materials. Article 6 of the London Protocol prohibits its so-called contracting parties from exporting waste to other countries for dumping at sea.

In 2008, a CCS legal and technical working group under the London Protocol determined that Article 6 prohibits the export of CO_2 from a contracting party to other countries for injection into sub-seabed geological formations. An amendment would thus be required to facilitate the development of CCS activities.[95] In 2009, the contracting parties to the London Protocol adopted a resolution amending Article 6 to provide for an exception for the export of CO_2 streams in certain specified circumstances (resolution LP. 3(4)).[96] However, more than ten years later, this amendment is still far from having been ratified by the required two-thirds of the contracting parties.

To overcome this problem, in October 2019 a resolution was passed for the provisional application of the 2009 amendment by those contracting parties which have deposited a declaration on provisional application of the 2009 amendment.[97] Further, the resolution clearly states that the export of CO_2 under the provisional application of Article 6 of the London Protocol will not be in breach of Article 6 as of the time of the export. Although it is uncertain just how robust this solution is from an international law point of view, the resolution seems to clear the way for a number of planned CCS projects in Europe, such as the Northern Lights project in Norway.[98]

7.6.2 CO_2 Transport by Ship and the EU ETS

No EU ETS permit is required for the transport of greenhouse gases by ship. At first sight, this seems to be advantageous to any entity looking to transport CO_2 for storage by ship since it does not have to abide by the provisions of the EU ETS Directive: the operator does not, for

---

[95]   IEA, 'Carbon Capture and Storage Legal and Regulatory Review (CCSReview)', www.cslforum.org/cslf/sites/default/files/documents/paris2012/IEALegalRegulatoryReview-WorkshopRiskLiability-Paris0712.pdf accessed 6 January 2021.

[96]   Ibid.

[97]   Text of the resolution available at: https://ablawg.ca/wp-content/uploads/2019/12/LC41wp.6_octore-2019_report-of-drafting-group.pdf accessed 6 January 2021.

[98]   https://northernlightsccs.com/en/about accessed 6 January 2021.

instance, have to monitor and report these emissions. Nevertheless, it also poses a financial problem to projects looking to have their CO_2 shipped to an offshore storage reservoir.[99]

This financial problem originates from Article 49(1) of Commission Regulation (EU) No 601/2012 on the monitoring and reporting of greenhouse gas emissions pursuant to EU ETS Directive 2003/87/EC (hereinafter the EU ETS Regulation). Article 49(1) states when a transfer of captured CO_2 to an installation gives the transferring operator the right to subtract the transferred CO_2 from his yearly emissions. The operator appears to have this right only when the captured CO_2 is transferred to one of the installations listed in the provision.

The limitative list of installations includes 'a transport network with the purpose of long-term geological storage in a storage site permitted under Directive 2009/31/EC'. Under the EU ETS Regulation the term 'CO_2 transport' is defined as 'the transport of CO_2 by pipelines for geological storage in a storage site permitted under Directive 2009/31/EC'. This seems to lead to the conclusion that the transfer of captured CO_2 to a ship (instead of a pipeline) does not render the transferring operator the right to subtract the transferred CO_2 from his yearly emissions.

If so, this would present an impediment to the development of offshore CO_2 transport and storage. For longer distances, CO_2 transport by ship could be a viable commercial option. However, any projects looking to capture their CO_2 and have it transported to an offshore storage reservoir by ship will lack the financial incentive under the EU ETS to do so. After all, the entity that wants to capture the CO_2 will not only bear the associated capital and operating expenditures, but will also have to add the transferred CO_2 to its yearly emissions for reporting purposes. In other words, its only prospect is an increase in cost.

Case C-460/15 Schaeffer Kalk

In 2017, the European Court of Justice (ECJ) ruled in the *Schaeffer Kalk* case, which sheds interesting light on the above.[100] *Schaeffer Kalk* revolved around a German producer of lime (Schaeffer Kalk) which operates an installation subject to the EU ETS. Schaeffer Kalk had applied to the German emissions authority for authorization to subtract from its emissions the CO_2 transferred for the production of precipitated calcium carbonate (PCC) to an installation not subject to the EU ETS. According to Schaeffer Kalk, since the transferred CO_2 is chemically bound in the PCC and thus not being emitted into the atmosphere, it should not be regarded as 'emissions' as defined in the EU ETS Directive. The German emissions authority had taken the view that such subtraction was not possible under, among other things, Article 49(1) of the EU ETS Regulation.

---

[99] In practice, this obviously particularly poses a problem for potential CO_2 capturers. However, since the CCS chain starts with them, any CCS project looking to have the captured CO_2 shipped to the storage location is confronted with this problem.

[100] Case C-460/15 *Schaefer Kalk GmbH*, 19-01-2017, ECLI:EU:C:2017:29. For an interesting case note in Dutch on this ruling see HvJ EU 19-01-2017, ECLI:EU:C:2017:29, m.nt. M.G.W.M. Peeters en M. Eliantonio (Schaefer Kalk), Milieu en Recht 2017/66.

The respective German administrative court asked whether the EU ETS Regulation is invalid and infringes the aims of the EU ETS Directive insofar as then Article 49(1) provided that CO_2 that is not transferred to one of the installations listed in the provision is to be considered emitted by the installation producing the CO_2.

The ECJ found Article 49(1) to be invalid insofar as it systematically includes the CO_2 transferred to another installation for the production of PCC in the emissions of the transferring installation, regardless of whether or not that CO_2 is released into the atmosphere. According to the ECJ, whether CO_2 transferred truly represents (atmospheric) emissions within the meaning of the EU ETS Directive should be decisive.

Of course, the Schaeffer Kalk ruling must be placed within the specific (factual) context of the case at hand. Nevertheless, it raises a number of interesting questions. The first is whether CO_2 transferred to a ship for geological storage should not be considered to fall outside the definition of emissions in the EU ETS Directive. After all, the captured CO_2 is not released into the atmosphere and is ultimately stored underground. If the former is to be answered positively, the second question is whether this means that the operator of a capture installation transferring its CO_2 to a ship for storage is likewise entitled to subtract the transferred CO_2 from its emissions, regardless of the wording in Article 49(1).

To provide clarity, CO_2 transport by ship with the purpose of geological storage could be added to Annex 1 of the EU ETS Directive and the definition of transport network in the EU ETS Regulation could be changed.[101] Alternatively, interested EU Member States could start a (so-called opt-in) procedure under Article 24 of the EU ETS Directive for unilateral inclusion of CO_2 shipping with a view to geological storage in the EU ETS. However, both options are likely to be time-consuming. As long as this issue remains unsolved, there exists a serious financial disincentive to CO_2 shipping for geological storage.

7.6.3   Offshore Storage of CO_2

Storing CO_2 onshore has faced significant (public) opposition in the past. The alternative is therefore to look to offshore storage of CO_2. There are various potential storage locations for CO_2 beneath the sea bed, the two most prominent candidates being aquifers and empty oil or gas fields, called 'depleted hydrocarbon reservoirs'.[102] Both have a porous layer of rock where the carbon dioxide is to be injected, and on top of this porous layer there is an impenetrable

---

[101]   In December 2018, the Commission followed up on the ECJ ruling in Schaeffer Kalk by adding to Article 49(1) of the EU ETS Regulation that the transfer of CO_2 out of an EU ETS installation to produce precipitated calcium carbonate, in which the CO_2 used is chemically bound, also gives the right to subtract the transferred CO_2 from the emissions of the installation in question. See Commission Implementing Regulation (EU) 2018/2066 of 19 December 2018 on the monitoring and reporting of greenhouse gas emissions pursuant to Directive 2003/87/EC of the European Parliament and of the Council and amending Commission Regulation (EU) No 601/2012 (Text with EEA relevance).

[102]   IEA Greenhouse Gas RD Programme (IEA GHG), 'CCS Site Characterisation Criteria', 2009/10, July 2009, p.1. Technically, coal beds are also available for carbon dioxide storage, but economically they are not feasible as coal bed storage is an immature technology and coal beds generally have small storage potential.

cap rock that prevents the injected CO_2 from leaking to the surface. As storage locations, aquifers and depleted hydrocarbon reservoirs have both advantages and disadvantages.

The aquifer is generally a larger storage location.[103] However, subsoil aquifers are filled with water and injecting CO_2 requires more energy. There is also less knowledge and data available on subsoil aquifers, because as yet there has been no economic incentive to study offshore subsoil aquifers.

Depleted hydrocarbons reservoirs are generally smaller compared to subsoil aquifers, but significantly more data is available on the suitability of the reservoir as a storage location. Because the reservoir is nearly depleted, it takes less energy to inject CO_2 into the reservoir. Another advantage of using hydrocarbon reservoirs is that CO_2 storage can start during the production of oil and gas. There is, however, also a downside to the use of (nearly) depleted hydrocarbon reservoirs. The decreased pressure within the reservoir is detrimental to the integrity of the cap rock as the cap rock may bend as a result of the pressure from the layers on top of it. This means that the cap rock decreases in strength and cannot contain much CO_2, or may even collapse.

Another important difference between aquifers and depleted hydrocarbon reservoirs is that with hydrocarbon reservoirs the infrastructure necessary for transporting and injecting CO_2 is already in place. This infrastructure consists of platforms, pipelines and wells. When CO_2 storage takes place in the context of enhanced recovery, two licences are required: one for the production of hydrocarbons and one for CO_2 storage. When CO_2 storage starts after the cessation of hydrocarbons production, the hydrocarbon production licence expires and the CO_2 storage licence takes its place. If the holder of the hydrocarbon production licence is not going to be involved in CO_2 storage activities, a new CO_2 storage licence holder takes over the infrastructure.

Ideally, the newly licensed CO_2 storage operator is able to begin CCS activities when the hydrocarbons production is finished. At the time of writing, only a limited number of offshore CO_2 storage projects involving reuse of existing hydrocarbon infrastructure are being developed. But given the available offshore infrastructure in the North Sea, it could be expected that more projects involving reuse of existing infrastructure will be developed in the future. In the offshore areas of the Netherlands, Norway and the United Kingdom, there are more than 5000 wells, around 1500 platforms and installations and 26000 km of subsea pipelines.[104]

However, if the emission allowance prices under the EU ETS are too low to incentivize CCS, then reusing depleted hydrocarbon reservoirs becomes more difficult. Operators of the offshore hydrocarbon installations and pipelines are under an obligation to remove these installations and pipelines (Art. 60(3) UNCLOS and Art. 5 OSPAR Convention Annex III).

---

[103]   Intergovernmental Panel on Climate Change (IPCC), *IPCC Special Report on Carbon Dioxide Capture and Storage. Prepared by Working Group III of the Intergovernmental Panel on Climate Change* (Cambridge University Press 2005), p.221.

[104]   See: www.nexstep.nl/decommissioning-landscape/ accessed 6 January 2021; Oil and Gas Authority, 'Decommissioning Strategy', 2016, p.5; D. Erlend Henriksen, 'Decommissioning Practice in Norway' in: M. Roggenkamp and C. Banet (eds), *European Energy Law Report* (Intersentia 2020), p.352.

Once decommissioned, the reservoirs can only be used for CCS activities after a new borehole is drilled. One of the challenges for the coming years is to find a way of preserving the existing hydrocarbon infrastructure for later reuse.

7.6.4   Reuse Strategy for Hydrocarbon Infrastructure for CO_2 Storage

When formulating a strategy for reusing existing hydrocarbon infrastructure for the permanent storage of CO_2, two factors have to be taken into account. First, the decommissioning regimes in the North Sea countries are based on the idea that when the production of hydrocarbons ceases, the infrastructure must be removed and the site has to be restored in its original state. Although the regime on decommissioning is strict on *which* infrastructure has to be removed and which can be left in place, the regime is not clear on *when* precisely the infrastructure has to be removed. Second, reuse of existing infrastructure is optional under the international regime on decommissioning and is sometimes mandatory under national law.

Bearing the two aforementioned factors in mind, three scenarios for hydrocarbon infrastructure reuse can be drawn, as in Figure 7.1.

In the first scenario, CO_2 storage is combined with hydrocarbon production in what is known as enhanced recovery. The hydrocarbon production licence overlaps with storage licence, and both licences are held by the same party.

In the second scenario, the hydrocarbon production licence is followed by the CO_2 storage licence. The two licences can be held by one party or by different parties. Should there be different parties, the infrastructure has to be transferred, together with the decommissioning obligations, from the hydrocarbon licence holder to the CO_2 storage licence holder.

In the third scenario, there is no direct follow-up between the two licences and there exists a temporal gap between the end of hydrocarbon production and storage of CO_2. During this period, the question is who is responsible for the infrastructure. Should nobody be responsible for the infrastructure, the responsibility shifts towards to the coastal state as the coastal state is, under international law, responsible for decommissioning. The holder of the hydrocarbon production licence who intends to stop production also wants to be relieved of the responsibility for the infrastructure. This dilemma requires a solution, and there are two possible ways of mitigating this problem.

The first possible solution is to extend the duration of the production licence, but without envisaging any production activities. National mining and petroleum laws generally allow for the extension of hydrocarbon production licences. Although production has ceased, the hypothesis would be that hydrocarbons are still being produced, whereas the actual goal is to maintain the existing infrastructure for future use. The advantage of this solution is that it does not require any drastic changes in the law. The drawback of this solution is that the licensing regime is used for another purpose than originally intended, namely for preservation instead of hydrocarbon production. This could be deemed as unintentional misuse of a licence. A hydrocarbon production licence holder would be willing to maintain the infrastructure when the licence holder is planning to start CO_2 storage in the future, or to pass the infrastructure on to a different storage operator who financially compensates the hydrocarbon licence holder.

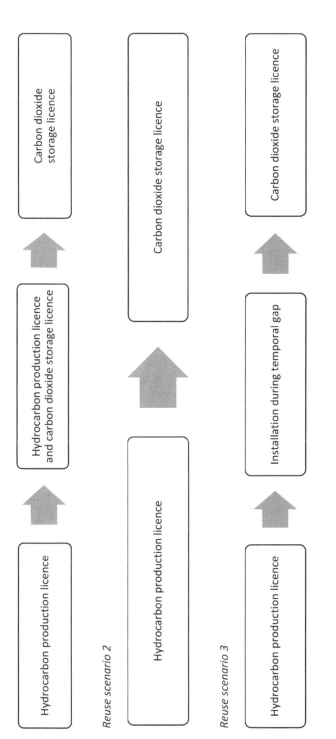

Figure 7.1 Hydrocarbon infrastructure reuse scenarios

The second possible solution is to introduce a new operator that maintains the infrastructure during the temporal gap. Because maintaining offshore hydrocarbon infrastructure for later reuse is not a commercial activity but rather a task performed in the general interest, it could be argued that the government should step in. The state can take over offshore infrastructure in order to preserve it, but it remains to be seen whether the state, in absence of a state oil or energy company, will have the personnel and equipment to maintain the infrastructure during the temporal gap. It therefore seems likely that in such a case the state would need to appoint an interim operator: an operator of last resort. This operator will then be given responsibility for maintaining the disused infrastructure until it is possible to begin awarding a CO_2 storage permit. This operator may be a separate licence holder or an offshore service company, hired by the state as an agent, through an Operation and Maintenance Agreement. Finally, transferring the infrastructure to the state also means that the decommissioning responsibility is transferred to the state. A careful selection procedure would be required to prevent a situation in which the state takes over infrastructure that is not going to be used for CO_2 storage in the future.

7.7　CONCLUSION

This chapter has introduced carbon capture and storage (CCS) as one of the possible instruments for reducing greenhouse gas emissions. At the time of writing, there still do not seem to be sufficient financial incentives for large-scale CCS to develop in Europe. Furthermore, offshore CCS project developers are confronted with several regulatory hindrances, such as CO_2 transport by ship not having been recognized under the EU ETS. If these hindrances are not removed, CCS deployment in the EU will likely exist on only a small scale in the near future.

The CCS Directive provides Member States and CCS market players with the opportunity to gain different experiences with the implementation of that legal framework. This is clearly visible in the general character and lack of detail of several of the Directive's provisions on CO_2 capture, transport and storage. In relation to a number of issues, exact standards will have to be set by the Member States, which have the Commission's legally non-binding guidelines at their disposal.

From a cost-effectiveness point of view, this regulatory approach has different effects. On the one hand, it allows for exact standards to be set in a way that better suits the characteristics of individual CCS projects. This, arguably, diminishes the chances of unnecessary (high) administrative burdens being imposed and allows for tailor-made regulatory solutions. On the other hand, the uncertainty caused by the general and imprecise character of the CCS Directive also raises transaction costs for CCS market players.

As for solidarity between the Member States, the CCS Directive's regulatory approach does not appear to be particularly helpful. The Directive does not prescribe a common EU approach for the selection and development of CO_2 transport and storage infrastructure. Likewise, more detailed rules regarding the accessibility of CO_2 transport and storage infrastructure are lacking. Moreover, the Directive hardly addresses the cross-border transport and storage of CO_2. The few provisions that deal with the latter only do so in a very rudimentary manner.

Ultimately, the lack of inter-Member State solidarity might affect Member States with little own storage capacity, since it reduces their chances of storing their CO_2 in other Member States. This needs to be addressed in the longer term.

CLASSROOM QUESTIONS

1. What is the biggest legal barrier to implementing CCS in the EU?
2. Which types of liability do CO_2 storage operators face in the EU?
3. What should be changed in EU climate mitigation law to better incentivize CCS?

SUGGESTED READING

Books

Havercroft I, Macrory R and Stewart RB (eds), *Carbon Capture and Storage: Emerging Legal and Regulatory Issues (Hart Publishing 2018).*

Roggenkamp MM and Woerdman E (eds), *Legal Design of Carbon Capture and Storage – Developments in the Netherlands from an International and EU Perspective (Intersentia 2009).*

Uwer D and Zimmer D, *Carbon Capture and Storage: The Legal Landscape of Climate Change Mitigation Technology (Globe Law and Business 2020).*

Articles and chapters

Brockett S, 'The EU Enabling Legal Framework for Carbon Capture and Storage and Geological Storage' (2009) 1 Energy Procedia 4433.

Haan-Kamminga A, 'Long-term Liability for Geological Carbon Storage in the EU' (2011) 29 Journal of Energy and Natural Resources Law 309.

Weber V, 'Uncertain Liability and Stagnating CCS Deployment in the European Union: Is It the Member States' Turn?' (2018) 27 RECIEL 153–61.

Policy documents

Global CCS Institute website on the Carbon Capture Use and Storage Legal Resource Net. www .globalccsinstitute.com/.

IEA, '20 Years of Carbon Capture and Storage' (IEA 2016). www.iea.org/reports/20-years-of-carbon -capture-and-storage.

IPCC, 'IPCC Special Report on Carbon Dioxide Capture and Storage. Prepared by Working Group III of the Intergovernmental Panel on Climate Change' (Bert Metz and others (eds), Cambridge University Press 2005).

8
Regulation of fluorinated gases

ABSTRACT

- The EU regulatory framework on fluorinated gases consists mainly of Directive 2006/40/EC (MAC Directive) and Regulation 517/2014 (F-gas Regulation);
- Fluorinated gases are powerful greenhouse gases used in refrigerators and air-conditioning units, for instance, or as blowing agents for foams and fire extinguishers;
- The Mobile Air-Conditioning (MAC) Directive prohibits the use of fluorinated gases with a global warming potential more than 150 times greater than CO_2 in new types of cars introduced from 2011 and in all new cars produced from 2017;
- The first F-gas Regulation (Regulation No 842/2006) aimed at (a) improving the prevention of leaks from equipment containing fluorinated gases and (b) avoiding the use of these gases in some applications where environmentally superior alternatives are more cost-effective;
- In 2012, the Commission proposed to improve the F-gas Regulation (COM (2012) 643 final) and to tighten its requirements as part of its 'Roadmap 2050' policy;
- A new F-gas Regulation (No 517/2014) has applied since January 2015 and is designed to achieve a reduction in the emissions of fluorinated gases. Its aim is to cut the EU's F-gas emissions by two-thirds in 2030 compared with 2014 levels. The Regulation is highly relevant in light of the global phase-down of the use of hydrofluorocarbons (HFCs), the most significant group of fluorinated gases, under the Montreal Protocol. The EU Green Deal may require amendments to this regulatory framework.

8.1 INTRODUCTION

This chapter explores the EU regulation of fluorinated gases (or: F-gases). Fluorinated gases are a group of chemicals containing fluorine and are mostly man-made. These F-gases are powerful greenhouse gases that trap heat in the atmosphere and contribute to global warming. They are used in a range of industrial applications. Because fluorinated greenhouse gases do not damage the atmospheric ozone layer, they are often used as substitutes for ozone-depleting substances. F-gases are used in several sectors for a wide range of applications, such as refrigeration, air-conditioning equipment, insulation foams and aerosol sprays (as solvents and fire-protection systems), as well as in the electronics sector and in the cosmetic and pharmaceutical industry. F-gases are mainly emitted due to leakages during use, maintenance and unsuccessful end-of-life recovery. Emissions of these fluorinated gases in the EU have risen by as much as 60 per cent since 1990, while all other greenhouse gases have been reduced. They account for approximately 3 per cent of overall greenhouse gas emissions in the EU.[1] That may not seem much, but action to reduce emissions of fluorinated gases is needed to combat climate change since these gases have relatively long lifetimes in the atmosphere and have a global warming effect up to 23,000 times stronger than that of CO_2.[2]

The EU regulatory framework on F-gases consists of two main legislative acts. Directive 2006/40/EC on Mobile Air-Conditioning (the MAC Directive) deals with emissions from air-conditioning systems in motor vehicles. Regulation 517/2014 (the F-gas Regulation) focuses on reducing emissions from certain equipment through a series of measures throughout their life cycle.[3]

In this chapter we examine the EU regulatory framework for fluorinated gases in light of the overarching EU climate policy principles of cost-effectiveness and solidarity. Section 8.2 explains what fluorinated gases are and how they contribute to global warming. In section 8.3 we discuss the MAC Directive by exploring its legislative approach and central provisions. In section 8.4 we discuss the F-gas Regulation (Regulation 517/2014), which is related to the Commission's policy document 'Roadmap 2050'. Section 8.5 concludes.

8.2 BASICS OF FLUORINATED GASES

The EU has had legislation on fluorinated gases in force since 2006. The overall objective of this regulation is to help fulfil the greenhouse gas reduction commitments of the EU and its Member States, first under the Kyoto Protocol of 1997 and now under the Paris Agreement of 2015. There are three main groups of gases containing fluorine: hydrofluorocarbons (HFCs), perfluorocarbons (PFCs) and sulphur hexafluoride (SF6), collectively known as fluorinated greenhouse gases.[4] Unlike many other greenhouse gases, these gases are man-made chemicals especially designed for industrial purposes.[5] HFCs, the most common fluorinated gases, are used in various sectors and applications, including refrigeration, air-conditioning and heat pump equipment. HFCs are also used as blowing agents for foams, fire extinguishers, aerosol propellants and solvents. PFCs are mostly used in the electronics sector as well as in the

---

[1] 'EEA Greenhouse Gas data viewer' www.eea.europa.eu/data-and-maps/indicators/emissions-and -consumption-of-fluorinated-2/assessment-2 accessed 12 November 2020.

[2] European Commission Press release, 7 November 2012, http://europa.eu/rapid/press-release_IP-12-1180_en .htm accessed 28 December 2018.

[3] Commission, 'Proposal for a regulation of the European Parliament and of the council on fluorinated greenhouse gases' COM (2012) 643 final.

[4] Annex A of the Kyoto Protocol to the United Nations framework convention in climate change. Approved by Council Decision 2002/358/EC of 25 April 2002 concerning the approval, on behalf of the European Community, of the Kyoto Protocol to the United Nations Framework Convention on Climate Change and the joint fulfilment of commitments thereunder [2002] OJ L130 and ratified by the EC and its Member States on 31 May 2002.

[5] W. Schwarz et al., *Preparatory Study for a Review of Regulation (EC) No 842/2006 on Certain Fluorinated Greenhouse Gases* (Öko-Recherche 2011 et al.) 1.

cosmetic and pharmaceutical industry. SF6 is primarily used as an insulation gas and can be found in high-voltage switchgear and as a cover gas in magnesium and aluminium production. Products containing fluorinated gases are used by millions of consumers, for instance through the use of domestic refrigerators and car air-conditioning.[6]

Emissions of F-gases occur during emissive uses, for example aerosol sprays, or due to leakage during the operation and disposal of products and equipment that contain fluorinated gases. In total, fluorinated gas emissions accounted for around 2.27 per cent of all greenhouse gas emissions in the EU in 2018.[7] Although this percentage is relatively low compared to other greenhouse gases, the effect of these gases in the atmosphere is powerful. The global warming potential of fluorinated gases can be up to 23,000 times higher than that of CO_2. F-gases also have very long lifetimes and can remain in the atmosphere for thousands of years.[8]

Fluorinated gases have become popular in industrial production since the 1990s, and since that time the use of these gases has more than doubled.[9] Most F-gases have been developed by industry to replace ozone-depleting substances (ODSs) that are being phased out under the Montreal Protocol on substances that deplete the ozone layer.[10] International action to reduce the production and consumption of F-gases, given that climate-friendly alternatives are available, led to the Kigali Amendment to the Montreal Protocol. This amendment, which came into effect in 2019, adds HFCs to the list of substances to be phased down. The EU ratified the Kigali Amendment in 2018.

The EU's regulation of fluorinated greenhouse gases is a key element of the first stage of the European Climate Change Programme (ECCP) established in 2000. The Commission's approach was to move forward in two stages.[11] The first stage was to establish a legal framework for fluorinated gases. This was followed by a period of monitoring and evaluation, after which the Commission would consider the option of strengthening the existing controls as

---

[6]   J. Harnisch and R. Gluckman, *Final Report on the European Climate Change Programme Working Group Industry Work Item Fluorinated Gases. Report Prepared on Behalf of the European Commission (DG Environment and DG Enterprise)* (2001) Ecofys and Enviros 2001, 5.

[7]   EEA Greenhouse Gas data viewer: www.eea.europa.eu/data-and-maps/data/data-viewers/greenhouse-gases -viewer accessed 1 December 2020.

[8]   A recently discovered fluorinated gas (Perfluorotributylamine) was found to be 7100 times more powerful at warming the Earth over a 100-year time span than CO_2. The former is estimated to remain in the atmosphere for about 500 years and has no natural 'sinks', such as forests and oceans, to absorb it. See Angela C. Hong et al., 'Perfluorotributylamine: A novel long-lived greenhouse gas', *Geophysical Research Letters* 2013, Vol. 40 (22), pp.6010–15.

[9]   European Commission Press release, 7 November 2012, https://ec.europa.eu/commission/presscorner/detail/ en/IP_12_1180 accessed 1 December 2020.

[10]   The Montreal Protocol on Substances that Deplete the Ozone Layer, adopted on 16 September 1987.

[11]   COM (2003) 492 final, 6.

well as the option of introducing additional measures to ensure that the policy objectives are achieved.[12]

In 2011 the Commission published a report on the application of the first F-gas Regulation.[13] One of the conclusions was that the 2006 F-gas Regulation assisted the EU to remain on track with the commitments under the Kyoto Protocol. Analysis showed that a full application of the former F-gas Regulation, together with the MAC Directive, has the potential to stabilize emissions of fluorinated gases.[14] However, the EU set more ambitious targets for the longer term. The Effort Sharing Decision (406/2009/EC), which was part of the so-called climate and energy package, laid down binding national greenhouse gas emission reduction targets for the period 2013–20 in sectors not covered by the EU Emissions Trading System (EU ETS) in order to contribute to the overall reduction objective for 2020.[15] Furthermore, the roadmap for moving to a low-carbon economy in 2050 confirmed the EU objective of reducing greenhouse gas emissions to at least 80–95 per cent by 2050 compared to 1990 levels.[16]

The 2030 framework on climate and energy constitutes the next step towards reaching the 2050 goal. Following the proposal by the Commission, the Council concluded in October 2014 to set a target for domestic EU greenhouse gas emissions of 40 per cent in 2030 relative to emissions in 1990. Emissions of the non-ETS sector, including emissions of F-gases, will have to be reduced by 30 per cent compared to 2005.[17] The methodology used to set national reduction targets for the non-ETS sectors on the basis of relative gross domestic product (GDP) per capita will be continued to reflect cost-effectiveness in a fair and balanced manner. Although there are no national reduction targets specifically for F-gases, the ambitions of the EU implied that it had to take further action to achieve cost-effective reductions of greenhouse gas emissions, including F-gases.[18] Therefore, in 2012 the Commission tabled a proposal to improve

---

[12] Reporting obligations are included in Article 10 of Regulation (EC) No 842/2006 of the European Parliament and of the Council of 17 May 2006 on certain fluorinated greenhouse gases [2006] OJ L161/1.

[13] Commission, 'Report from the Commission – On the application, effects and adequacy of the Regulation on certain fluorinated greenhouse gases (Regulation (EC) No 842/2006)' COM (2011) 581 final.

[14] Schwarz et al., n 5, p.1.

[15] Decision No 406/2009/EC of the European Parliament and of the Council of 23 April 2009 on the effort of Member States to reduce their greenhouse gas emissions to meet the Community's greenhouse gas emission reduction commitments up to 2020 [2009] OJ L140/136.

[16] Commission, 'Communication from the Commission to the European Parliament, the Council, the European Economic and Social Committee and the Committee of the Regions: A roadmap for moving to a low carbon economy in 2050' COM (2011) 112 final.

[17] Commission, 'Communication from the Commission to the European Parliament, the Council, the European Economic and Social Committee and the Committee of the Regions: A policy framework for climate and energy in the period from 2020 to 2030' COM (2014) 015 final and European Council, Conclusions on 2030 Climate and Energy Policy Framework, Brussels, 23 October 2014 (OR. en) SN 79/14.

[18] COM (2011) 581 final, 11.

and strengthen the F-gas Regulation.[19] In January 2015 the new F-gas Regulation (Regulation 517/2014) came into force; it will be discussed in section 8.4.

Since the introduction of the F-gas Regulation in 2015, EU policy on reducing greenhouse gas emissions has evolved even further. The Effort Sharing Regulation has replaced the ESD for the period 2021–30.[20] In December 2019 the European Green Deal was presented by the European Commission. It aims to achieve a climate-neutral EU by 2050 and sets an intermediate target of at least 55 per cent net reduction in greenhouse gas emissions by 2030.[21] This new policy will lead the Commission to revise the Effort Sharing Regulation. The current F-gas Regulation will be reviewed and updated in view of the European Green Deal and climate law, international obligations on HFCs (Montreal Protocol) and progress made and lessons learned. A proposal for an updated F-gas Regulation is expected by the end of 2021.[22]

8.3   THE MOBILE AIR-CONDITIONING DIRECTIVE

Directive 2006/40/EC, the MAC Directive, aims to reduce emissions of specific fluorinated gases in air-conditioning systems fitted to passenger cars and light commercial cars.[23] In this section we explore the legal implications of the MAC Directive by discussing its legislative approach and its central provisions.

8.3.1   Legislative Approach

During the legislative procedure for the MAC Directive, two essential changes were introduced to the Commission's proposal to reduce emissions from motor air-conditioning.[24]

First, there was discussion about the Commission's approach to reach the objectives. The Commission's original proposal included a gradual, flexible system based on transferable

---

[19]   COM (2012) 643 final.

[20]   Regulation (EU) 2018/842 of the European Parliament and of the Council of 30 May 2018 on binding annual greenhouse gas emission reductions by Member States from 2021 to 2030 contributing to climate action to meet commitments under the Paris Agreement and amending Regulation (EU) No 525/2013 (Text with EEA relevance).

[21]   The European Green Deal COM/2019/640 final.

[22]   See https://ec.europa.eu/info/law/better-regulation/have-your-say/initiatives/12479-Review-of-EU-rules-on-fluorinated-greenhouse-gases accessed 6 January 2021.

[23]   According to Article 2 the MAC Directive shall apply to motor vehicles of categories M1 and N1 as defined in Annex II of Council Directive 70/156/EEC of 6 February 1970 on the approximation of the laws of the Member States relating to the type-approval of motor vehicles and their trailers [1970] OJ L0156.

[24]   European Parliament, 'I Report on the proposal for a European Parliament and Council regulation on certain fluorinated greenhouse gases (COM (2003) 492 – C5-0397/03- 2003/0189 (COD))' 18 March 2004, Final A5-0172/2004, 40 and 46. Opinion of the European Economic and Social Committee on the proposal for a Regulation of the European Parliament and of the Council on certain fluorinated greenhouse gases (COM (2003) 492 final, 0189/2003 COD), OJ C108/62.

quotas to gradually phase out the use of HFC-134a in vehicle air-conditioning systems.[25] According to the Commission, a phased approach would allow car manufacturers and importers enough time to introduce the changes to the platforms of cars in a cost-effective manner.[26] According to the proposal (Article 10), the possibility of selling passenger cars and light commercial vehicles with an air-conditioning system containing HFC-134a would be phased out gradually. It stipulated that between 1 January and 31 December 2009 only 80 per cent of a predetermined quota of such cars and vehicles may be placed on the market. Over the following years this level would be reduced to 60 per cent, 40 per cent, 20 per cent and eventually 10 per cent (in 2013). Quota holders would be allowed to transfer quotas, without any restrictions, to other quota holders and special provisions were made for new entrants on the market.

The Parliament's amendments, however, considered that the proposed quota system was not the most practical way to achieve the goal of reducing emissions from air-conditioning systems. The tradable quota system was bureaucratic and entailed relatively high administrative costs. The Parliament was of the opinion that the quota system was complicated and hard to implement.[27] The Council agreed with the Parliament and therefore the proposed quota system was deleted during the legislative process.

Instead of the tradable quota system, the Parliament intended to use the so-called type-approval system (pursuant to Framework Directive 70/156/EEC). Its purpose is to control the way in which vehicles are equipped with eco-friendly air-conditioning systems (Council Directive No 92/53/EEC).[28] The type-approval system aims at ensuring that new vehicles placed on the market in the EU are in line with safety and environmental performance requirements. It is a procedure whereby a Member State certifies that certain types of vehicles, systems, components or separate technical units satisfy the relevant administrative provisions and technical requirements.

The Council agreed with the use of a type-approval approach and divided the single legislative proposal into two pieces of legislation. The mobile air-conditioning part of the Commission's proposal was now moved to a separate Directive: the MAC Directive. However, the Council underlined that, although there are two elements (a Regulation and a Directive), the Council and the Commission were in agreement that there is still only one proposal.[29]

The Framework Directive 70/156/EEC, which had been the main legal instrument for the type-approval system since 1970, was replaced by the Framework Directive 2007/46/EC for the approval of motor vehicles and their trailers and of technical systems intended

---

[25]   COM (2003) 492 final, 4.

[26]   COM (2003) 492 final, 11.

[27]   European Parliament, 'I Report', n 24.

[28]   Council Directive 70/156/EEC of 6 February 1970 on the approximation of the laws of the Member States relating to the type-approval of motor vehicles and their trailers [1970] OJ L42/1.

[29]   Council of the European Union, 'Draft statement of the Council's Reasons' 22 March 2005, 2003/0189B (COD), 9.

for such vehicles.[30] In Article 3 of this current Framework Directive, 'EC Type-Approval' is defined as the procedure whereby a Member State certifies that a type of vehicle, system, component or separate technical unit satisfies the relevant administrative provisions and technical requirements of this Directive and the regulatory Acts in Annex IV or XI. The MAC Directive is one of the regulatory Acts listed in Annex IV and XI in which harmonized technical requirements applicable to individual parts and characteristics of a vehicle are specified. 'National type-approval' is also defined: it refers to a type-approval procedure laid down by the national law of a Member State, the validity of such approval being restricted to the territory of that Member State. The MAC Directive establishes the fundamental provisions on mobile air-conditioning systems in vehicles. Technical specifications and administrative provisions are laid down by the following implementing measures: Commission Regulation (EC) No 706/2007 and Commission Directive 2007/37/EC.[31]

The two main objectives of the MAC Directive are (a) to control leakage of fluorinated greenhouse gases with a global warming potential higher than 150 from air-conditioning systems in motor vehicles and (b) a prohibition on using these gases in such air-conditioning systems. The Directive provides for a phase-out of air-conditioning systems designed to contain fluorinated gases with a global warming potential higher than 150 in the period 2011–17. Global warming potential (GWP) refers to the climatic warming potential of a (for instance: fluorinated) greenhouse gas relative to that of CO_2. The global warming potential is calculated in terms of the 100-year warming potential of one kilogram of a gas relative to one kilogram of CO_2.[32] The GWP of CO_2, therefore, is 1.

In practice, this approach comes down to a phase-out of hydrofluorocarbon-134a (HFC-134a), which has a global warming potential of 1300. This used to be the only identified fluorinated gas with a global warming potential higher than 150. The Directive does not prescribe the use of another refrigerant or particular system, but states that cost-effective and safe alternatives to HFC-134a are expected to be available in the near future. Nowadays more climate-friendly alternatives are available, in particular R1234yf, which is used almost exclusively. Another alternative is CO_2, which is expected to be more widespread in the future.[33]

The application of the MAC Directive has led to an infringement procedure against Germany. In September 2012 the Commission was informed by Daimler, a German car

---

[30]   Directive 2007/46/EC of the European Parliament and of the Council of 5 September 2007 establishing a framework for the approval of motor vehicles and their trailers and of systems, components and separate technical units intended for such vehicles [2007] OJ L263/1.

[31]   Commission Regulation (EC) No 706/2007 of 21 June 2007 laying down, pursuant to Directive 2006/40/EC of the European Parliament and of the Council, administrative provisions for the EC type-approval of vehicles, and a harmonized test for measuring leakages from certain air conditioning systems [2007], OJ L 161/33, 22.6.2007. Commission Directive 2007/37/EC of 21 June 2007 amending Annexes I and III to Council Directive 70/156/EEC on the approximation of the laws of the Member States relating to the type approval of motor vehicles and their trailers.

[32]   Article 3(8) of the MAC Directive.

[33]   Recital 3 of the Preamble to the MAC Directive. See https://ec.europa.eu/clima/policies/f-gas/alternatives_en accessed 1 December 2020.

manufacturer, that it had safety concerns regarding the use of HFO-1234yf as a refrigerant and that it would continue to use HFC-134a.[34] The Commission declared that no evidence existed that there were no technical solutions to mitigate the flammability risks associated with HFO-1234yf and pointed out that each Member State was responsible for implementation of the MAC Directive.[35] The CJEU declared in 2018 that Germany had infringed EU law by allowing Daimler to place vehicles on the market that were not in conformity with the MAC Directive, and by failing to take remedial action.[36]

8.3.2   Key Legal Instruments and Provisions

In this section we discuss the key legal instruments, laid down in Article 5 and Article 6, of the MAC Directive.

Article 5 of the MAC Directive requires vehicles that are fitted with air-conditioning systems designed to contain fluorinated greenhouse gases with a global warming potential higher than 150 to be type-approved. It also requires the establishment of limit values for leakage rates from such systems. Articles 5(2) and 5(3) of the MAC Directive require EC type-approval or national type-approval to be granted by Member States only if the rate of leakage from the air-conditioning system does not exceed certain thresholds (namely 40 grams of fluorinated greenhouse gases per year for a single evaporator system or 60 grams of fluorinated greenhouse gases per year for a dual evaporator system). The control of leakage is mandatory from 21 June 2009 for all new vehicles, including new vehicles for existing types. The harmonized leakage detection test to evaluate the release into the atmosphere of refrigerant fluid (referred to in Articles 5(2) and 5(3) of the MAC Directive) is laid down in Annex II to Commission Regulation (EC) No 706/2007. This Annex addresses the equipment requirements, test conditions, test procedures and data requirements. The Regulation also contains some administrative provisions for EC type-approval to implement the obligations of the Member States stipulated in Article 4. For example, Member States have to ensure that manufacturers supply information on the type of refrigerant used in air-conditioning systems fitted to new motor vehicles.

The phase-out of HFC-134a started in 2011. From 1 January 2011, Member States were no longer allowed to grant EC type-approval or national type-approval for vehicles with an air-conditioning system designed to contain fluorinated gases with a global warming potential higher than 150 (Article 5(4) of the MAC Directive). Since 1 January 2017, 'certificates of conformity' are no longer valid for new vehicles which are fitted with an air-conditioning system designed to contain F-gases with a global warming potential higher than 150. A 'certificate of conformity' is a document, issued by the manufacturer, that certifies that a vehicle belongs

---

[34]   Press release issued by Daimler, 25 September 2012, see https://vasa.org.au/major-car-maker-renounces -r1234yf/ accessed 5 January 2021.

[35]   Commission, 'Declaration by the European Commission regarding Point 9 of the agenda of the 31st meeting of the "Technical Committee – Motor vehicles" (TCMV): State of Play of the EU Mobile Air-Conditioning directive (2006/40/EC)' Brussels, 19 December 2012.

[36]   CJEU 4 October 2018, *Commission v Germany*, ECLI:EU:C:2018:802.

to the series of the type approved in accordance with the legal framework of type-approval.[37] Member States have to refuse registration and prohibit sale and entry into service (Article 5(5) MAC Directive) when such a certificate is no longer valid.

To ensure that the prohibition of certain fluorinated gases is effective, the possibilities of installing an air-conditioning system in a vehicle after it has been registered are limited. Therefore, by virtue of Article 6(1) MAC Directive, air-conditioning systems designed to contain fluorinated greenhouse gases with a global warming potential higher than 150 may not be 'retrofitted' to vehicles that have been type-approved from 1 January 2011. From 1 January 2017, such air-conditioning systems may not be retrofitted in any vehicle.

8.4 THE F-GAS REGULATION

Regulation (EC) No 842/2006 on certain fluorinated gases was the first F-gas Regulation and aimed to contain, prevent and thereby reduce emissions of fluorinated greenhouse gases through a series of measures or actions taken throughout their life cycle. In 2012 the Commission tabled a proposal for a revision of the EU regulation on F-gases.[38] In 2014 the European Parliament and the Council formally approved a slightly amended text that maintained the ambition level of the proposal by the Commission. The current F-gas Regulation (EU) No 517/2014 has applied since January 2015.[39] In this section we will explore the legal implications of regulating F-gases in the EU. We will do so by discussing the legislative approach to both the first and the new F-gas Regulation (section 8.4.1) and by describing key legal instruments and provisions of the F-gas Regulation No 517/2014 (section 8.4.2).

8.4.1 Legislative Approach

In this first F-gas Regulation, the Commission chose to follow two lines of action.[40]

First, Regulation No 842/2006 aimed at avoiding the use of fluorinated gases where environmentally superior alternatives were already available at a sufficiently low cost at the time of adoption, by restricting the use and marketing of specific applications. The provisions on fluorinated gases covered the key applications in which fluorinated gases are used, such as air-conditioning and refrigeration systems, heat pumps, fire-protection systems and switchgears. Furthermore, it introduced a prohibition on placing on the market products and equipment that contain fluorinated gases, such as footwear, windows for domestic use and fire-protection systems.[41] These prohibitions were technically feasible and cost-effective

---

[37] 'Certificate of conformity' is defined in Article 3 of the Framework Directive 2007/46/EC.

[38] COM (2012) 643 final.

[39] Regulation (EU) No 517/2014 of the European Parliament and the Council of 16 April 2014 on fluorinated greenhouse gases and repealing Regulation (EC) No 842/2006, OJ L 150/95.

[40] Commission, 'Report from the Commission on the application, effects and adequacy of the Regulation on certain fluorinated greenhouse gases (Regulation (EC) No 842/2006)' COM (2011) 581 final, 2.

[41] See Article 9 and Annex II of the first F-gas Regulation.

because alternatives to the use of F-gases were available.[42] Alternative technologies based on low-GWP fluids were deemed feasible in most relevant fields of application. For example, the chemical industry has developed new 'fluorocarbons' which as early as 2011 were considered to be potential substitutes to high-GWP fluorinated gases in several sectors (including refrigeration and air-conditioning, heat pumps and foams).[43]

Second, Regulation No 842/2006 aimed at reducing leakage from equipment where fluorinated gases could not be replaced by alternatives. It contains a series of measures targeting the proper handling of equipment throughout its lifetime, such as mandatory inspections for leakages by certified personnel, mandatory record-keeping by the operator and mandatory end-of-life recovery. Regular checks for leakages are considered one of the most effective ways to reduce emissions from equipment; this followed from the experiences with the implementation of the emission control provisions of Regulation (EC) 2037/2000 on substances that deplete the ozone layer.[44] The focus on containment is underpinned by the success of the so-called STEK scheme, a scheme developed by the Dutch Foundation for Emission Prevention in Refrigeration (*Stichting Emissiepreventie Koudetechniek*; STEK). This scheme has been used in the Netherlands since 1992 to promote and achieve careful and skilful handling of refrigerant gases during use, maintenance and end-of-life recovery to avoid leakages.[45]

Member States that applied stricter national policy measures addressing fluorinated gases prior to the introduction of Regulation No 842/2006 were allowed to maintain their stricter provisions until 2012.[46,47]

According to the Commission's Roadmap for moving to a competitive low-carbon economy in 2050,[48] the revised F-gas Regulation (No 517/2014) sets out a cost-effective pathway for a low-carbon society. The new F-gas Regulation aims to reduce fluorinated gas emissions in the EU by two-thirds by 2030 compared to 2014 emissions by introducing a phase-down mechanism, which will reduce the total amount of HFCs sold in the EU in steps to one-fifth by 2030. It establishes that non-CO_2 emissions (including fluorinated greenhouse gases) should be reduced by about 72–73 per cent by 2030 and by 70–78 per cent by 2050, compared to 1990 levels. The intended reductions represent – according to research by the Commission – a fair

[42]   COM (2003) 492 final, 8.

[43]   COM (2011) 581 final, 8. Visit www.fluorocarbons.org for more information last accessed 1 December 2020.

[44]   COM (2003) 492 final, 7.

[45]   See for a review of the STEK scheme and a comparison to other Member States' policy: Enviros Consulting Ltd, *Assessment of the Costs and Implication on Emission of Potential Regulatory Frameworks for Reducing Emissions of HFCs, PFCs & SF6*, Enviros Consulting Ltd. Report CAN EC002 5008, 2003.

[46]   See further Schwarz et al., n 5, pp.49–51. The deadline followed from Decision 2002/358/EC concerning the approval of the Kyoto Protocol.

[47]   Council Decision of 25 April 2002 concerning the approval, on behalf of the European Community, of the Kyoto Protocol to the United Nations Framework Convention on Climate Change and the joint fulfilment of commitments thereunder (2002/358/CE), OJ L 130/1. Also see Recital 11 of the preamble to Regulation No 842/2006.

[48]   COM (2011) 112 final.

and cost-efficient contribution by the fluorinated gas sector to the EU's climate objective for 2050 at the time of introduction of the new F-gas Regulation.

Emerging opportunities from available and developing technologies in the relevant sectors offer significant scope for additional cost-effective reductions of F-gas emissions in the EU. The Roadmap stipulates that all sectors of industry should be involved in the reduction efforts and that without action on the reduction of fluorinated gases emissions, other economic sectors would have to pick up the bill for reducing greenhouse gas emissions, at a potentially higher cost. Furthermore, it is anticipated that the reduction targets will stimulate innovation, green jobs and growth by encouraging the use of green technologies based on more climate-friendly refrigerants. The revised F-gas Regulation was also intended to facilitate a global agreement to phase down the use of F-gases under the Montreal Protocol.

The targets set for 2030 in the revised F-gas Regulation are higher than the targets set out in the Communication from the Commission on a policy framework for climate and energy in the period from 2020 to 2030 at that time and the following Council conclusion of October 2014.[49] However, neither the Commission's proposal nor the Council's conclusion referred specifically to the reduction of F-gas emissions, which belongs to the non-ETS sector. The targets set in the current F-gas Regulation were deemed necessary and appropriate to achieve the overall goals of the policy framework. The European Green Deal that was presented by the European Commission in December 2019 introduces an even more ambitious goal for 2050. The F-gas Regulation will be reviewed in order to assess whether more stringent targets should be introduced in light of the new overall reduction targets for greenhouse gas emissions. No amendments have been proposed at the time of writing (December 2020).

The current F-gas Regulation aims to improve the measures of containment and end-of-life treatment of products and equipment that contain fluorinated gases that were laid down in Regulation No 842/2006. However, in light of the Roadmap for moving towards a competitive low-carbon economy, the Commission believed that the EU should take additional action on F-gases given the potential to further reduce emissions at relatively low costs.[50] Research prepared for the Commission showed that emerging low-GWP technologies were already available or technically feasible and could be cost-effective in almost all sectors in which F-gases were used.[51] The F-gas Regulation replaced Regulation (EC) No 842/2006 in order to ensure a more cost-efficient contribution to achieving the EU's climate objections by discouraging the use of F-gases in favour of more energy-efficient and safer alternatives.[52]

The most important new measure introduced in 2015 is the phase-down mechanism, which involves a gradually declining cap in the form of quantitative limits on the supply of bulk

---

[49]   Commission, 'Communication from the Commission to the European Parliament, the Council, the European Economic and Social Committee and the Committee of the Regions: A policy framework for climate and energy in the period from 2020 to 2030' COM (2014) 015 final and European Council, Conclusions on 2030 Climate and Energy Policy Framework, Brussels, 23 October 2014 (OR. en) SN 79/14.

[50]   COM (2011) 581 final, 11.

[51]   Schwarz et al., n 5.

[52]   COM (2012) 643 final, 2.

HFC substances in the EU that decreases over time. The Commission performed an impact assessment in which four policy options were assessed in detail: (a) voluntary agreements; (b) extended scope for containment and recovery measures; (c) quantitative limits on the supply of HFCs (phase-down); and (d) a ban on placing certain products and equipment that contain fluorinated gases on the EU market.[53] Current legislation was used as the baseline against which the supplementary policy options where assessed. From the impact assessment it followed that a phase-down would allow for the highest additional environmental benefit, would stimulate innovation to the highest degree and would come at a relatively low cost to society as a whole.[54] This mechanism was therefore deemed to be cost-effective. Furthermore, it is intended to support the system established at EU level for the phase-down of ozone-depleting substances.

8.4.2   Key Legal Instruments and Provisions

In this section we will discuss the key legal provisions of the F-gas Regulation. Several of the key provisions of Regulation No 517/2014 are based on the 2006 Regulation, with adjustments to ensure better implementation and more effective enforcement by national authorities.[55]

According to Article 1 of the F-gas Regulation, its objectives are: (a) to establish rules on containment, use, recovery and destruction of F-gases; (b) to impose conditions on the placing on the market of specific products and equipment that contain, or whose functioning relies upon, F-gases; (c) to impose conditions on specific uses of fluorinated greenhouse gases; and (d) to establish quantitative limits for the placing on the market of hydrofluorocarbons. The key legal instruments to achieve these objectives concern the proper handling of equipment throughout its lifetime: containment, training, certification, labelling and reporting.

Article 3 is concerned with containment. It first stipulates that the *intentional* release of F-gases is prohibited if the release is not technically necessary for the intended use (Art. 3(1)).

Operators of equipment are obliged to prevent leakage (*unintentional* release) from equipment by taking all measures which are technically and economically feasible to minimize leakage and to repair any detected leakage as soon as possible (Arts 3(2) and 3(3)). Equipment that contains F-gases in quantities of 5 tonnes of CO_2-equivalent and more in: (a) stationary refrigeration equipment; (b) stationary air-conditioning equipment; (c) stationary heat pumps; (d) stationary fire-protection equipment; (e) refrigeration units of refrigerated trucks and trailers; (f) electrical switchgear; and (g) organic Rankine cycles, shall be subject to regular checks for leakage by a certified natural person and with a prescribed frequency (Art. 4). Pursuant to Article 5 of Regulation 517/2014, leakage detection systems are mandatory under specific circumstances. Operators of equipment that is required to be checked pursuant to Article 4 have to maintain relevant records, including the quantity and type of F-gases installed, iden-

---

[53] COM (2012) 643 final, 2.

[54] SWD (2012) 364 final, 51.

[55] COM (2012) 643 final, 7. Annex VIII of the proposal provides a correlation table. 64 Recital 3 of the preamble and Article 3 of Regulation No 517/2014.

tification of the company or technician who performed the servicing or maintenance and the dates and results of the checks (Art. 6).

Article 8 of the F-gas Regulation deals with 'recovery', which means the collection and storage of F-gases from products, including containers, and equipment during maintenance or servicing (or prior to the disposal of the products or equipment).[56] Operators of certain equipment are responsible for putting in place arrangements for proper recovery of F-gases by certified personnel, to ensure their recycling, reclamation or destruction.[57] Training and certification requirements follow from Article 10. Member States are obliged to set up and notify their training and certification systems on the basis of the Commission's minimum requirements. In 2011, these training and certification requirements applied to approximately 600,000 persons and 66,000 companies.[58] Reporting requirements follow from Article 19: each producer, importer and exporter of fluorinated gases has to annually report data on production, import and export to the Commission. Member States have to establish the reporting systems for the relevant sectors covered by the F-gas Regulation. Article 12 deals with labelling rules that apply to certain products and equipment that are placed on the market.

The provisions discussed above are complemented by Commission Regulations concerning the format for reports, the form of labels and additional labelling requirements, standard requirements for checking leakage, requirements for training and certification programmes and the format for notifying them.[59]

Articles 11 and 13 regulate the placing on the market and control of the use, respectively, of products and equipment containing F-gases. Specific uses of F-gases are prohibited in Article 13, for example the use of SF6 for filling car tyres. The prohibition on placing on the market specific products and equipment follows from Article 11. These specific products and equipment (e.g. windows for domestic use, footwear and tyres that contain fluorinated greenhouse gases), including the date from which the prohibition applies, are listed in Annex III.

The current F-gas Regulation also introduces additional bans on the placing on the market of new equipment for refrigeration, air-conditioning and fire protection that operates using specific F-gases, where suitable alternatives for the use of those substances are available.[60] Article 11(2) instructs that such bans can only be introduced where they will result in lower overall greenhouse gas emissions, in particular from the leakage of any fluorinated gases as well as the CO_2 emissions resulting from their energy consumption. This means that equipment containing F-gases is allowed if its overall greenhouse gas emissions are less than those that would result from an equivalent equipment without F-gases (which should nevertheless

---

[56]   Article 2(14) of the F-gas Regulation.

[57]   Member States shall also encourage producer responsibility schemes for the recovery of F-gases pursuant to Article 9.

[58]   COM (2011) 581 final, 4.

[59]   Find all Commission Implementing acts on the European Commission website on fluorinated greenhouse gases at http://ec.europa.eu/clima/policies/f-gas/index_en.htm. Regulation (EC) No 307/2008 of 2 April 2008, OJ L92/25; Commission Regulation (EC) No 308/2008 of 2 April 2008, OJ L92/28.

[60]   Recital 11 of the preamble of Regulation No 517/2014.

respect the maximum allowed energy consumption set out in relevant measures adopted under Directive 2009/125/EC on ecodesign requirements for energy-related products).[61]

The key innovation of F-gas Regulation No 517/2014 compared to the previous F-gas Regulation is the introduction of a gradually declining cap of HFCs that are allowed to be placed on the market, regulated in Chapter IV. Companies must have rights if they want to place HFCs on the EU market for the first time. In 2015, the Commission allocated free quotas for each producer and importer, based on reported data in the past (Articles 16(5) and 19). By 31 October 2014 the Commission had to determine for each producer or importer – by means of implementing decisions – a reference value that was based on the annual average of the quantities of HFCs the producer or importer reported to have placed on the market from 2009 to 2012 (Art. 16(1)). This reference value was recalculated as of 31 October 2017, to be updated every three years after that (Art. 16(3)). Each producer or importer for which a reference value has been established receives a quota. The Commission will use the allocation mechanism laid down in Annex VI of the Regulation. In order not to exclude small operators, 11 per cent of the overall quantitative limit is reserved to importers and producers that have not imported or produced more than 1 tonne of fluorinated gases in the reference period (Recital 15 of the preamble). The maximum quantity for each year is calculated by applying percentages to the annual average of the total quantity produced and imported into the EU during the period 2009–12. The phase-down mechanism starts in 2015 with a freeze, followed by a first reduction in 2016, reaching 21 per cent of the levels sold in 2009–12 by 2030 (Annex V). Transferring quotas is permitted in order to maintain flexibility in the market (Recital 19) and to allow for a cost-effective pathway to decrease the use of HFCs and emissions reduction. Any producer or importer for whom a reference value has been determined and who has been allocated a quota may transfer that quota for all or any quantities to an undertaking represented in the EU, and also to producers and importers that have not been active in the sector before (Art. 18). A central electronic registry is used to manage the quotas (Art. 17).[62] As was discussed regarding the regulatory choice not to allow transferable quotas in the MAC Directive, the question remains whether the administrative burdens of the use of the registry will outweigh the economic advantages of the possibility of transferring quotas allocated on the basis of reference values to another producer or importer in the Union. It remains somewhat unclear why the idea of transferable quotas was dropped in the negotiations for the MAC Directive but survived in those for the current F-gas Regulation.

In July 2017 the Commission reported that it was too early for an in-depth assessment of the functioning of the phase-down mechanism and to thoroughly appreciate all possible impacts of the chosen quota allocation method. Nevertheless, the Commission stated that the phase-down was functioning as it should and that quota price development, although causing

---

[61] Directive 2009/125/EC of the European Parliament and of the Council of 21 October establishing a framework for the setting of ecodesign requirements for energy-related products, OJ L285/10. See also Chapter 6 of this book.

[62] See Recital 19 of the preamble. This central electronic registry is based on the system for licensing trade under Regulation (EC) No 1005/2009 of the European Parliament and the Council of 16 September 2009 on substances that deplete the ozone layer, OJ L286/1.

some concern among stakeholders, was in line with expectations and remains far below the 50-euro level (per tonne of CO_2 emitted) considered reasonable. Moreover, the Commission concluded that there was good compliance with the EU's total HFC limit. The Commission therefore has no intention to amend the quota allocation method.[63] A new review procedure of the F-gas Regulation was initiated in 2020 in view of the European Green Deal and could lead to a proposal for an amended F-gas Regulation in 2021.

8.5   CONCLUSION

The MAC Directive (2006/40/EC) prohibits the use of fluorinated gases (F-gases) in new types of cars introduced from 2011 and in all new cars produced from 2017. The first F-gas Regulation (No 842/2006) aimed at: (a) improving the prevention of leaks from equipment containing F-gases; and (b) avoiding the use of F-gases in some applications where environmentally superior and cost-effective alternatives exist.

In accordance with the EU's greenhouse gas emission reduction goal for 2050, a revised (current) F-gas Regulation (No 517/2014) has applied since 2015 and introduces a gradually declining cap for HFCs. It is designed to achieve a reduction in the emissions of fluorinated gases as a cost-efficient contribution to the overall economic effort needed to avoid more costly consequences of climate change in the future. Without reducing F-gases, other economic sectors would have to pick up the bill for reducing greenhouse gas emissions, at potentially higher cost. The revised F-gas Regulation ensures flexibility in the market for bulk HFCs by introducing quota and by allowing the transfer to another producer or importer in the EU. This flexibility is likely to ensure a cost-effective decrease in the use of HFCs.

From the above analysis in light of the EU climate policy principles of cost-effectiveness and solidarity, it appears that the revised F-gas Regulation (No 517/2014) is mainly based on the desire to achieve the reduction targets in a cost-effective way. A review of the current F-gas Regulation is formally required by December 2022 but the new EU policy stipulated by the European Green Deal may lead to proposals for amendments to the F-gas Regulation before that time in order to achieve the even more ambitious goal of carbon neutrality by 2050.

CLASSROOM QUESTIONS

1. Why does the MAC Directive prescribe a type approval system instead of a quota system?
2. What is the key innovation in F-gas Regulation 517/2014 compared to F-gas Regulation 842/2006?
3. What is de quota allocation method prescribed by the F-gas Regulation?

---

[63]   See COM(2017) 377 final, p.9. For more information visit the website for the F-gas Consultation Forum in 2018 (and 2020) at https://ec.europa.eu/clima/events/articles/0106_en accessed 1 December 2020.

SUGGESTED READING

Books

No specific books available (to our knowledge).

Articles and chapters

Glaubitz P, Siebert C, Zuber K, 'SF$_6$, Its Handling Procedures and Regulations' in Krieg T and Finn J (eds), Substations (CIGRE Green Books/Springer 2019).

Gonçalves F and Cavique M, 'F-gas Regulation: Possible Solutions for the Retrofit Dead End' in Slatineanu L et al, IManE&E 2018, MATEC Web of Conferences Volume 178, 09023 (2018).

Lindley AA and McCulloch A, 'Regulating to Reduce Emissions of Fluorinated Greenhouse Gases' (2005) 126 Journal of Fluorine Chemistry 1457–62.

Policy documents

European Commission website on fluorinated greenhouse gases. http://ec.europa.eu/clima/ policies/f-gas/index_en.htm.

European Environment Agency, Fluorinated Greenhouse Gases 2017. Data Reported by Companies on the Production, Import, Export and Destruction of Fluorinated Greenhouse Gases in the European Union, 2007–2016, EEA report 20/2017 (ISSN 1977-8449).

European Commission, 'Commission staff working paper. Impact assessment. Review of Regulation (EC) No 842/2006 on certain fluorinated greenhouse gases', SWD (2012) 364 final.

PART III
OVERARCHING ISSUES IN EU CLIMATE REGULATION

9

EU climate law and energy network regulation

ABSTRACT

- Promoting the consumption of renewable energy sources is one of the key means to combat climate change (see also Chapter 5 above);
- Network operators need to re-design their networks following the requirement to promote the use of renewable energy sources and to promote energy efficiency (also referred to as 'decarbonizing' energy networks);
- The feed-in of large volumes of renewable energy sources to the grid has an impact on both the upper level (transmission) and the lower level (distribution) of the grid. It requires investments in network expansion, the construction of new networks and a new approach to network balancing;
- The feed-in of renewable energy sources on the lower level of the grid (distribution network) has led to a new category of energy producers, namely consumers who act as producers at the same time (prosumers). The new 2019 Electricity Directive acknowledges these prosumers but refers to them as active consumers;
- Smart grids entail that network operators increasingly rely on ICT solutions and on the introduction of smart meters, which is crucial for balancing the grid;
- The development of offshore wind energy requires the establishment of new transport infrastructure offshore. In the absence of an EU legal framework Member States may opt for a variety of solutions for bringing the electricity to shore, involving either the transmission system operator (TSO), the wind park developer or a third party;
- The development of offshore grids and smart grids as well as the upgrading and/or expansion of existing networks involves a wide range of planning and licensing procedures. EU policy on trans-European energy networks requires Member States to accelerate these procedures.

9.1   INTRODUCTION

Since the industrial revolution, energy production and supply in Europe has relied on the use of fossil fuels, often situated at a considerable distance from the energy consumers. However, the design of the energy sector is changing due to EU climate policy and subsequent EU targets for renewable energy consumption. The EU's ambitions regarding use of renewable energy sources have moved from a goal of 20 per cent renewable energy consumption in 2020 to a goal of 32 per cent renewable energy consumption by 2030 and even 100 per cent by 2050 (see

Chapter 5). Such a large-scale introduction of renewable energy sources, like wind and solar, will impact energy transportation.

Although the mode of transportation used to bring the energy source to the end-consumer may differ, cables and pipelines dominate as the means of transporting electricity and natural gas. The electricity and gas sectors are therefore referred to as network-bound energy sectors. The design of the network-bound energy sectors has been based on a 'top-down' approach, meaning that upon production, electricity and gas is fed into the national grid. The national grid is again connected to several regional – distribution – grids to which most end-consumers are connected. However, renewable energy sources are increasingly fed into the lower level of the grid. The operators of these distribution grids are thus faced with new operational challenges.

In contrast to the transportation of other commodities, electricity and gas do not really 'travel' from A to B. As a general rule the transport of electricity and gas is a matter of keeping the grid in balance, which means that any feed into the grid needs to be matched by a similar quantity taken from the grid. Since gas can to some extent be stored in the pipeline (referred to as 'line pack'), the operator of the gas grid has some flexibility when operating the grid. Electricity differs from gas as it cannot easily be stored (at least not in an economically viable fashion) and therefore the issue of balancing is of utmost importance for 'transporting' electricity, especially if based on intermittent renewable resources.

In this chapter we examine the process of decarbonizing the energy networks. Section 9.2 presents the main challenges affecting energy networks. Thereafter section 9.3 provides a brief overview of EU network regulation following the energy market liberalization process. It presents in particular how the decarbonization process influences the use of existing networks and the development of new networks. Section 9.4 will then focus on the electricity chain, as the decarbonization process directly impacts the upper and lower levels of the grid and has led to a new approach of network regulation. Section 9.5 concludes.

9.2   CHALLENGES OF DECARBONIZING ENERGY NETWORKS

9.2.1   Main Challenges

Network operators are required to meet two fundamental obligations when managing the grid. First, they need to ensure that the grid is safe and reliable; second, they need to guarantee all users (producers and consumers) non-discriminatory access to the grid. The large-scale introduction of renewable energy sources raises several challenges for network operators. It requires them to rethink how to comply with these obligations, while at the same time facilitating the feed-in of production from renewable energy sources.

The obligation that network operators must ensure that the grids are safe and reliable entails, for example, that their networks are in balance and certain quality specifications are being met. Here you can think of voltage and frequency levels for electricity networks, but also of the way in which users of gas networks are bound by certain gas quality specifications. The latter is of specific relevance for the use of biogas and green hydrogen, as a producer of these 'green gases' who wishes to make use of the existing natural gas grid needs to meet the required quality specifications of the gas network.[1]

Meeting the above-mentioned renewable energy goals depends not only on an increase of levels of renewable energy production but also on the renewable energy reaching the final customer, that is, that the producers will get access to the grid. Therefore, EU law provided until recently that network operators should guarantee renewable energy sources access to their networks or even prioritize them. Since the entry into force of the new Renewable Energy Sources Directive or Directive 2018/2001/EU (hereafter 2018 RES Directive) the rules on guaranteed or priority access have been abolished, as a result of which the general non-discrimination obligation placed on network operators also applies to producers and suppliers of renewable energy sources.

Another challenge resulting from the large-scale introduction of renewables is the need to develop new networks or replace existing networks with more innovative systems. The large-scale production of offshore wind also has an impact on the transmission system that needs to transport large volumes of electricity to end-consumers. The existing networks may not have sufficient transport capacity; network operators are thus faced with congestion management rules and/or the need to invest in extra grid capacity. The decarbonization of the gas sector is facing some other challenges too, especially given the possibility to convert electricity from renewable energy sources into green hydrogen. It is not yet clear how to qualify pipelines dedicated to transporting green hydrogen, and which rules apply to their development.

9.2.2 Specific Challenges Relating to Electricity Networks

The electricity sector is facing some particular decarbonization challenges at both the upper and lower level of the grid. These challenges are not only technical in scope but also require a reassessment of the existing legal framework. They also necessitate additional attention due to the increasing electrification of our society. The share of electricity in final energy demand will be around 30 per cent in 2030 and around 40 per cent in 2050.[2]

One of the challenges relates to the development of offshore wind energy, that is, electricity generated by offshore wind turbines. Offshore wind energy developers need to bring the electricity produced to shore. Who is entitled to construct these offshore cables and how should these cables be regulated? In the absence of a proper EU legal framework, Member States are currently developing their own regimes, which in the longer run could create new challenges involving harmonization of laws. Moreover, these offshore wind parks are usually large production facilities, which are directly or indirectly connected to the transmission grid. The

---

[1]   M. Roggenkamp, J. Sandholt, D.G. Tempelman, 'Innovation in the EU Gas Sector: Injection of Biomethane into the Natural Gas System', in S. Zillman, M. Roggenkamp, L. Paddock and L. Godden (eds), *Innovation in Energy Law and Technology: Dynamic Solutions for Energy Transitions*, OUP, 2018, pp.262–83.

[2]   European Commission, Joint Research Centre, 'Global Energy and Climate Outlook 2019: Electrification for the low-carbon transition: The role of electrification in low-carbon pathways, with a global and regional focus on EU and China', Publications Office of the European Union, 2020, doi:10.2760/350805, p.79.

final consumers, however, are often not located close to these facilities but are further inland. Therefore there may be a need for additional grid capacity.

Another challenge can be found on the level of the distribution grid to which (smaller) renewable energy facilities are increasingly connected. This may result in a need to adapt and invest in the existing grid unless better or smarter use is made of the grid. The need to develop smart grids is recognized but so far, the term has not yet been clearly defined. Be that as it may, the development of smart grids will be indispensable to deal with the gradually increasing production of renewable energy and thus with challenges involving network balancing. In addition, it can be noted that a new category of producers is emerging; that is, small consumers (such as household consumers) are now entitled to generate and, to some extent, to supply electricity. The 2018 RES Directive and the 2019 Electricity Directive recognize the development of 'energy communities', which may consist of final consumers actively engaged in the production and supply of electricity and also grid management. This poses new legal challenges for network management.

9.3   ENERGY NETWORKS AND NETWORK GOVERNANCE IN THE EU

9.3.1   Types of Networks and Regulatory Approaches

The networks connecting energy producers and consumers are usually divided into three main categories:

- networks connecting production facilities to the national grid;
- national grid (transmission line);
- distribution grids supplying small consumers (mainly households and small- and medium sized companies).[3]

The first category includes the pipelines and cables connecting production facilities (gas fields and major electricity generators) with the main grid. Whereas cables connecting electricity generators with the main grid are basically considered part of the production installation, a separate category of pipelines exists for those pipelines connecting gas production facilities to the main pipeline system. These are generally referred to as upstream pipelines. The next category involves the main grid, which includes high pressure and high voltage networks. These so-called transmission (pipe)lines usually cover the entire territory of a state and are

---

[3]   In addition to these three main categories, EU law also refers to 'direct lines' (e.g. an electricity line linking an isolated generation site/a producer with an isolated customer or an electricity supply undertaking to supply directly their own premises, subsidiaries and customers) and 'interconnectors' (e.g. a cross-border transmission line connecting two national transmission systems).

considered the 'national energy highways'.[4] Transmission lines are again connected to another – third – category of infrastructure consisting of distribution lines, which operate under lower pressure and/or voltage and have a direct connection to (small) end-consumers.[5] Historically the picture of stages of energy supply appears as in Figure 9.1.

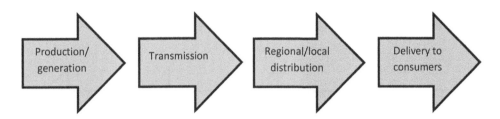

Figure 9.1   Overview of traditional stages of energy supply

Following the process of energy market liberalization that started in the EU in the late 1980s, the legislation applying to these networks is now primarily based on EU law.[6] Energy market liberalization means that all market parties have freedom of choice. Hence, energy production is based on authorizations awarded in a competitive manner and electricity generators are free to decide about the primary fuel to be used. As electricity generation from renewable energy sources is usually more expensive than fossil fuels, the 2018 RES Directive allows Member States some financial support (see further Chapter 5). In addition, all consumers are given freedom of choice, that is, they have the right to choose their supplier. In such a liberalized market, it is a prerequisite that producers and consumers will have non-discriminatory and transparent access to the grid.

In order to facilitate such access, the EU legislator envisaged that the (*de facto* or *de jure*) vertically integrated energy companies had to be unbundled, that is, that the network activities would be separated from production and supply. Such unbundling establishes independent network operators and ensures that decisions by network operators cannot be influenced by the production and supply companies. The level of unbundling has increased since the entry into force of the first directives in the 1990s and differs between types of networks. Since 2009 several unbundling models have applied to transmission networks.[7] Examples are (a) the most far-reaching model of ownership unbundling, where the shareholders of the network company are different from the shareholders in the production/supply company; and (b) the

---

[4]   An exception can be found in some federal states, such as Germany, where four electricity and fourteen gas transmission systems operate in different parts of the country.

[5]   Please note that sometimes large consumers can also be directly connected to the transmission grid.

[6]   COM 1988 (238) final.

[7]   See Directive 2009/72/EC, OJ 2009, L.211/55 and Directive 2009/73/EC, 2009, OJ L.211/94. The entry into force of Directive 2018/2001/EU, Directive 2019/944/EU and Directive (EU) 2019/692 has not led to any changes in the unbundling regime.

regime of an Independent Transmission Operator (ITO), where the transmission network operator remains a separate legal entity within the vertically integrated energy company but is subject to strict administrative and regulatory control by the EU Commission.[8] Energy distribution networks, however, need only to be legally unbundled, which means that each distribution company has to designate a distribution system operator (DSO) that needs to be a separate legal entity.[9] Member States are however entitled to apply stricter unbundling rules to DSOs. The new energy chain is presented in Figure 9.2.

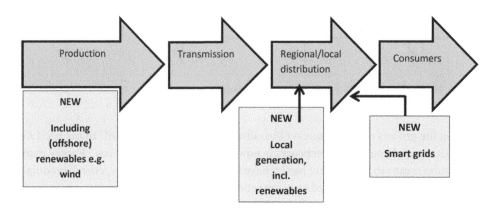

Figure 9.2   New energy chain

In the EU the process of market liberalization is accompanied by a process of market integration, that is, the need to create one internal energy market while taking into account climate change goals. In this regard, the following four developments can be noted. First is the need to connect the national grids in order to create one European grid and to secure long-term energy supply, as energy resources are often located further away from final consumers. Second is the process of market liberalization, which has led to an unbundling of vertically integrated energy companies and the establishment of independent network operators. Third is the need to make use of more renewable energy sources in order to safeguard energy supply and to meet climate mitigation obligations. Finally, the EU aims at 'putting consumers at the heart of the energy transition' by giving them more choice, strengthening their rights and enabling everyone to participate in the transition themselves by producing their own renewable energy

---

[8]   A third model involves the Independent System Operator (ISO), which entails that a third party is appointed to operate the transmission system. This is the least popular model. See further N. Bel and R. Vermeeren, 'Unbundling in the EU Energy Sector: The Commission's Practice in Assessing the Independence of Transmission System Operators for Electricity and Gas', in M.M. Roggenkamp and H. Bjørnebye, *European Energy Law Report X*, Intersentia, 2014, pp.49–64.

[9]   Chapter IV Directive 2019/944/EU amending Directive 2012/27/EU and Chapter V Directive 2009/73/EC.

and feeding it into the grid.[10] The latter trend has a serious impact on grid design as renewable energy is increasingly fed into the distribution grid instead of the transmission grid.

9.3.2   Network Operations

Third party access

For proper functioning of a liberalized energy market it is a prerequisite that all market parties have non-discriminatory and transparent access to the grids.[11] Whereas the requirement of transparency entails that access requirements should be publicly available and known before-hand, non-discrimination means that network operators need to treat similar user groups in a similar way but may treat different user groups in a different manner.[12] Although the term 'access' can be interpreted quite broadly, the EU court ruled in the *Sabatauskas* case that the term 'access' refers to the right to use (in this case) the electricity system and should be distin-guished from the term 'connection' that corresponds to a physical connection to the system.[13] Hence, 'access' is the ability to have energy 'transported' or 'transmitted' through networks, which requires that network capacities are available and can technically facilitate transporta-tion, including the necessary system services required for this.

In practice different access regimes apply, depending on the type of infrastructure involved. Most importantly, a regime of regulated third party access applies to the transmission and distribution grids. Whereas transport conditions can be set by the network operators, the transport tariffs, or the methodologies used to calculate them, are regulated, and this means that they need to be approved by an independent energy regulator and should take into account the need for system integrity and reflect the actual costs incurred, while including an appropriate return on investments. In contrast, access to upstream pipelines is based on negotiations between pipeline operators and third parties who wish to gain access, and occurs without any *ex ante* involvement of the energy regulator.[14] As a rule network operators may refuse access but only on some limited grounds, primarily if there is no capacity available or in the case of incompatibility of technical specifications that cannot be reasonably overcome when it involves upstream pipelines. Anyway, a refusal to provide access has to be based on duly substantiated reasons and on objective and technically and economically justified criteria. In addition, TSOs and DSOs are required to indicate which measures are needed to reinforce the network, that is, how to increase the available capacity.[15]

---

[10]   European Commission, 'Clean Energy Package for All Europeans', March 2019. doi:10.2833/9937.

[11]   Cf Article 6 of Directive 2019/944/EU and Article 32 of Directive 2009/73/EC as amended by Directive 2019/692/EU.

[12]   H.K. Kruimer, *The Non-Discrimination Obligation of Energy Network Operators: European Rules and Regulatory Practice*, Intersentia, 2014.

[13]   C-239/07 *Julius Sabatauskas and Others* [2007] ECR II-7523, para 42.

[14]   Article 34 Directive 2009/73/EC as amended by Directive 2019/692/EU. Please note that the competition authority can play a role *ex post* if third parties feel that the pipeline operator has abused its (monopoly) position and has, for example, set higher tariffs than necessary or denied access without proper reasons.

[15]   Article 6 para 2 of Directive 2019/944/EU.

In order to promote the use of renewable energy sources, the first RES Directives provided that electricity produced from renewable energy sources should be granted '*guaranteed access*' or even '*priority access*' to the grid.[16] With the entry into force of the 2018 RES Directive, the scope of the directive was extended to renewable gases, but at the same time this favourable access regime for electricity from renewable energy sources was abolished. Although energy from renewable energy sources will be treated in the same manner as 'conventional' electricity and natural gas, the 2018 RES Directive acknowledges that priority dispatch still may apply for renewable energy sources, and in particular electricity from renewable energy sources.[17] The term dispatch refers to the situation where the energy producers decide when and how much electricity they will feed in to the grid. In a liberalized market, such decision will usually be based on an economic assessment. Dispatching is thus market-based, and this means that the available energy sources are ranked according to a merit order based on price signals. As renewable energy sources have very low marginal costs, they will often be first in the merit order. Moreover, the amounts of electricity fed in to the grid need to match the volumes consumed in order to avoid an imbalance of the system (as discussed in the next sub-section). In case of oversupply or congestion, the network operator needs to curtail the volumes injected, including electricity from renewable energy sources. In practice, the TSOs are responsible for balancing the overall system and the DSOs for controlling distribution until the points of consumption.[18] As decentralized energy production increases, Member States are still entitled to require DSOs, when dispatching generating installations, to give priority to generating installations using renewable sources.[19]

Network balancing

Balancing is at the core of the operation of the energy systems and in particular the electricity system. Generally, it entails that each offtake of the grid needs to be compensated directly with a similar quantity of input to the grid. As gas can be stored in (underground or above ground) storage facilities and/or in the pipeline itself (line pack), the operator of the gas grid has more flexibility when operating then electricity network operators have. If the volumes of electricity taken from the grid and fed into the grid do not match, there may be a voltage dip, causing a brownout or blackout. Whereas brownouts refer to smaller voltage dips and can cause poor performance of equipment or even incorrect operation, blackouts involve a total loss of power in an area. Network operators (usually TSOs) thus need to know on a daily basis how much electricity (or gas) will be fed into the grid and how much will be taken from the grid. In case of a possible imbalance, network operators can organize extra production or demand reduction. All Member States have therefore introduced some sort of a market for

---

[16]   Article 16 of Directive 2009/28/EC. See also Chapter 10, section 3 in the first edition of this book.

[17]   Please note that this provision does not apply to renewable gases. See recital 60 Directive 2018/2001/EU.

[18]   F. Elskamp, 'Renewable Energy Sources and the Impact on Security of Supply and Dispatching', in M.M. Roggenkamp et al., *Energy Law, Climate Change and the Environment*, Edward Elgar Publishing, 2021, pp.377–87.

[19]   Article 4 para 2 of Directive 2019/944/EU. The same applies for using high-efficiency cogeneration, in accordance with Article 12 of Regulation (EU) 2019/943.

balancing services, in other words, a marketplace for acquiring extra production capacity or demand reduction. Household consumers are exempted from this specific market and the responsibility for balancing will usually be held by their supply companies or DSOs. However, increasing scales of distributed generation and growing numbers of small consumers also producing electricity[20] will make it more difficult to operate the distribution system on the basis of existing predictions of demand patterns. Moreover, the capacities of the distribution grids are becoming insufficient to accommodate their increased use due to increased amounts of generation connected to the distribution grid, but also increasing demand as a result of the objective of further electrification, for example by electric vehicles. At the same time, new technologies have been introduced to facilitate the balancing of supply and demand at the distribution system level. This development is known as smart grids or smart electricity systems (see further section 9.4.3).

9.3.3 Developing Decarbonized Energy Networks

The process of decarbonizing EU energy networks also requires investments in existing and/or new infrastructure. Investments can be needed to expand/upgrade existing networks in order to accommodate higher volumes of renewable energy sources, especially if these resources are situated away from traditional generators and end-consumers. In addition, it may be necessary to construct new infrastructure given the absence of dedicated networks, for example in the case of offshore wind. As networks are natural monopolies and thus regulated, investments in the upgrade of existing networks and investments in new networks are subject to EU regulation.

As discussed above in 9.2.2, transmission and distribution tariffs are regulated, which means that either the tariffs or the methodology underlying their calculation must be approved by the national regulatory authority (NRA). These tariffs must be objective and non-discriminatory but also cost-reflective.[21] The latter means that the applicable tariffs will ensure the network operator a (normal) profit and covers network investments, but only those that are deemed necessary to operate the network efficiently. The decarbonization of energy networks will require considerable additional investment, which needs to be taken into account by the network operators and NRAs when establishing the network tariffs.[22]

---

[20] Please note that the possibility of decentralized renewable gas products will be limited. In practice it involves, for example, farmers producing biogas from manure. See M.M. Roggenkamp and D.G. Tempelman, 'Gas Sector Developments in the Netherlands and the EU: From Manufactured Gas via Natural Gas to Biogas', *Journal of Energy and Natural Resources Law*, 2012, vol. 30, no. 4, 523–37.

[21] See, inter alia, Article 18 para 1 Regulation (EU) 2019/943 on the internal market for electricity.

[22] Machiel Mulder and Edwin Woerdman, 'Energy Networks, Natural Monopolies and Tariff Regulation', in M.M. Roggenkamp et al., *Energy Law, Climate Change and the Environment*, Edward Elgar Publishing, 2021, pp.563–72.

Network development plans

In order to identify where investments are needed, TSOs and DSOs are required to present transparent network development plans on a regular basis. Whereas both the electricity and gas TSOs are required to present a network development plan, such a requirement does not (yet) apply to DSOs in the gas sector; it will most likely be included in the next EU Gas Directive.

DSOs in the electricity sector are required to submit a network development plan to the NRA at least every two years. The plan will be developed in consultation with the TSOs and all relevant network users and needs, *inter alia,* to present for a period of five to ten years the investments that are needed in the main distribution infrastructure to connect new generation capacity, that is, production from renewable energy sources, as well as alternatives that can be used to avoid network expansion, that is, energy storage facilities. The NRA may request amendments to the plan submitted to them.[23]

Albeit more detailed, a similar provision applies to all TSOs.[24] After having consulted all relevant stakeholders, they also have to submit a ten-year network development plan, which is based on existing and forecast supply and demand, to the NRA at least every two years. The plan has to show the main transmission infrastructure that needs to be built or upgraded over the next ten years, the decided and planned investments for the next three years and the time frame for these investments. The plan needs to take into account alternatives to network expansion but also developments in cross-border trade and regional networks. Most importantly, these ten-year network development plans also need to take into account the decarbonization of energy networks as the NRAs are explicitly required to examine the consistency of the ten-year network development plan with the national energy and climate plan submitted in accordance with the 2018 Governance Regulation[25] (see Chapter 5). When drafting these ten-year network development plans the NRAs have to consult all network users and publish the result of the consultation process. They will in particular need to examine whether the ten-year network development plan covers all investment needs identified during the consultation process and whether it is consistent with the non-binding Union-wide ten-year network development plan.[26] If any doubt arises as to consistency with the Union-wide network development plan, the NRA shall consult ACER and may require the TSO to amend its ten-year network development plan. Last but not least, the NRA will monitor and evaluate the implementation of the ten-year network development plan and if a TSO is not investing according to the plan, and there are no overriding reasons beyond its control, the NRA may (i) require the TSO to execute the investments in question, or (ii) organize a tender procedure open to any investors for the investment in question, or (iii) oblige the TSO to accept a capital increase to finance the necessary investments and allow independent investors to participate in

---

[23]   Article 32 paras 3 and 4 Electricity Directive 2019/944/EU.

[24]   Article 51 Electricity Directive 2019/944/EU and Article 22 Gas Directive 2009/73/EU.

[25]   Regulation (EU) 2018/1999.

[26]   See Article 30(1) of Regulation (EU) 2019/943.

the capital. Be that as it may, the network tariffs – that is, the applicable tariff regulation – need to cover the costs of the investments.

Trans-European networks and projects of common interest

In practice there may be networks that are relevant for the decarbonization process, but which need additional investment. Such financial support is envisaged by EU law governing the development of trans-European networks for energy (TEN-E), but only if the specific energy infrastructure is classified as a project of common interest ('PCI').[27] Since Regulation 347/2013[28] these PCIs need to be part of the 'priority corridors and areas of TEN-E' and be approved by the Regional Group specifically established for that specific corridor or area.[29] On the basis of some general and some specific selection criteria, each Group will select those projects that can be included in a Union list of projects of common interest ('Union list'), which is established every two years and annexed to the Regulation.[30] Each of the listed PCIs will then need an implementation plan, which includes a timetable for feasibility and design studies as well as the award of permits and authorizations.[31]

Although the 2013 Regulation confirms that the costs for developing, constructing and operating a PCI project need to be borne by the users of the infrastructure,[32] it also acknowledges that certain projects will require some EU financial assistance. In order to assess such a need, the Regulation introduces a cost–benefit analysis, which needs to be included in the above-mentioned ten-year network development plans.[33] In practice, the requirements for financial support may differ per Regional Group but in essence, financial assistance in the form of grants is possible if the developers can demonstrate a lack of commercial viability and that significant positive externalities – such as supply security, solidarity and innovation – will be provided by the project. If so, financing will be based on the Connecting Europe Facility 2014–2020 and is capped at €5.35 billion for energy PCIs.[34] In October 2019 the European Commission published the fourth list of PCIs, which consists of some 100 electricity transmis-

---

[27]   See Articles 170–172 TFEU on trans-European networks (TEN).

[28]   This Regulation replacing Decision 1364/2006/EC directly binds Member States as of 1 June 2013. Council Decision No. 1364/2006/EC was replaced by Regulation (EU) No. 347/2013.

[29]   See Article 3 of the Regulation and Annex III. On the basis of this Regulation 12 Regional Groups are established consisting of the relevant Member States, TSOs, NRAs and the Commission, ACER and ENTSO-E or ENTSOG.

[30]   See Article 4 of Regulation 347/2013.

[31]   See Article 5 of Regulation 347/2013. Article 7(10) of Regulation 347/2013 provides detailed requirements for permit granting and public participation. The aim is to reduce permitting to 3.5 years. See also the first edition of this book.

[32]   Recital 35 and Article 12 Regulation 347/2013.

[33]   Article 11 of Regulation 347/2013.

[34]   Article 15 of Regulation 347/2013. See https://ec.europa.eu/inea/en/connecting-europe-facility accessed 16 June 2021.

sion and 32 gas projects.[35] This also shows that such financial support can be relevant for the decarbonization of the energy network as it includes several energy storage projects (primarily hydro-electricity pumped storage), some cross-border carbon dioxide (CCS) projects, six smart grid projects and the development of some offshore wind energy hubs in the North Sea (such as the 'North Sea Wind Power Hub').

9.4   DECARBONIZING ELECTRICITY NETWORKS

9.4.1   Changes in the Electricity Chain: From Offshore Wind to Local Energy Communities

In the above we discussed in general terms the EU legal framework applying to the energy sector, and in particular electricity and gas networks, and how this impacts the decarbonization process as regards both the use of energy networks and the upgrading and development of new networks. In this section we will go one step further and focus on the electricity system only, as in this area some specific new developments promoting the use of renewables directly influence network design and thus also regulation. Each development impacts a particular section of the electricity chain and involves a different challenge for network operators and energy producers when developing new and renewable energy sources. These challenges involve:

– the development of electricity from offshore wind and the construction of an offshore electricity grid;
– increasing volumes of decentralized production of renewable electricity and the introduction of smart grids.

The first development takes place at the upper part of the grid and causes challenges primarily for the electricity transmission network operators. Although the production of offshore wind energy, like 'traditional' energy supply developments, takes place at the upper end of the grid, the challenge here is the way in which offshore wind energy is transported to shore, and determining who is responsible for doing so. The second development primarily causes challenges for distribution systems operators. As argued above, the latter were traditionally not faced with the direct feed-in of energy, but this has changed as a consequence of the promotion of renewable energy consumption. Besides, the second development may also involve a new type of producer: household consumers generating electricity by themselves. This again has led to new market designs and grid concepts. In the following sections we will discuss these two developments in more detail.

---

[35]   Commission Delegated Regulation (EU) 2020/389 of 31 October 2019 amending Regulation (EU) No. 347/2013 of the European Parliament and of the Council as regards the Union list of projects of common interest.

9.4.2   Offshore Wind Energy and Cable Design

Offshore wind energy production is increasing, especially in the North Sea, the Baltic Sea and the Irish Sea. As the majority of wind energy production is developed in the North Sea[36] we will take the developments in this region as an example and specifically focus on developments in five countries which have some of the largest offshore wind capacity: the UK (45 per cent), Germany (34 per cent), Denmark (8 per cent), Belgium (7 per cent) and the Netherlands (5 per cent). The interest in developing offshore wind energy is the result of the national renewable targets and the need to combat climate change, the available techniques to construct wind turbines at sea and the fact that several Member States do not have sufficient possibilities to develop renewable energy sources onshore due to scarcity of land and local opposition to the construction of large wind turbines. By contrast to the situation onshore, where it is possible to construct stand-alone wind turbines for personal use, wind energy offshore is always a matter of grouping several turbines together in a wind park, which can consist of some 20–60 wind turbines.

Legal basis for offshore wind energy production

Offshore wind parks can be situated in the territorial sea (12 nautical miles from the coast) or beyond in the exclusive economic zone (EEZ). Constructing wind parks in the territorial waters is more economical but has as a disadvantage that these wind turbines remain visible from land and thus often lead to severe local opposition delaying construction. In the North Sea area we therefore see different approaches. Whereas the Netherlands originally decided to limit the development of wind turbines in the territorial sea to one demonstration project, most wind farms in Denmark and the UK were initially situated in the territorial sea. Another advantage of developing these wind parks in the territorial sea is the fact that there is no doubt about the legal basis for this development. As the territorial sea is considered part of the territory of a state, coastal states have full jurisdiction in that area.[37] Outside the 12-mile zone, coastal states have a limited or functional jurisdiction, but only if the coastal state has a continental shelf or an EEZ. In contrast to a continental shelf, which applies automatically, the EEZ has to be established explicitly by the coastal state. If a coastal state has declared an EEZ, it has '(i) sovereign rights for the purpose […] of the production of energy from the water, currents and winds, and (ii) jurisdiction […] with regard to the establishment and use of artificial islands, installations and structures'.[38]

Consequently, these coastal states have exclusive jurisdiction over such installations and structures and may issue laws and regulations they deem necessary for developing offshore

---

[36]   In 2019 the following figures applied: 77 per cent North Sea, 13 per cent Irish Sea, 10 per cent Baltic Sea and <1 per cent Atlantic Ocean. See *The European Offshore Wind Industry: Key Trends and Statistics 2019*, Wind Europe, 2019, p.15. Available at www.windeurope.org accessed 16 June 2021.

[37]   Article 2 UNCLOS.

[38]   Article 56 UNCLOS.

wind parks and bringing the energy to shore.[39] All North Sea states have established an EEZ or a similar zone like the Renewable Energy Zone (REZ) in the UK and thus have the exclusive right to produce wind energy within the territorial sea and the EEZ. However, an express decision needs to be made as to the laws applying to the EEZ. Hence, coastal states may decide to extend (parts of) national laws to the EEZ or issue new legislation to regulate specific activities on the EEZ. Moreover, EU law applies to these maritime zones insofar as Member States have offshore jurisdiction. Consequently, EU energy laws (relevant Treaty provisions and the electricity market and renewable energy sources directives and/or regulations) apply to those coastal states that are EU Member States as well as to Norway via the Agreement on the European Economic Area.[40] The extent to which EU law or principles of EU law will continue to apply to the UK from 2021 onwards remains to be seen.[41]

Regulating electricity production from offshore wind

EU-level guidance on the development of offshore wind energy production has so far been limited to several policy declarations issued in 2004, 2005 and 2007 and a Communication from the European Commission of 2008.[42] These documents provide some policy goals but are not binding. Consequently, coastal states have established different legal regimes for developing offshore wind energy and some of these legal regimes have been amended if deemed necessary. Most North Sea States apply the relevant onshore Electricity Act and/or Renewable Energy Sources Act to the EEZ in order to ensure that the financial support schemes (e.g. feed-in tariff, premium tariff or certificate regime as discussed above in Chapter 5) also apply to wind energy projects offshore.[43] Some other countries, such as the Netherlands and Germany, have opted to issue dedicated legislation for offshore wind energy in addition to the above-mentioned energy laws.[44] In general, the development of offshore wind energy depends on a prior authorization. The award procedure may differ but it can be noted that regimes based on a 'first come, first

---

[39]   Article 60 UNCLOS.

[40]   H.K. Müller, *A Legal Framework for a Transnational Offshore Grid in the North Sea*, Intersentia, 2015, chapter 2.

[41]   S. Goldberg, 'UK De-Coupled – Brexit and the Energy Market', in M.M. Roggenkamp and C. Banet, *European Energy Law Report XIV*, Intersentia, 2021.

[42]   See EWEA, *Delivering Offshore Wind Power in Europe: Policy Recommendations for Large-Scale Deployment of Offshore Wind Power in Europe by 2020*, at www.ewea.org/fileadmin/files/library/publications/reports/Delivering _Offshore_Wind_Power_in_Europe.pdf accessed 16 June 2021 and COM(2008) 768 final.

[43]   See for a recent overview of the legal developments and applicable laws: C.T. Nieuwenhout, *Regulating Offshore Electricity Infrastructure in the North Sea – Towards a New Legal Framework*, PhD, Groningen, 16 November 2020, available at www.rug.nl/research/portal/nl/publications/regulating-offshore-electricity-infrastructure-in-the-north -sea accessed 16 June 2021.

[44]   The Netherlands: Wet Wind op Zee (Wind at Sea Act) of 2015. Germany: Windenergie auf See Gesetz (Offshore Wind Energy Act) of 2016.

served' approach[45] as initially applied in the Netherlands and Germany have been replaced by a regime based on competitive bidding. Currently, the UK, Denmark, Germany, Belgium and the Netherlands apply a regime of tendering offshore locations. The way in which these areas are selected and tendered may differ. First, the tender may include a specific reference to the financial support required. The award of such financial support will often result in a requirement that the energy produced offshore needs to be fed into the national grid. Second, the tender may refer to some specific preselected locations. Such preselection is usually based on a regime of spatial planning on the basis of which the government assesses and balances the interests offshore, including the interests of the hydrocarbons sector and fishery but also environmental interests. Sometimes the authorizing authorities will even apply an environmental impact assessment of the selected areas before the tendering process begins. By doing so, they facilitate and speed up the licensing process and thus the development of offshore wind energy production, and the deadlines for meeting the EU renewable energy goals.

An offshore wind energy permit is usually awarded for a limited period of time and gives the wind park developer the right to construct a specific number of wind turbines in the given location based on some specific technical standards and subject to some offshore safety obligations, such as the establishment of a 500-metre safety zone around the offshore installation (e.g. the wind park) and the obligation that any installation that is no longer in use shall be removed in order to ensure safety of navigation.[46] In addition to establishing the wind turbines, the wind-park developer will lay a number of submarine cables linking these offshore turbines with a converter station from which the electricity will be brought to shore. Sometimes the wind park developers can be charged with the construction and operation of these cables as well.

Initially all North Sea states opted for a regime where the park-to-shore cable was considered to be part of the installation and thus should be constructed and exploited by the wind park developer.[47] However, a distinction can be made between the developments in the UK and Denmark, on the one hand, and those in Germany, Belgium and the Netherlands, on the other. By contrast to Denmark and the UK, where the first wind parks were situated close to shore (i.e. the territorial waters), the Netherlands, Belgium and Germany opted for wind parks further from shore, that is, in the EEZ. When looking at the situation in Germany and the Netherlands, it can be noted that both countries made use of a regime where a permit allowing for the construction of an offshore installation was awarded on a 'first come, first served' basis.[48] In both countries this led to high interest in offshore permits but also difficulties in

---

[45]   This means that the first person to apply for a permit and meet the minimum requirements regarding financial and technical capability will be awarded the permit/licence.

[46]   See Article 60 UNCLOS.

[47]   See for the early regimes applied in Belgium, Denmark, Germany, the Netherlands and the United Kingdom, M.M. Roggenkamp and Ulf Hammer (eds), *European Energy Law Report I*, Intersentia, 2004, pp.93–173.

[48]   Section 5 of the Seeanlagenverordnung of 23 January 1997, BGBL. I Section 57, as amended in March 2002 by BGBL I Section 1193 (1216), and the *Wet beheer rijkswaterstaatswerken* (Public Works and Water Management Act) as amended in 2000 (Staatsblad 510) and the accompanying policy rules (Staatscourant 6 May 2002, No. 85).

financially supporting the projects for which a permit had been awarded. An extra compli-
cation existed in the German part of the North Sea as all cables had to cross the wetland area
Wadden Sea, which is a designated nature protection area. Hence, the legal regime in both
countries was amended, as a result of which the onshore TSO is now exclusively responsible
for connecting the wind parks offshore and thus for constructing and exploiting the offshore
cable (see further below).

Denmark is the only coastal state in which a wind park developer will be able to construct
the park-to-shore cable. The Promotion of Renewable Energy Act establishes two regimes for
offshore wind energy development: a tendering regime and a regime where the wind park
developers apply directly for a wind energy permit, but only for those areas which had not
been preselected for tendering. This regime was only applied with regard to small wind parks
close to shore.[49] However, following the Energy Agreement of 2018 a new tender model has
been introduced for nearshore wind parks (at least 4km from shore), based on the assumption
that nearshore wind parks will be cheaper to build as they are located in shallower waters and
require shorter cable connections. The tendering requirements are similar to those applying to
onshore wind parks, which entails that the wind park developer has to offer local residents and
companies a share of at least 20 per cent in the project and has to develop the cable itself. The
Thor Offshore Wind Park is a typical example of this new approach.[50]

The role of TSOs in developing park-to-shore cables

Denmark was also the first North Sea country to provide that the TSO could be charged with
the obligation to connect the wind park offshore. For wind energy projects authorized under
the above-mentioned tender regime, the TSO 'Energinet' is obliged to construct the cable
connection and to cover the costs of the cable, the offshore converter station (transforming
the electricity to a lower voltage) and the onshore connection. If Energinet fails to deliver the
connection in time, it is subject to strict liability as the wind park developer will not be able to
market the electricity as scheduled.[51] The amount of compensation is included in the tender
conditions.[52] The converter station that connects the offshore wind park and the subsea cable
is partly financed by the wind park developer. Energinet, however, owns the substation and
underwater cables, which are considered part of the onshore grid. As this offshore cable is
part of the onshore grid, most of the costs are socialized among the electricity consumers.
Currently, Energinet owns and operates two cable connections to the wind parks at Horns
Rev, a cable connection at Anholt and another at Kriegers flak that also has a connection with

---

[49] H.K. Müller, *A Legal Framework for a Transnational Grid in the North Sea*, Intersentia, 2015, p.162.

[50] Danish Energy Agency at https://en.energinet.dk/Infrastructure-Projects/Projektliste/Thor-Offshore-Wind
-Farm accessed 16 June 2021.

[51] Promotion of Renewable Energy Act, Article 31(2).

[52] Ibid., Article 31(3).

an offshore wind park in Germany. Pursuant to the above-mentioned 2018 Energy Agreement the number of offshore wind parks will increase, and thus also the number of offshore cables.[53]

The legal regimes in Belgium, Germany and the Netherlands have changed drastically over the past 10–15 years. Apart from involving the TSO in developing the offshore cables, these countries have also opted to organize the tendering regime in such a way that several wind parks can be connected to one cable in order to make transport to shore more cost-efficient. The regime that was introduced in the Netherlands in 2015 follows to a large extent the German regime, although some lessons have been learned from the German experience. Whereas in Germany the TSOs were given the responsibility to construct and own the offshore cables as early as 2006,[54] the Renewable Energy Sources Act was amended in 2012 in order to explicitly oblige the TSO to connect offshore electricity producers to the grid at a point as close as possible to the production facility (joined connections).[55] A similar obligation was included in the Dutch Electricity Act in 2016.[56] So far TenneT has developed one offshore grid connection in Germany and two offshore connections in the Netherlands.[57] As a result, TenneT is the TSO charged with development of the offshore grid in the North Sea in both countries. In both countries this obligation is accompanied by some specific legal measures facilitating offshore planning. In Germany, for example, the TSO is required to draft an offshore grid development plan,[58] which will enable the TSO to design the offshore infrastructure in a more efficient manner and to establish the expected completion date of the grid.[59] The expected completion date is crucial as it provides the TSO with an incentive to establish the offshore grid in time. If the TSO is unable to provide the wind farm developer with a grid connection at the expected completion date, it is obliged to pay damages to the wind-farm developer.[60] This rule aims to give wind-farm developers certainty that when the construction of the wind farm is completed, the transport of electricity may commence directly. As this connection regime was proven to be complex and costly, the Dutch legislator opted for a standardized approach as this would limit some of these risks. Pursuant to the Water Act, the government drafts a National Water

---

[53]   https://ens.dk/en/our-responsibilities/wind-power/ongoing-offshore-wind-tenders/nearshore-wind-tender accessed 16 June 2021.

[54]   Section 7 of the Act facilitating planning procedures for infrastructure projects (Gesetz zur Beschleunigung von Planungsverfahren für Infrastrukturvorhaben, Infrastrukturbeschleunigungsgesetz) of 9 December 2006 BGBl. I S. 2833, Nr. 59, 2007 I S. 691 amending the Energy Industries Act (EnWG).

[55]   Section 5(1).

[56]   Act of 23 March 2016 amending the 1998 Electricity Act, Staatsblad 116.

[57]   In Germany the offshore cable has a length of 200 km and was constructed in 2015 to connect BorWin I and 12 other wind parks. In the Netherlands, each offshore cable connects two offshore wind parks at the location Borssele and Hollandse Kust.

[58]   Section 17b German Energy Act.

[59]   The date will become fixed 24 months in advance of the expected completion of the grid connection. This date can only be changed by the federal network agency (Bundesnetzagentur).

[60]   If not, the TSO is not liable until 11 days after the expected completion date has expired. It is up to the court to determine the amount payable.

Plan that, among other things, allocates those areas in the North Sea to be used for developing offshore wind energy. Subsequently, the Minister of Economic Affairs and Climate designates within those areas the sites to be tendered on the basis of the Offshore Wind Energy Act.[61] By tendering sites that are closely situated to each other and including the basic requirements in the tender (including the requirement to connect the wind park to the offshore converter station), a high level of standardization is achieved. As in Germany, the TSO can be held liable, but only if the offshore cable is not completed when the offshore wind turbines are able to start producing electricity or if the offshore transmission system cannot be used and this unavailability is not due to maintenance.[62]

Alternative approach: third parties operating the offshore cable

Although wind-park developers in the UK were initially responsible for constructing and maintaining the cables connecting the offshore wind parks to the onshore grid, this situation changed in 2009 when the UK decided to opt for an alternative approach by giving third parties the responsibility to operate the offshore cable. For this purpose it created the concept of an Offshore Transmission Owner or OFTO. The reason for this change was the assumption that the development of offshore wind energy would require massive investments which would only be feasible if costs were as low as possible. Involving new investors (third parties) was considered the most cost-efficient approach. In addition, the UK government anticipated the concept of ownership unbundling introduced by the third energy package of the EU (see above section 9.2.2). The OFTOs are thus considered TSOs under UK law and subject to a strict (ownership) unbundling regime.[63]

Consequently, the Energy Act 2004 was amended in 2009 and supplemented by the Electricity (Competitive Tenders for Offshore Transmission Licences) Regulations 2013. The new regime provides for a specific licensing regime for offshore transmission activities and entails that an offshore transmission licence can only be obtained through a competitive tendering process.[64] This tendering process is organized by the energy regulator, Ofgem, and runs in parallel with the tendering offshore wind parks. Hence, when initiating a tender for an offshore wind park, another tender procedure will be opened for developing the cable connecting the wind park to shore by an OFTO. This new OFTO tendering regime applies to projects which meet the qualifying requirements after 31 March 2012.[65] Six tender rounds under the new OFTO regime have been opened since February 2014[66] and approximately 20 OFTO licences were awarded in 2020. Such licence entitles the licensee a regulated rate of return on the costs of building and operating the networks.[67] The advantage of the OFTO tendering

---

[61]   Act of 24 June 2015, Staatsblad 261.

[62]   Article 16f(1)(a) and (b) Dutch Electricity Act 1998.

[63]   Section 6(2A) Electricity Act.

[64]   Section 6(1)(b) in conjunction with 6C(1) Electricity Act.

[65]   A transitional regime applied for projects meeting the qualifying requirements by 31 March 2012.

[66]   This tender ran in parallel with the 3rd round tendering the zones for developing offshore wind parks.

[67]   Section 6(6A) Electricity Act.

model is that the investment provides for a fixed 20–25-year revenue, which is independent of the generator activities and provides incentives for those OFTOs who manage costs savings. Investment in an OFTO project is thus a low-risk investment with higher returns. Wind-park developers, however, run a risk as this model does not guarantee that the transmission system and the wind park are ready to operate at the same time. Another concern involves the price to be paid to the generator for the transmission assets.[68]

Assessment

It can be concluded that the large-scale development of offshore wind energy poses challenges to the electricity transport infrastructure. Although most North Sea states originally focussed on wind energy production only, we gradually notice a shift towards a need to consider transportation to shore as well. Whereas most coastal states started with a regime considering the transportation cable as part of the installation or the wind park, we now see a gradual switch towards treating the cable as a separate object requiring separate legal treatment. Most North Sea states have changed the law so that the TSO is charged with the wind-park connection offshore. In doing so the coastal states meet another challenge: streamlining the development of the wind park with the construction of the cable in order to avoid that the wind park is ready to operate but the cable is not yet there (or vice versa). Moreover, there is a trend towards clustering several wind parks to one offshore cable. This increases cost-efficiency. For reasons of cost-efficiency, the UK has chosen a slightly different approach, with the offshore grid licensed to an OFTO. Consequently, investments are done by third parties but coordination problems are avoided as the development and operation of the grid is usually still taken care of by the wind park developers. A next step would be further integration of the national grids into a North Sea grid. This requires further cooperation between the North Sea States and the removal of several legal barriers.[69]

9.4.3   Decarbonizing and Smartening the Electricity Distribution System

The electricity distribution network is also gaining importance with regard to facilitating energy transition. This is the result of increasing amounts of renewable energy sources connected to the distribution grid, also known as decentral, or distributed, generation. Although renewable energy sources (mainly onshore wind) have been connected to the distribution grid since the energy crisis in the 1970s, the scale is now much larger as it is no longer limited to solitary wind turbines or solar panels. Currently it also involves large, onshore wind parks and/or photovoltaic solar projects as well as active involvement of small (household) consumers who, in addition to being energy consumers, have increasingly become 'prosumers'. Commonly,

---

[68]   See also PWC, 'Offshore Transmission Market Update', 2018, p.15, available at https://www.ofgem.gov.uk/system/files/docs/2018/10/pwc_ofto_tr6_market_update.pdf accessed 16 June 2021.

[69]   See also C.T. Nieuwenhout, *Regulating Offshore Electricity Infrastructure in the North Sea: Towards a New Legal Framework*, PhD, Groningen, 16 November 2020, available at www.rug.nl/research/portal/nl/publications/regulating-offshore-electricity-infrastructure-in-the-north-sea accessed 16 June 2016.

prosumers are understood as energy consumers who start generating energy primarily for their own use and on their own premises. Prosumers often generate surplus electricity due to the variable character of RES, which does not coincide with the demand patterns of prosumers. One of the resulting regulatory questions is whether these prosumers are allowed to sell any surplus electricity and whether network tariff structures should take into account the difference between conventional consumers and prosumers as the latter use the grid in a different way: less consumption combined with a feed-in of surplus electricity.[70] Both developments require a new focus on the distribution system. Whereas the integration of the distributed generation on the basis of RES poses a technical challenge due to limited grid capacities and inadequate information on electricity flows, from a legal perspective the roles of the distribution system operators and the system users need to be redefined and further specified. These challenges have been addressed in the new 2019 Electricity Directive (discussed above in section 9.1.2) and will be discussed further below.

Prosumers and the need for smart grids

When the EU Commission published the proposal for reforming the EU legal framework of the electricity sector in November 2016, the proposal was especially promoted as paving the way for a 'consumer-centred energy transition'.[71] While all consumers have been free to choose their supplier since July 2004, the legal reform of 2018/19 further pushes consumers, and especially small consumers, into the market by extending their role beyond 'consuming'. The rationale behind this explicit focus on consumers is not only consumer empowerment, but also the need to integrate demand flexibilities of consumers in the electricity system. Increasing amounts of decentral generation on the basis of RES require flexibilities at the distribution grid level as alternative to expanding grid capacities. The electricity market directive 2019 states this objective in a recital as follows:

> Consumers should be able to consume, to store and to sell self-generated electricity to the market and to participate in all electricity markets by providing flexibility to the system, for instance through energy storage, such as storage using electric vehicles [...] or through energy efficiency schemes. New technology developments will facilitate those activities in the future.[72]

This objective potentially opens a new dimension for consumers in the market as they are empowered not only by the freedom of choice of supplier, but also by actively adjusting their demand and possibly offering flexibility services, which are relevant for system operation. This not only changes the role of consumers, but also requires a change in the way the system is operated. Whereas TSOs are responsible for balancing the system, DSOs can assist to manage the grid by integrating the flexibilities of connected system users, that is, small consumers.

---

[70]   Lea Diestelmeier, 'Regulating Residential Prosumers', in M.M. Roggenkamp, K. de Graaf and R. Fleming (eds), *Energy Law, Climate Change and the Environment*, Edward Elgar Publishing, 2021, pp.729–37.

[71]   EU Commission, 'Clean Energy For All Europeans' COM(2016) 860 final Brussels, 30.11.2016.

[72]   Recital 42 Directive 2019/944/EU.

A precondition for such a system is a constant information exchange between consumers and DSOs on available grid capacities, demand and supply. On the basis of this information, varying prices for grid use and energy consumption can be determined and ideally provide an incentive for consumers to adjust their demand and system use. Only then can consumers engage in the market by offering flexibilities for system operation and, vice versa, can DSOs make use of the potential flexibilities. A 'consumer-centred energy transition' thus requires adding information and communication technologies (ICT) to the electricity system, including smart meters which enable consumers to access exact prices of their network use and consumption. These systems are often referred to as 'smart grids' or 'smart electricity systems'.

The concept of smart grids is not entirely new: the European Commission stated in Directive 2009/72/EC that 'Member States should encourage the modernization of distribution networks, such as through the introduction of smart grids, which should be built in a way that encourages decentralized generation and energy efficiency'.[73] The 2019 Electricity Directive kept the same approach.[74] Smart grids are thus considered to be essential for modernizing distribution grids with the aim of facilitating decentral generation and improving energy efficiency. Perhaps surprisingly, a definition of smart grids is not included in the Electricity Directive, but in the Regulation for Trans-European Energy Infrastructure the smart grid is defined as

> an electricity network that can integrate in a cost efficient manner the behaviour and actions of all users connected to it, including generators, consumers and those that both generate and consume, in order to ensure an economically efficient and sustainable power system with low losses and high levels of quality, security of supply and safety.[75]

Smart grids, therefore, are more than cables supplying electricity; they include ICTs in order to better plan and operate existing distributions systems, to intelligently control (decentral) generation and to enable new energy services and energy efficiency improvements. For this purpose, smart grids incorporate three main technologies into the current electricity system. First is the consumer's need to have access to flexibility technologies, such as storage or other devices which allow them to shift their demand to different times (these are often referred to as 'smart home technologies').[76] Second, all consumers (or prosumers) and DSOs need access to communication networks. Third, this also requires data exchanges between consumers, prosumers and DSOs, ideally in real time on generation, consumption and flexibilities within one system, that is, the smart grid. All three technical components of smart grids bear a variety

---

[73] Recital 36 Directive 2009/72/EC.

[74] Recital 51 Directive 2019/944/EU.

[75] Article 2(7) Regulation (EU) No. 2013/347 Guidelines for Trans-European Energy Infrastructure and repealing Decision 2006/1364/EC [2013] OJ L 115/39.

[76] Charlie Wilson, Tom Hargreaves and Richard Hauxwell-Baldwin, 'Benefits and Risks of Smart Home Technologies' (2017) 103 *Energy Policy* 72–83.

of legal questions relating, inter alia, to the use, operation and/or security of smart grids.[77] The relevant legal framework for smart grids is therefore very broad and cannot merely be related to energy-specific legislation, but also relates to legislation governing data protection and privacy, standardization and ICTs. The following sections of this chapter focus on the provisions provided by the 2019 Electricity Directive and on the question of how smart grids potentially contribute to decarbonizing the electricity distribution system and what they imply for the role of system users and system operation.

Smart meters as a prerequisite for smart grids

One of the essential preconditions for a functioning smart grid is the use of smart meters. The 2019 Electricity Directive regards smart meters as part of a 'smart metering system', which is defined as 'an electronic system that is capable of measuring electricity fed into the grid or electricity consumed from the grid, providing more information than a conventional meter, and that is capable of transmitting and receiving data for information, monitoring and control purposes, using a form of electronic communication'.[78]

Smart meters are, so to speak, the 'gateway' to the market for consumers and also allow for more precise operation of the distribution system. Smart meters are installed at the premises of the consumer and, like conventional meters, serve the purpose of measuring electricity consumption for billing. In contrast to conventional or analogue meters, however, smart meters do not work in one direction and measure consumption only, but work digitally and thus enable the two-directional communication of the real-time price generation, loads and flexibilities within one system.[79] The 2019 Electricity Directive affirms that smart meters have a twofold rationale: they will empower consumers by allowing them to receive accurate information on consumption and generation, to participate in demand response[80] and subsequently to lower their bills, but also will enable DSOs to gain more insight in the use of their networks and thus allow them to reduce their operation and maintenance costs, which ideally leads to lower distribution tariffs.[81]

Given the role these smart meters play in the establishment of an internal electricity market, Member States also need to ensure the interoperability of these metering systems[82] and to have due regard to the use of appropriate standards, best practice and any interoperability require-

---

[77]   Lea Diestelmeier, 'A Legal Framework for Smart Grids', in M.M. Roggenkamp, K. de Graaf and R. Fleming (eds), *Energy Law, Climate Change and the Environment*, Edward Elgar Publishing, 2021, pp.645–55.

[78]   Article 2(23) Directive 2019/944/EU.

[79]   Sabine Erlinghagen, Bill Lichtensteiger, and Jochen Markard, 'Smart Meter Communication Standards in Europe: A Comparison' (2015) 43 *Renewable and Sustainable Energy Reviews* 1249–62, 1250.

[80]   Demand-response refers to the situation where final consumers will amend their normal consumption pattern in response to market signals (including time-variable prices or incentive payments).

[81]   Recital 52 of the preamble to Directive 2019/944/EU.

[82]   Article 2(24) defines interoperability as 'the ability of two or more energy or communication networks, systems, devices, applications or components to interwork to exchange and use information in order to perform required functions'.

ments issued by the EU Commission.[83] Article 20 of the 2019 Electricity Directive specifies in more detail the functionalities of smart metering systems. Data provided by smart meters has to include actual consumption on the basis of actual time of use; validated historical data and non-validated near real-time data have to be made available through a standardized interface or also through remote access. Regarding the frequency of data availability, the article specifies that smart meters have to 'enable final customers to be metered and settled at the same time resolution as the imbalance settlement period in the national market'.[84] Indeed, this complies with the aim of enabling all consumers to participate in the (entire) electricity market. Moreover, if so requested by the final customer, the smart meter has to be able to account for electricity fed into the grid from the final customer's premises and ensure that the metering data on their electricity input and off-take is made available to them or to a third party in an easily understandable format that they can use to compare deals on a like-for-like basis. The latter requirement is of special importance for prosumers who are engaged in feed-in and take-off from the grid on a daily basis. Furthermore, these smart meters have to comply with EU rules and best available techniques regarding cybersecurity and privacy and the protection of data has also to comply with relevant EU data protection and privacy rules.[85] This exemplifies how the transition towards a 'smart, consumer-centred' energy system necessarily includes legislation which is not explicitly linked to the energy sector. The legal framework for the operation of such smart electricity systems thus becomes much broader and more complex.[86]

Further guidance on how to put these smart meters into operation is included in both the 2019 Electricity Directive and the 2018 Energy Efficiency Directive (Directive 2018/2002/EU). This illustrates that smart meters are not only crucial for better positioning consumers in the electricity market but are also key to achieve further energy savings (see Chapter 6). Although Member States are required to promote the introduction of smart meters, they have been provided some choice in their approach. First and foremost, a competitively priced individual meter or smart meter should always be provided when an existing meter has to be replaced and when a new connection is made in a new building or a building undergoes major renovations.[87] With regard to the need to replace existing conventional meters, Member States have the possibility to make use of a prior economic assessment, which should take into account the long-term costs and benefits of smart meters to the market and the individual consumer or the type of intelligent metering that is economically reasonable and cost-effective and the

---

[83]   Article 24(2) Directive 2019/944/EU.

[84]   Article 20(g) Directive 2019/944/EU.

[85]   J. Milaj and J.P. Mifsud Bonnici, 'Privacy Issues in the Use of Smart Meters—Law Enforcement Use of Smart Meter Data', in A. Beaulieu et al (eds), Smart Grids from a Global Perspective: Bridging Old and New Energy Systems, Springer, 2016, pp.179–96.

[86]   L. Diestelmeier, *Unlocking Flexibility with Law: developing a Legal Framework for Smart Electricity Systems*, PhD, RUG, 2019.

[87]   See Article 9 of Directive 2018/2002/EU (on energy efficiency) and Directive 2018/844/EU on energy performance of buildings.

timeframe that is feasible for their distribution.[88] Subject to the outcome of that assessment, Member States have to prepare a timetable for the implementation of these smart meters and, in case of a positive assessment, at least 80 per cent of all consumers have to be equipped with a smart meter either within seven years of the date of the positive assessment or by 2024 for those Member States that have initiated the systematic deployment of smart metering systems before 4 July 2019.[89] However, if the outcome of the cost–benefit analysis is negative, final consumers are, upon a specific request, still entitled to have a smart meter installed which complies with the defined functionalities. In that case, though, final customers have to bear the associated costs. In order to ensure transparency of the costs, the competent regulatory authority has to review and make publicly available regularly, and at least every two years, the associated costs, and trace the evolution of those costs as a result of technology developments and potential metering system upgrades.[90] Should a final customer require such an installation, the competent regulatory authority has to ensure that the installation is done in a reasonable time-frame, but no later than four months after the request.[91]

According to the EU Commission, some 123 million smart electricity meters or 43 per cent of electricity metering points (households and small and medium-sized enterprises) were expected to have been installed and in operation in 2020.[92] Moreover, the roll-out of smart meters, or plans for their roll-out, varies considerably among the Member States. According to the Agency for Cooperation of Energy Regulators (ACER), which monitors implementation and publishes the current status on a yearly basis, several Member States were expected to complete the national roll-out of smart meters in 2018 and others planned to do so in 2020 or 2021, but seven Member States have decided not to implement the roll-out of smart meters based on a cost–benefit analysis.[93] Even if a smart meter has been installed it is not guaranteed that consumers actually make use of it, for example by concluding a dynamic supply price contract or accurately measuring electricity fed back into the grid. In practice, the extent to which smart meters are being deployed differs among EU countries.

Smart grid operation and the role of DSOs

Installing smart meters is key to the development of smart grids. In the EU, smart grid developments still remain on pilot project scales and depend on private investments or specific regulatory schemes for grid innovation (see section 9.2.2).[94] Nevertheless, with the development of smart grids and smart meters the operation of the distribution system changes. The

---

[88]   See Annex II of Directive 2019/944/EU.

[89]   Article 19 Directive 2019/944/EU.

[90]   Article 21(2 c) Directive 2019/944/EU.

[91]   Article 21(2 b) Directive 2019/944/EU.

[92]   ACER Market Monitoring Report 2019 – Energy Retail and Consumer Protection Volume, p.65.

[93]   ACER/CEER, Annual Report on the Results of Monitoring the Internal Electricity and Natural Gas Markets in 2018 – Consumer Empowerment Volume, October 2019, available at acer.europe.eu accessed 16 June 2021.

[94]   Commission of the EU, Joint Research Centre, 'Smart Grid Projects Outlook 2017: Facts, Figures and Trends in Europe', Publications Office of the European Union, 2017, p.28.

operation becomes much more complex when compared to the current approach of distribution system operation, as many more variables have to be considered. The instant availability and transparency of information which is relevant for 'smart' system operation also requires re-thinking the role of the DSOs.

As explained above, smart grids are electricity systems which are upgraded with ICT and which facilitate data exchanges among system users and system operators. In a liberalized energy market, it is relevant to assess whether and to what extent DSOs could engage in the operation of communication networks and could manage the flexibility sources of system users by processing their data.[95] The principle of independent network operation entails that system operators need to provide non-discriminatory access to the system, so that any potential producer or consumer can participate in the market. As communication networks and data are an integral part of smart grids, access to ICT becomes a precondition for participating in the market in the smart grid scenario.[96] Even though communication networks and data are an integral part of smart grids, the operation of these communication networks and the processing of data do not necessarily need to be carried out by the DSO. As early as 2011, the EU Commission recognized the potential risk to a competitive market setting caused by assigning the DSO new tasks of smart grid operation and stated that 'DSOs would obtain access to detailed information about consumers' consumption patterns, which could give DSOs a considerable competitive advantage over other market actors in offering tailor-made services to consumers. The regulatory setting will need to ensure that these risks are properly addressed.'[97]

Ensuring a liberalized market setting for smart grids thus requires the legislator to carefully assess whether new tasks regarding smart grid operations should be carried out by the DSOs or whether this would foreclose potential markets, or whether a new additional regulated actor needs to be assigned this task, especially with regard to the management of data (see also further below). Essentially, this requires striking a balance between the need for coordination for the purpose of system operation and the aim to enable competition for flexibility services offered by system users.[98]

The new 2019 Electricity Directive gives some further insight into this matter as it provides some relevant and initial provisions regarding the operation of smart grids and the role of DSOs. Article 23 provides for a new provision on data management, but it leaves a considerable degree of discretion to the Member States for the exact implementation of data management by requiring them to specify the rules on eligible parties' access to the data of the final customer. Although the directive does not specify who these eligible parties are, it does require that access to data is provided simultaneously and in a non-discriminatory manner to all

---

[95]   Marius Buchmann, 'Governance of Data and Information Management in Smart Distribution Grids: Increase Efficiency by Balancing Coordination and Competition' (2017) 44 *Utilities Policy* 63–72, 64.

[96]   Lea Diestelmeier and Dirk Kuiken, 'Smart Electricity Systems: Access Conditions for Household Customers under EU Law (2017) 1(1) *European Competition and Regulatory Law Review* 1–11, 9.

[97]   EU Commission, 'Smart Grids: From Innovation to Deployment' (Communication) COM(2011) 202 final, 10.

[98]   Lea Diestelmeier, 'A Legal Framework for Smart Grids', in M.M. Roggenkamp, K. de Graaf and R. Fleming (eds) *Energy Law, Climate Change and the Environment*, Edward Elgar Publishing, 2021.

eligible parties, and that procedures for obtaining access to data are publicly available.[99] Some extra provisions relate to situations in which DSOs are involved in data management, and in particular if a DSO is still a part of an vertically integrated undertaking (that is, involved in production and/or supply), by requiring Member States to take 'all necessary measures to ensure that vertically integrated undertakings do not have privileged access to data for the conduct of their supply activities'.[100] This shows that the development of smart grids requires an increased focus on market liberalization at the distribution system level and on the regulation of DSOs. The differences between new tasks and existing regulatory tasks must be clearly defined and delineated.

Citizen energy communities and distribution network operation

The 2019 Electricity Directive introduced a new actor that could contribute to decarbonizing the distribution grid and possibly also to take on system operational tasks. This actor is called 'citizen energy community' (CEC) and is defined as a legal entity if:

(a) [it is] based on a voluntary participation and effectively controlled by members or shareholders that are natural persons, small enterprises or local authorities, including municipalities,

(b) it has as its primary purpose to provide environmental, economic or social community benefits to its members or shareholders or to the local areas where it operates rather than to generate financial profits, and

(c) [it] engages in, *inter alia*, generation, including from renewable sources, distribution, supply, consumption, energy storage, energy efficiency services or charging services for electric vehicles or provide other energy services to its members or shareholders.[101]

While local community initiatives in the energy sector already exist across Member States,[102] EU legislation has never previously explicitly defined the role of such initiatives. Generally, CECs are expected to facilitate organizational structures among local actors (private and public) which jointly engage in the energy sector. This could create institutional support for creating decentral solutions for energy transition which are locally tailored. The EU legal definition is very broad and offers a variety of potential activities that could be carried out by a CEC, inter alia system operational tasks. The preamble of the directive clearly sees CECs as 'competitors' of DSOs, as it acknowledges that 'This Directive empowers Member States to allow citizen energy communities to become DSOs [...]. Once a citizen energy community is granted the status of a DSO, it should be treated as, and be subject to the same obligations

---

[99]   Article 23(2) Directive 2019/944/EU.

[100]   Article 34 Directive 2019/944/EU.

[101]   Article 2(11) Directive 2019/944/EU. This section does not exhaustively discuss the concept of CEC, but focuses on the potential role of CEC with regard to network operation and their relation vis-à-vis DSOs.

[102]   See also Bridge Horizon 2020, Task Force Energy Communities. See www.h2020-bridge.eu/wp-content/uploads/2020/01/D3.12.d_BRIDGE_Energy-Communities-in-the-EU-2.pdf accessed 18 June 2021.

as, a DSO.'[103] Potentially, this bears a legal conflict with the unbundling requirements as CECs may carry out a variety of activities, including production and network management. However, as a general rule Member States may exclude undertakings from the unbundling requirements if they serve fewer than 100,000 customers.[104] This exemption rule might become of greater relevance with the further uptake of CECs. In the case that CECs are allowed to carry out distribution system operation, it will be necessary to clarify the relation between a CEC and the respective DSO when transposing the directive into national law. Article 16 of the 2019 Electricity Directive provides several options for Member States regarding how to organize this relation as well as some minimum requirements these CECs need to meet.

These minimum requirements oblige Member States to ensure that DSOs cooperate with CECs. Beyond this obligation, this provision seems to indicate that the members of CECs do not have to be in geographical proximity. The provision states that DSOs are obliged to 'facilitate the transfer of electricity within CECs'. As the definition of CEC does not contain an element which refers to a confined location, this could be understood as meaning 'among the members'. This would then further imply that members of a CEC can be connected to the same distribution system but do not need to be located in the same area, that is, a confined part of the grid. Furthermore, this also implies that in the area where the CEC operates, other system users who are not members of that CEC can be connected to the distribution grid. In this case, the DSO needs to cooperate with the CEC and provide electricity transfer services for 'fair compensation' from the CEC.[105]

Member States have the option to allow CECs to autonomously manage distribution grids. This option implies obligatory conditions and may also be extended by further exemptions. As the CEC would then have the right to manage distribution systems in 'their areas of operation', the proximity condition has to be fulfilled. This further implies conditions which mainly refer to the regulation of connection points with neighbouring networks.[106] Moreover, the CEC may not discriminate against connected customers who are not participating in the CEC.

Furthermore, where CECs are allowed to operate (a part of) a distribution grid, Member States may also decide to grant them specific exemptions. These are the same that already apply to so-called closed distribution systems (CDS).[107] These exemptions relieve the operator of a CDS and thus CEC from a number of relevant obligations such as the requirements (i) that tariffs, or their methodologies, need prior approval; (ii) to procure flexibility services and to develop the operator's system on the basis of network development plans; and (iii) not to own, develop, manage or operate recharging points for electric vehicles and energy storage facilities. These exemptions basically give the operator of the CDS/CEC considerable leniency with regard to the operation of the grid and the charging of network tariffs compared to the standard regulatory regime applicable to DSOs (see section 9.3.3).

---

[103]   Recital 47 Directive 2019/944/EU.

[104]   Article 35(4) Directive 2019/944/EU.

[105]   Article 16(1 d) Directive 2019/944/EU.

[106]   Article 16(4 a–b) Directive 2019/944/EU.

[107]   Article 38(2) Directive 2019/944/EU.

While the 2019 Electricity Directive aims at providing minimum requirements which are relevant for establishing a level playing field for CECs, overall the provisions leave a rather large degree of leniency to the Member States in determining the relation between CECs and DSOs. Some might exclude system operation from the potential task package of CECs and others might allow CECs to autonomously operate systems and possibly also grant them special exemptions. The implementation is not only relevant for the role of CECs; it is just as important for DSOs. All DSOs will have to prepare to at least cooperate with CECs.

Assessment

Technical innovation and profitable support schemes, for example in the form of feed-in tariffs, have incentivized the development of small-scale RES generation connected to the distribution network in the past decades. While the increase in decentral generation contributes positively to the overall share of RES, it also requires changes to the way in which distribution networks are operated. Smart meters used in combination with smart grids facilitate upgrading the distribution network by integrating ICT for data exchange between system users and system operators. Smart meters and smart grids, however, are not only a technical innovation, but also enable new organizational forms for operating the distribution networks. Consumers become prosumers and could offer demand flexibility for system operation, and local communities could operate their own distribution grid from new organizational entities in the electricity sector. These developments are included in the 2019 Electricity Directive. It remains to be seen how Member States will transpose these provisions into national law and which new developments will emerge at distribution network level.

9.4.4   Legal Innovations to Network Designs

The process of decarbonizing the electricity system has led to some important changes to the existing network design. At the transmission level it is noteworthy that TSOs are gradually moving towards offshore activities, as many EU Member States require the TSO to make an offshore connection for new offshore wind parks. At the same time it can be noted that a new trend may be emerging for wind parks close to shore. However, it can be assumed that new legislative initiatives will be required as a result of the need to also connect offshore consumers (e.g. oil and gas production facilities) to the offshore grid and to make cross-border connections between two wind parks. Similarly, changes are taking place on the distribution level as a result of a more active role played by consumers. Here we can note a trend towards a new type of network design, which seems to be based on an integrated approach and thus challenges the unbundling principles, as in the case of energy communities discussed above. It shows that climate change also challenges market and network design.

9.5   CONCLUSION

The need to combat climate change and the associated requirement to increase the use of renewable energy sources has a considerable impact on energy networks. For several decades the design of these networks has been based on a top-down approach, which entails that the energy is fed into the upper part of the grid (the transmission grid) and taken out at the lower part of the grid (the distribution grid). Renewables, however, can be fed into the grid at all levels, which poses new challenges to all network operators and thus requires a bottom-up regulatory approach.

The EU's climate goal to promote renewable energy sources needs to be balanced with the energy market goal and, thus, network regulation. The decarbonization process may affect the use of existing networks and possibly requires the development of new networks, but can also directly affect network design. This is illustrated by two important developments in the electricity sector: (i) the use of the EEZ for clean energy production (e.g. offshore wind energy) and (ii) the active involvement of small, household consumers in renewable energy production. Whereas the first development required legislators to consider the qualification and thus regulation of the offshore cables, the latter involved the need to introduce smart grids and potentially a new approach to electricity networks by energy communities. Such examples do not exclusively apply to the electricity sector but can also be found in the gas sector, as natural gas may be replaced by green hydrogen and biogas upgraded to biomethane. Both would require a reassessment of current network design.

New technologies are often introduced and applied with little or no guidance at EU level and see Member States taking the lead and opting for national solutions. This can result in a variety of legal choices being taken towards the organization and use of some specific networks and a need for subsequent harmonization of laws. Be that as it may, the first steps have been taken, and more will follow, in the process of energy transition and decarbonization of energy networks.

CLASSROOM QUESTIONS

1. What are the main tasks of system operators assigned by EU energy law and are they compatible with each other?
2. What are the main legal issues for developing offshore wind energy and which alternatives exist for developing and operating subsea cables bringing the electricity to shore?
3. Do you think that Citizen Energy Communities will play a significant role in decarbonizing energy networks in the near future? What legal obstacles might there be?

SUGGESTED READING

Books

Müller HK, A Legal Framework for a Transnational Offshore Grid in the North Sea (Intersentia 2015).
Roggenkamp MM, de Graaf KJ, Fleming RC (eds), Energy Law, Climate Change and the Environment (Edward Elgar Publishing 2021).
Zillmann D et al (eds), Innovation in Energy Law and Technology: Dynamic Solutions for Energy Transitions (Oxford University Press 2018).

Articles and book chapters

Diestelmeier L, 'A Legal Framework for Smart Grids' in MM Roggenkamp, KJ de Graaf and RC Fleming (eds), *Energy Law, Climate Change and the Environment* (Edward Elgar Publishing 2021).

Nieuwenhout C, 'Offshore Hybrid Grid Infrastructure: The Kriegers Flak Combined Grid Solution' in MN Roggenkamp and C Banet, *European Energy Law Report XII* (Intersentia 2017).

Savaresi A, 'The Rise of Community Energy from Grassroots to Mainstream: The Role of Law and Policy' (2019) 31(3) Journal of Environmental Law 487–510.

Policy documents

European Commission, DG ENERGY 'Impact Assessment Study on Downstream Flexibility, Price Flexibility, Demand Response & Smart Metering' https://ec.europa.eu/energy/sites/ener/files/documents/demand_response_ia_study_final_report_12-08-2016.pdf.

Promotion (Progress on Meshed HVDC Offshore Transmission Networks), Deliverable 7.2: Designing the Target Legal Framework for a Meshed Offshore Grid, April 2019 www.promotion-offshore.net/results/deliverables/.

Bridge Horizon 2020, 'Energy Communities in the EU Task Force Energy Communities' (2019) www.h2020-bridge.eu/wp-content/uploads/2020/01/D3.12.d_BRIDGE_Energy-Communities-in-the-EU-2.pdf.

10

Multi-level governance in EU climate law

ABSTRACT

- Multi-level governance refers to the following four levels of governance: industry, national, EU and international;
- Implementing a climate change policy requires action primarily at the level of the Member States and industry, given that the EU itself has a very limited carbon footprint;
- As the Member States rely on industry and citizens to actually reduce greenhouse gas emissions, governance within a Member State can be complex as well, particularly with local governments sometimes taking the lead in combating climate change;
- This entails significant problems relating to the resulting multi-level governance scheme that involves the international, EU and national level insofar as public authorities are concerned;
- The resulting multi-level governance complexity is partly shaped by the fact that the industries involved exert influence and experience the effects of EU climate law at all four levels identified above;
- EU climate law addresses the multi-level governance issue only indirectly, and to a large extent general EU law has shaped this system of multi-level governance;
- The resulting governance structure is (overly) complicated as EU Member States resist a clear governance structure in which the Member States are legally bound to individual targets;
- Solidarity and cost-effectiveness are major drivers in shaping this system of multi-level governance;
- Solidarity needs to be enforced because not all Member States and non-EU countries may share the same commitment to curbing climate change;
- Cost-effectiveness explicitly involves regulated competition at lower levels of governance (Member States and industries), and thus presumes a market regulator, because cost-effectiveness is attained by the use of market mechanisms such as the EU ETS and these markets require supervision.

10.1 INTRODUCTION

Talking of 'EU climate law' as such may be legally correct, but in a way it is strange: the overwhelming majority of greenhouse gas emissions and (renewable) energy production does not involve any European Union institution. EU climate law ultimately involves citizens, consumers and companies established in the Member States and needs to take into account the positions of the industry as well as the positions of governments and companies in third countries. A frequently expressed fear is that the competitive position of the industry will be jeopardised when the costs of climate action increase. This situation is complex because of the multitude of actors and the various levels of governance involved.

The relations between these actors that make up the system of multi-level governance are shaped by the (to a lesser or greater extent) shared desire to combat climate change in mutual solidarity as well as by the need to preserve a level playing field. This implies that the forces of solidarity and cost-effectiveness shape EU climate law.

This chapter will analyse this multi-level or 'polycentric' system of governance. We will identify the forces that have resulted in the current system of governance and extrapolate it to predict what the system may look like in the future. Multi-level governance refers to the following four levels of governance: international, EU, national[1] and industry. In the preamble to the EU ETS, for instance, compliance with the UNFCCC is reiterated as well as the need to review the EU ETS in the light of international developments.[2] We also read that 'emission allowance trading should form part of a comprehensive and coherent package of policies and measures implemented at Member State and [Union] level'.[3] In other words: the EU ETS forms a part of a system of multi-level governance that aims to reduce greenhouse gas emissions.[4] The more recent Energy Union and Climate Action Governance Regulation (hereafter the Governance Regulation[5]) further highlights that EU climate action is now an integral part of the Energy Union, so that its governance interacts with the other four dimensions of the Energy Union.[6] This also holds true for the renewable energy and energy efficiency pillars of EU climate law.

Section 10.2 of this chapter sketches the multi-level governance of EU climate law. Section 10.3 focuses on intra-EU multi-level governance, whereas section 10.4 considers its relation with international multi-level governance. Section 10.5 concludes.

---

[1]   The sub-national level, e.g. regional and city governments, could be added as well, as these are increasingly active in climate mitigation and adaptation: cf. E. Scanu and G. Cloutier, 'Why Do Cities Get Involved in Climate Governance? Insights from Canada and Italy' 2015 *Environnement Urbain/Urban Environment* 9, available at: http://journals.openedition.org/eue/635. See further recital 62 of the preamble to the 2018 RES directive, Directive 2018/2001, OJ 2018 L 328/82.

[2]   Directive 2003/87, OJ 2003 L 275/32, recital 22 of the preamble.

[3]   Ibid., recital 23 of the preamble.

[4]   Ibid., recital 26 of the preamble.

[5]   Regulation 2018/1999 on the governance of the Energy Union and Climate Action, OJ 2018 L 328/1.

[6]   The five dimensions of the Energy Union are: energy security; internal energy market; energy efficiency; decarbonisation; and research, innovation and competitiveness. cf. Article 4 of the Governance Regulation.

10.2   MULTI-LEVEL GOVERNANCE AND EU CLIMATE LAW

Analysis of multi-level governance matters because EU climate law has shown itself to be a rapidly developing area of law. Moreover, it is an area of law that has developed in response to the multi-level nature of the problem involved. Analysing this development from the perspective of multi-level governance intuitively starts with an identification of the levels, and – importantly – the hierarchy between them. In particular in a legal analysis, hierarchy between rules and governance levels is used to overcome inconsistencies between the rules emanating from the various levels. In the EU, for example, the principle of primacy of EU law over national law is used to solve legal problems that arise when there is a discrepancy between EU law and national law.[7] Translating this situation into the terminology of multi-level governance, a clear hierarchy can be identified, placing the EU in a hierarchically superior position compared to its Member States. This reflects the true nature of multi-level governance in the EU only partially.

10.2.1   Multi-level Governance, Integration and Energy Sovereignty in the EU

On the one hand, this application of the principle of primacy simultaneously results in more European integration, in line with the EU's objective to create 'an ever closer Union' (Article 1 TEU). This reinforces the position of the EU as a legitimate source of governance. Connecting this general process of EU integration to climate change, we notice that the EU has defined several aims and values that are very closely connected to combating climate change.[8] Building on this, the EU has set itself a climate change policy and manifests itself as an international negotiator in the talks leading to a new climate accord.

On the other hand, at the same time the Treaties show that the Member States are reluctant to completely hand over all powers to the European Union. In relation to EU climate law we see this when the Treaty on the Functioning of the European Union provides for an energy sovereignty clause that dictates a different decision-making procedure that increases the powers of the Member States in EU decision-making whenever the EU adopts 'measures significantly affecting a Member State's choice between different energy sources and the general structure of its energy supply'.[9] In terms of multi-level governance, therefore, the EU cannot be seen as a unitary actor that exists independent of the Member States. In certain areas, the EU has 'a life of its own' and can act in relative independence of the interests of the Member State(s) involved. We see this notably where the Commission is granted powers to apply the

---

[7]   The applicable EU law then sets aside the national law, enabling the (private) actors at the national level to enjoy the benefits given to them by EU law.

[8]   Notably, Article 3(3) and (5) and 21(2)(f) TEU.

[9]   Article 192(2)(c) TFEU. This is confirmed in Article 194(2), last sentence, TFEU.

competition provisions.[10] We see it to a lesser extent when the normal voting procedure relies on qualified majority voting among the Member States in the Council,[11] whereas measures significantly affecting the choice between energy sources require unanimity in the Council.

This reluctance to confer powers on the EU in the energy field can, however, be contrasted with the almost entire transfer of powers in the area of the internal market. Whenever, for example, Member State climate action involves state aids, EU law applies to the effect that all national measures require the Commission's prior approval.[12] So, governance in the EU as a whole and EU climate law in particular sometimes takes shape more at the EU level, and at other times more at the Member State level.

10.2.2 Governance of Solidarity and Cost-effectiveness

Still, the EU law that emanates from this overwhelmingly needs to be applied at the Member State level in order to have effects at the industry level. Ultimately, producers and consumers need to change their decisions in the market if we are to reduce greenhouse gas emissions. These consumers may do their shopping with a company that falls within the scope of the EU ETS or do business with a competitor from a third country to which the EU ETS does not apply. Obviously, being in or out of the EU ETS and having or not having to pay a carbon price affects the competitive position of these companies. The inclusion of carbon costs in production will affect either prices or profitability and thus impact competition on the market for capital,[13] or the market on which the products are sold. This is one of the reasons why cost-effectiveness is so central to the EU ETS. Only by keeping carbon costs as low as possible is the risk of distorting the level playing field minimised.

At the level of the EU and its Member States, we see that cost-effectiveness and solidarity play an equally prominent role in the Effort Sharing Decision[14] and the Effort Sharing Regulation.[15] Internationally, cost-effectiveness underlies the Kyoto flexible mechanisms[16] and solidarity is embodied in the central principle of the 'common but differentiated responsibility'.[17]

The centrality of cost-effectiveness and solidarity invariably triggers governance problems. The idea of solidarity inherently requires a higher-level governance framework or compulsion. Solidarity, certainly when the group size increases and becomes more differentiated,

---

[10]   See e.g., Articles 101, 102, 106 and 107 TFEU. Notably, the latter has been used by the Commission to adopt a competition policy that is solely in the interest of the European Union, as Article 17(1) TEU requires.

[11]   Article 238 TFEU; this boils down to roughly three quarters of the votes in favour, offering a possibility to outvote a minority of Member States that represent one quarter of the votes.

[12]   Cf. Article 107 in connection with 108(3) TFEU.

[13]   Publicly traded companies will see their share value drop when they report lower profitability.

[14]   European Council Conclusions 23 and 24 October, SN 79/14, pp.1 and 4.

[15]   Regulation 2018/842, OJ 2018 L 156/26, notably recital 2.

[16]   In the Paris Agreement Article 6(4) envisages a mechanism similar to that in place until 2020 on the basis of the Kyoto Protocol.

[17]   Enshrined in Article 2(2) of the Paris Agreement. See further T. Honkonen, 'The Principle of Common But Differentiated Responsibility in Post-2012 Climate Negotiations' 2009 RECIEL 18/3, at p.264.

does not come about by itself. This is why states force their citizens to engage in solidarity, for example, by contributing to mandatory mutual insurance schemes. Similarly, solidarity between the Member States is unlikely to be realised in the absence of some form of coercion by a higher level. The judgment in the *OPAL* case provides a clear example of the enforcement upon Member States of the energy solidarity principle enshrined in Article 194(1) TFEU.[18] Cost-effectiveness, by contrast, involves lower levels of governance in defining the norms as establishing cost-effectiveness invariably requires knowledge of facts that are only known accurately at the lower levels. The fact that cost-effectiveness as well as solidarity shape EU climate law results in, respectively, both bottom-up and top-down influences in governance.

10.2.3 Integration, Solidarity, Cost-effectiveness and the Shaping of Governance

Whereas solidarity and cost-effectiveness have opposing impacts on governance regimes, the integration that is a prominent objective of the EU as a whole also affects the governance structure. Integration exerts an upward effect on governance structures, meaning that decisions are increasingly taken at the higher levels. In the framework of the EU, this means that the role of the Member States as loci of governance decreases while the importance of the EU increases.

Energy market integration, for example, obviously involves more integration in that prices will be aligned in a bigger market that may encompass more than one Member State. It is enabled by having more interconnection, thus necessitating coordination of decisions concerning network investments and development, but also of decisions relating to network management. That in turn requires coordination of the interventions taken by the National Regulatory Authorities in all the Member States. More fundamentally, such market integration enables purchasing and production decisions to be taken in a transboundary manner; for example, enabling consumers to choose between expensive indigenous electricity and cheap imported electricity.[19] Obviously, such purchasing decisions may also 'significantly affect a Member State's choice between different energy sources' within the meaning of Article 194(2) TFEU. This adds a dynamic dimension to the governance, meaning that, for example, even when the law reflects a governance scheme with considerable national influence, the ongoing market integration may actually result in a shift of governance towards the EU level.

---

[18] Case T-883/16 *Poland v Commission* ECLI:EU:T:2019:567, paras 67–85. Note that the Commission, which sanctioned a decision by the German regulatory authority, was found to have failed to comply with the principle of energy solidarity. The result is that from now on the Commission will have to test member states' plans for their compatibility with the principle of energy solidarity and thus enforce compliance with this principle.

[19] A similar issue was at hand in the legal proceeding against a Spanish plan that granted preferential access to indigenous coal for electricity production: see Case T-57/11 *Castelnou Energía, SL v Commission* ECLI:EU:T:2014: 1021.

10.3   EU CLIMATE LAW AND INTRA-EU MULTI-LEVEL GOVERNANCE

The previous chapters have revealed that the bulk of EU climate law requires implementation at the national level. This means that programmes to encourage, for example, consumption of renewable energy or increases in energy efficiency are developed and implemented by the Member States in a way that is intended to have effects in the industry. We have also seen how the role of the Member States in implementing the EU ETS has diminished in particular with the entry into force of the third trading phase. As a result, we can ask to what extent the centralisation and concomitant Europeanisation that has taken place in the EU ETS can also be expected with regard to these other elements that constitute EU climate law. To answer this question, we need to identify the initial multi-level nature of EU climate law governance as well as its development into its current form.[20] This requires us to start with an overview of the constitutional framework within which EU climate law is set. On the basis of this framework, the various legal acts that make up EU climate law are analysed with a view to identifying and comparing the role of the EU, the Member States, sub-national authorities and private parties such as industry and environmental NGOs.

10.3.1   The Constitutional Framework for EU Climate Law

The current constitutional framework for all EU policies is set in the Treaties.[21] Since the entry into force of the Treaty of Lisbon, the framework for EU climate law can be found in Article 194 TFEU. However, this provision to a large extent only codifies a pre-existing policy.[22] As a result, a significant part of EU climate law was adopted within the framework of Article 192 and 114 TFEU. Moreover, both Article 192 and 114 TFEU remain relevant today. In appraising this legal framework, it is important to take into account Article 4 TFEU, according to which 'the environment' and 'energy' as well as the 'internal market' are so-called shared competences. Climate change as such is not mentioned in that list of shared competences, but we see it mentioned in the provisions that elaborate the environmental and energy areas.[23] As a result, both the EU and the Member States are competent in these fields. In practice this means that as long and insofar as the EU has not enacted legislation in a certain area, regu-

---

[20]   See further on this issue J. van Zeben, *The Allocation of Regulatory Competence in the EU Emissions Trading Scheme*, Cambridge University Press 2014, p.18 et seq., who reviews theories of multi-level and polycentric governance.

[21]   See, by analogy, Case 294/83 *Parti écologiste 'Les Verts' v European Parliament* ECLI:EU:C:1986:166, para. 23.

[22]   For an overview see H.H.B. Vedder, 'The Treaty of Lisbon and European Environmental Law', 2010 *JEL* 22/2, pp.285–99.

[23]   Article 191(1), fourth indent, TFEU. Article 194 TFEU: climate change can be said to be implicit in the inclusion of energy efficiency and renewable energy in that provision.

lating this area remains for the Member States.[24] The shared competence means that there is ample room for true multi-level governance, but the results of such governance at the Member State level have to comply with EU law. The concept of true multi-level governance intends to distinguish between those areas where the Member States have no governance powers and essentially only act to give effect to EU measures and situations where the Member States have room for autonomous decisions and can thus 'govern' in this area.

The constraints set by EU law to such national initiatives follow from essentially two sets of rules: (a) the internal market rules in the Treaty and (b) the detailed rules laid down in EU secondary law, such as the ETS Directive and the RES Directive. The framework for multi-level governance scheme laid down in EU secondary law is also shaped by the Treaties through the dual principles of:

- subsidiarity, and
- proportionality.[25]

The exercise of all shared competences must comply with these two principles that essentially seek to curb the EU's use of its competences.[26]

Subsidiarity

In a nutshell, the principle of subsidiarity constrains the EU's exercise of its competences to those instances where there is a transnational aspect to the matter that is regulated or where the EU's action would result in economies of scale or scope compared to Member State action. In practice it means that the EU, and notably the Commission, is under a duty to explain why EU action complies with the subsidiarity principle.[27] In relation to EU climate law, both elements of the subsidiarity test would seem to present little difficulty in view of the truly global nature of climate change and the obvious impossibility of dealing with this effectively at the Member State level. Still, we see that the principle of subsidiarity is used by the Member States to challenge proposed EU action. In its proposal for a European Soil Framework Directive, for instance, the Commission argues that EU action is necessary as, for example, drainage of peat soils causes massive emissions of greenhouse gases, which is a problem of a clearly transnational nature.[28] The fact of the matter is that the Commission has withdrawn the 2006 proposal, but remains committed to soil protection.[29] As a result, there is to this day no

---

[24] Article 5(2) TFEU. See further Protocol No. 25 and Declaration No. 18 attached to the Treaty of Lisbon on shared competences. Moreover, when the EU has regulated a certain area, there is a possibility for the EU to 'withdraw itself from this area and thus hand back the competence to regulate these matters to the Member States.

[25] Article 5(3) and (4) TEU.

[26] For an analysis of subsidiarity and proportionality T. Tridimas, *The General Principles of EU Law*, Oxford: Oxford University Press 2006.

[27] Protocol No. 2 contains more detailed rules.

[28] COM (2006) 232. Note that the European Parliament had adopted a favourable opinion following the first reading, OJ 2008 C 282 E/281.

[29] OJ 2014 C 153/3.

specific EU-level governance in this field. More fundamentally, there is no clear governance structure that allows for the identification of a clear division of competences and the corresponding assignment of responsibilities. Soil protection is thus a clear example of the failure of multi-level governance, because of the principle of subsidiarity, in the sense that a clearly transboundary problem is not effectively addressed at the EU level, despite the fact that the EU is competent in this field.[30]

Proportionality

On a similar note, the principle of proportionality seeks to protect national competences. This principle means that in its regulations the EU should leave the Member States as much room as is possible. It thus requires the EU to continuously reflect upon the objectives it seeks to attain with the regulation at hand and to seek to ensure the means chosen do not go beyond what is necessary. In view of the room left for national implementing measures, this results in a preference for a framework directive over a detailed directive and a preference for directives over the use of regulations. Moreover, the substance of such directives and regulations should leave the Member States as much freedom as possible, which could involve freedom to set national standards that go beyond the European standards and thus protect the climate to a greater extent. This invigorates the multi-level governance in these matters.

The proportionality principle raises the question: to what extent has the EU exhaustively and completely harmonised certain areas of climate law in the EU? This is a question of the scope and degree of harmonisation.[31]

Essentially, the basic reasons underlying harmonisation dictate that the degree of harmonisation will be higher, leading to full or total harmonisation, whenever the impact on the internal market is greater. In terms of EU climate law this means that those measures that are more directed at the environmental side of affairs are likely to be of a minimum harmonising character, whereas the measures that are closely linked to trade between Member States are often of a fully harmonising nature. The fully harmonising nature of these rules negates the room for national rules that are in any way different from those in the EU rules.[32] In terms of multi-level governance, full harmonisation entails a clear shift of governance to the EU level.[33]

10.3.2 Multi-level Governance of Renewable Energy Sources

We see harmonisation's impact on the multi-level governance scheme very prominently in the RES Directive.[34] While the majority of the rules in this Directive are of a minimum harmo-

---

[30]   This competence could result from the environmental impact (Article 192 TFEU), the agricultural impact (Art. 43 TFEU) or the climate impact (Art. 194 TFEU).

[31]   For an overview of this general doctrine of EU law see J.H. Jans and H.H.B. Vedder, *European Environmental Law*, Groningen: Europa Law Publishing 2013, pp.97–104.

[32]   Case 148/78 *Criminal proceedings against T. Ratti*, ECLI:EU:C:1979:110, paras 25–27.

[33]   For an application of this in the field of renewables see Case C-242/17 *Legatoria Editoriale Giovanni Olivotto (L.E.G.O.) SpA v Gestore dei servizi energetici (GSE) SpA* ECLI:EU:C:2018:804, paras 29, 28 and 52–55.

[34]   See Chapter 5.

nising nature, thus allowing the Member States to impose national targets on the percentage of renewable energy that go beyond those laid down in the Directive, some of the provisions are fully harmonising, such as the rules on the sustainability criteria of biofuels. This is clearly indicated in the text of the Directive.[35] As a result, any multi-level governance concerning these sustainability criteria is completely ruled out. This has resulted in very extensive and all-encompassing rules on the sustainability of biofuels and bioliquids that are only becoming more comprehensive at the EU level with new iterations of the Directive.[36] Multi-level governance concerning the rest of the RES Directive is entirely possible and indeed happening.

This emanates clearly from two recent cases, *Essent Belgium* and *Ålands Vindkraft*, as well as the older *PreussenElektra* case.[37] All three deal with essentially the same question: to what extent can a Member State encourage renewable energy production by means of a measure that restricts the incentive for renewable energy that is produced in that Member State? In this regard the multi-level nature of the governance starts from the general objective set by the EU to ensure that in 2020 at least 20 per cent of energy is renewable. This is then translated into national goals as well as reporting obligations and a duty to draw up National Renewable Energy Plans. Decisions on how the renewable energy targets are to be achieved are thus taken on the national level (and at times sub-national levels).[38] In view of the fact that encouraging and incentivising renewable energy production still involves overcoming market failures, most of the decisions concerning renewables involve subsidisation. Decisions on such subsidies are preferably taken by the Member States in a way that ensures that such funds benefit the national economy so that 'leakage' to industries in other Member States and third countries is minimised.

The German scheme in *PreussenElektra*, for example, required German electricity utilities to buy the renewable electricity that was produced in the area where they were active. This means that this obligation attached only to green electricity produced in Germany. The Swedish scheme at hand in *Ålands Vindkraft* required Swedish electricity supply companies to surrender green certificates at the end of the year corresponding to a certain percentage of the energy they supplied in that year. Here the catch was that such certificates were only provided to renewable electricity generators that are located in Sweden, whereas the Åland archipelago – although interconnected to Sweden – is part of Finland. In both schemes the purchasing obligation, whether this involves the green electricity itself that is to be bought at a premium or the green certificates that have a certain value, ensures a financial incentive for national renewable

---

[35] 2018 RES directive, Directive 2018/2001, recital 94 of the preamble and Article 29(12).

[36] The 2018 RES Directive, for example, contains more elaborate annexes, compared to Directive 2009/28, that set rules on, for example, the transport distance of biofuels to be taken into account for determining their contribution.

[37] Respectively joined cases C-204/12 to C-208/12 *Essent Belgium* ECLI:EU:C:2014:2192, case C-573/12 *Alands Vindkraft* ECLI: EU:C:2014:2037 and C-379/98 *PreussenElektra* ECLI: EU:C:2001:160.

[38] See M. Dreyfus, 'A Bottom-up Approach to Energy Transition in Europe: The Case of "Local Climate Energy Plans" in France', in L Squintani and H. Vedder (eds), *Sustainable Energy United in Diversity: Challenges and Approaches in Energy Transition in the European Union*, e-book available at www.eelf.info/home-7.html last accessed 16 June 2021.

energy production. In *ENEA*, finally, Polish electricity supply companies were required to buy minimum amounts of renewable energy from Polish generators.[39]

Whenever such a decision is taken at the Member State level, it is bound to lead to legal opposition from the private companies being regulated. In *PreussenElektra*, the (staged) proceedings resulted from the national energy utilities fearing the costs resulting from this feed-in tariff that would have to be borne by them and passed on to the final consumers.[40] In *Ålands Vindkraft*, the opposition came from a company in a neighbouring Member State that wanted to benefit from the incentive scheme.[41]

The fact that private companies, the subject of regulation, are involved indicates yet another layer in the multi-level governance scheme of EU climate law. Such private involvement in shaping EU climate law follows from the direct effect of the internal market rules. The internal market consists of the free movement of goods, services, people and capital as well as a system ensuring undistorted conditions of competition.[42] In relation to national incentive schemes such as those in the cases mentioned above, the direct effect of the internal market rules results in the inclusion of private actors as well as the EU judiciary in the multi-level decision-making system. The former have a right to challenge the legality of such national schemes in the light of the free movement of goods (Article 34 TFEU) as well as the provisions on state aids (Article 107 and 108 TFEU). This right may be enacted before a national court of the Member States who then may (be obliged to) make a preliminary reference to the Court of Justice of the European Union.[43]

As a result, the internal market rules form an integral part of the constitutional framework for the multi-level governance of EU climate law. All national measures that restrict the free movement of goods[44] or that distort competition must comply with the Treaty rules. In keeping with the basic approach in EU law, the basic provisions are construed broadly, encompassing all measures that may affect trade or competition. This is then connected to a possibility to objectively justify such restrictions. This requires the Member State to put forward an objective in the common interest and to demonstrate that the national measure

---

[39]   Case C-329/15 *ENEA S.A. v Prezes Urzędu Regulacji Energetyk* ECLI:EU:C:2017:671, notably para 3 where the national law is set out.

[40]   This also drove the ENEA case, *ibid*.

[41]   For a more detailed analysis see H. Vedder, 'Good Neighbourliness in a Sustainable European Internal Electricity Market: A Tale of Communities and Uncommunautaire Thinking', in E. Basheska and D. Kochenov (eds), *The Principle of Good Neighbourliness in EU Law*, Martinus Nijhoff 2014, pp. 94–113.

[42]   Article 26(2) TFEU in connection with Protocol No. 27 on the Internal Market and Competition.

[43]   According to Article 267 TFEU any national court may make a preliminary reference, whereas the highest national court is under an obligation to do so unless there is an acte clair or acte éclaire: case 283/81 *CILFIT*, ECLI:EU:C:1982:335, paras 13–20. Note that the Court is reluctant to grant the highest national courts too much discretion in this regard. Moreover, the Court has also accepted *Francovich* Member State liability for highest national courts that fail to make a preliminary reference in Case C-224/01 *Köbler*, ECLI:EU:C:2003:513.

[44]   Perhaps somewhat counterintuitively, the Court has classified electricity as a good within the meaning of Article 34 TFEU.

does not go beyond what is necessary to attain this objective. This proportionality test allows the Court to exercise considerable influence over the national measures, potentially firmly establishing itself as an actor in the multi-level governance system. We see the practical impact of this in the *E.On Biofor* case. This case essentially arose because of Swedish rules that made it impossible for E.On Biofor, a Swedish subsidiary of E.On, to use biogas produced in Germany by another company in the E.On group, in compliance with the Swedish renewables regulations when that biogas is transported though gas pipelines.[45] The Court notes that biogas fed into the Swedish grid could be taken into account, whereas gas fed into an interconnected gas grid in another Member State could not. This inconsistency, the Court held, contributed to the incompatibility of the Swedish rules with Article 34 TFEU.[46] As a result of the appeal by E.On Biofor, Sweden had to amend its rules and to allow imported gas to be taken into account.

The actual impact of this tandem made up of private companies and the EU judiciary in the multi-level system of governance depends on the intrusiveness of the Court's review.[47] In *PreussenElektra* the Court showed very considerable deference and thus reduced the practical impact of the internal market rules as a framework for the multi-level governance of EU climate law. One of the reasons for this was the fact that electricity market liberalisation was only just underway in the EU. This resulted in a clearly temporally limited acquiescence on the part of the Court.[48]

Expectations were therefore high when the Court had to rule on similar national incentive schemes well over a decade, and two energy market liberalisation packages, later. The two later cases, *Essent Belgium* and in particular *Ålands Vindkraft*, highlight the connection between the governance of climate law and the relevance of governance at the industry level. The applicant in that case, a Finnish company producing renewable energy from the Finish Åland archipelago, in effect could only sell that wind power on the Swedish market, since the only significant power cable connected the islands to Sweden.[49] It was, therefore, in competition with Swedish electricity generators and in that sense 'governed' by the local market conditions.[50] However, the regulatory regime resulting from the Swedish rules meant that it could not qualify for the incentive, affecting its position on the market.

---

[45]   Case C-549/15 *E.ON Biofor Sverige AB v Statens energimyndighet* ECLI:EU:C:2017:490. Actually, the imported gas could be used if it was transported in a (rail)road-based vessel, an utterly economically unattractive means of transportation.

[46]   Case C-549/15 *E.ON Biofor Sverige AB v Statens energimyndighet* ECLI:EU:C:2017:490, paras 94–99.

[47]   It may be noted that the Court shows considerable deference in this regard: see H. Vedder, 'Good Neighbourliness in a Sustainable European Internal Electricity Market: A Tale of Communities and Uncommunautaire Thinking', in E. Basheska and D. Kochenov (eds), *The Principle of Good Neighbourliness in EU Law*, Martinus Nijhoff 2014, pp.94–113.

[48]   Case C-379/98 *PreussenElektra* ECLI: EU:C:2001:160, para. 81, where the Court finds the German scheme justified 'in the current state of Community law concerning the electricity *market* [emphasis added]'.

[49]   See for a fuller analysis: H.H.B. Vedder, 'EU Law and the Financing of New Energy Infrastructure', in M.M. Roggenkamp and U. Hammer (eds), *European Energy Law Report X*, Intersentia 2014, at pp.27–29.

[50]   In effect, it would sell its electricity on the pan-Scandinavian and Baltic Nord Pool spot market.

In much more elaborate and reasoned judgments, the Court found the Swedish and Belgian schemes compatible with the EU law on the free movement of goods. This compatibility had its roots in the way in which the RES Directive barely regulates such national incentive schemes and clearly puts the Member States in the driving seat. We see this when the Court refers extensively to the provisions in the 2009 RES Directive to find that the Member States are free to organise their national incentive schemes, whereas the Directive limits itself to stating that the Member States may conclude cooperation agreements.[51] In this regard, the progress made in market liberalisation since *PreussenElektra*[52] only resulted in a more elaborate, but not a more intrusive, proportionality test. In terms of multi-level governance, the private companies and EU level clearly matters less than the national level in relation to national incentive schemes. This becomes clear when the relevant provisions in the 2009 RES Directive are analysed and this finding also influences the Court's findings on compatibility with the internal market rules. In other words: the governance scheme provided for in the 2009 RES Directive is one that envisages a minimal role for the EU and fundamentally chooses sovereign Member States as the drivers for policy and action in this field. The 2018 RES Directive by and large respects this national sovereignty, but adds a provision on the opening of support schemes for imported renewable electricity.[53] This provision is clearly intended to allow Member States to gather experience with such cross-border schemes.

For the moment, these sovereign Member States engage in a limited amount of solidarity. We see this when we find that the incentive schemes are essentially restricted to national production of renewable energy only. This limited amount of solidarity comes at the cost of employing the market mechanism and all the potential offered by the comparative advantages that the market harnesses. Indeed, the reduced amount of 'EU enforced' solidarity limits the role of markets to the situation *within* a Member State, if a market-based instrument is chosen. The companies', Commission's and Courts' roles in this multi-level governance scheme are very limited. *E.On Biofor*, however, shows that the Court will test the proportionality of national measures that limit the importation of renewable energy increasingly strictly. The proportionality of national measures is also reviewed by the Commission, subject to Court review, as part of the state aid rules. By and large, this means that national measures to subsidise renewable energy generation will be reviewed by the Commission for their compatibility with the internal market. Increasing market integration has resulted in increasingly tight review by the Commission. This is also clearly visible in relation to the state aid supervision

---

[51]   C-573/12 *Ålands Vindkraft* ECLI: EU:C:2014:2037, paras 84–119 and notably paras 97–100.

[52]   The Court explicitly noted this in paras 84 and 85 of the judgment in case C-573/12 *Ålands Vindkraft*, ECLI: EU:C:2014:2037.

[53]   Directive 2018/2001, Article 5.

of national (renewable) energy policies,[54] as well as ancillary schemes such as capacity and generation adequacy schemes.[55]

The same holds *a fortiori* for the rules on energy efficiency. From Chapter 6 of this book we know that the rules on energy efficiency lack binding targets at the Member State level and do not provide for any market-based mechanism whatsoever. Where a market is envisaged, the EED confines itself to reiterating the applicability of the general EU rules on competition law.[56] This centrality of the Member States comes with an overwhelming focus on cost-effectiveness and reduces the potential for solidarity as well as (market) integration, negating the European Union's role.[57]

This is different when the market-based rules are more central to the EU regulation in place, as is the case in the EU ETS, to be discussed hereafter.

10.3.3 Multi-level Governance of the EU ETS

The EU ETS has experienced a marked evolution at an astonishing pace. In little more than 15 years the EU ETS has developed from a scheme that fundamentally put the Member States in the driving seat to one that acknowledges the need to have effective market supervision for markets to really work.[58] Discussions on so-called windfall profits accrued by companies due to free allowances triggered a major overhaul of the scheme, with a greater emphasis on auctioning as the main mechanism for allocating allowances. This firmly establishes markets as a main driver for change in greenhouse gas emissions, in addition to the trading that takes place once allowances have been distributed. More fundamentally, the 2009 overhaul of the ETS Directive has recognised that there is obvious regulatory competition between the Member States. Whereas regulatory competition could result in regulatory experimentation and thus sustain efficiencies in the EU ETS, in reality it has resulted in a classic 'race to the bottom'. Over-allocation of allowances by the Member States is the foremost reason for the absolute lack of scarcity on the allowance market and the resulting absence of a serious incentive for companies to invest in low-carbon technology and innovation.

The solution has been to put the Commission, as a supervisor of the market for regulatory competition, in a central position from the third trading phase onwards. To be clear, as of

---

[54]   Cf. A. Haak and M. Bruggemann, 'Compatibility of Germany's Renewable Energy Support Scheme with European State Aid Law – Recent Developments and Political Background' 2016 *Eur. St. Aid L.Q.* 91, at 100.

[55]   The increased uptake of intermittent renewable energy in the grid can jeopardize grid stability and generation adequacy: see Chapter 5 in this book and e.g. Case T-793/14 *Tempus Energy Ltd and Tempus Energy Technology Ltd v Commission* ECLI:EU:T:2018:790, on the UK capacity and generation adequacy scheme and the state aid supervision of that.

[56]   Directive 2012/27, OJ 2012 L 315/1, Article 18(3).

[57]   See further European Council Conclusions 23/24 October, SN 79/14, p. 6, where an EU target of 27 per cent for improving energy efficiency is set, which will not be translated into nationally binding targets and even omits the 'need to deliver collectively the EU target' that guides the Member States for the EU renewables target.

[58]   For a fuller overview of this evolution see Chapter 3 in this book.

2013 the EU ETS remained a system of multi-level governance, but had shifted the bulk of competences to the EU level.[59]

Part of this shift also came from the private company level, as is evidenced by the *Arcelor* case.[60] This case was started by a producer of ferrous metals that was subject to the EU ETS as part of the implementation in France. It considered that this implementation, and the ETS Directive that underlies it, was illegal as it deteriorated its competitive position compared to that of the aluminium and plastics industry. The latter industries were outside the scope of the ETS Directive and French legislation, but they were in competition with Arcelor, for example where packaging is concerned. Under the heading of the general principles of equality and proportionality, the ECJ investigated whether the choice by the EU legislator to include the ferrous metal industry but to leave out the aluminium and chemicals industries could be objectively justified. In what is again a deferential judgment that acknowledges the discretionary room for the EU legislature, the Court ultimately accepted the different treatment of the steel industry and the aluminium and plastics industry for reasons of administration and enforcement costs.[61] At the basis of this analysis, however, is the observation that from a climate change perspective, all greenhouse gas emissions are comparable.[62] In this regard, the fact that the industries involved are (potential) competitors is not a decisive criterion, according to the Court.[63] This shows that the Court is primarily thinking in terms of markets, but works from an assumption of equality of polluters when analysing the EU ETS.

The implications of the Court's vision of the EU ETS for the governance of the EU ETS are significant, as it implies that all sectors that contribute to climate change and have a potential for greenhouse gas abatement are included. As a result, it becomes an issue for the EU legislature to explain why – as an exception – a sector is not included. The reasoning accepted in this regard is one based on enforcement and administration costs in the light of the novelty and complexity of the scheme. Based on this there would be an expectation for the scope of the EU ETS to expand over the years, as the novelty wears off, experience is gained and administration and enforcement costs decrease. This is indeed what has happened, as recital 10 in the preamble to Directive 2009/29 clearly shows in its statement that other industries have been included in the EU ETS. The steps taken towards the inclusion of aviation and maritime shipping – albeit in an imperfect manner – in the EU ETS are further evidence of this expansion to include all sectors of the economy.[64]

Interestingly – and this is also alluded to in *Arcelor*[65] – the wider scope of the EU ETS including the allowance market as an exchange mechanism allows for comparative advantages to be

---

[59]   J. van Zeben, *The Allocation of Regulatory Competence in the EU Emissions Trading Scheme*, Cambridge University Press 2014, p.144 et seq.

[60]   Case C-127/07 *Arcelor* ECLI:EU:C:2008:728.

[61]   Case C-127/07 *Arcelor* ECLI:EU:C:2008:728, paras 46–73.

[62]   Case C-127/07 *Arcelor* ECLI:EU:C:2008:728, para. 32–38.

[63]   Case C-127/07 *Arcelor* ECLI:EU:C:2008:728, para. 36.

[64]   Recital 4 of the preamble to Directive 2018/410 OJ 2019 L 76/3.

[65]   Case C-127/07 *Arcelor* ECLI:EU:C:2008:728, para. 33.

reaped. Indeed, allowances can be traded freely in the EU and possibly even, in the future, with third countries (so-called linking of emissions trading markets).[66] This inherent expansion of the scope of emissions trading brings an ever greater group of companies as well as states into the multi-level governance scheme.

The reduced role for the Member States also can be seen in the EU legislature's reasoning in curbing the Member States' freedom to enact more stringent protection legislation. The replacement of the Directive on Integrated Pollution Prevention and Control (IPPC) with the 2010 Directive on Emissions from Industrial Installations (IED) came with the introduction of a clause that banned the Member States from introducing emissions limit values in the permits on the basis of this Directive.[67] The idea underlying this provision was to ensure the undistorted functioning of the EU ETS by means of largely ruling out the possibility of national environmental regulation. This seems to be at odds with Article 193 TFEU that allows for such more stringent protection measures,[68] but such measures must be compatible with the Treaties. Above, we have already seen how climate law at both the EU and the Member State level interacts with the constitutional framework of the EU that is laid down in the Treaties. In relation to more stringent measures, however, the catch is that the Treaties do not stand in the way of a Member State putting its own industry at a competitive disadvantage. Essentially, therefore, the self-restraint of the Member States, which will undoubtedly be lobbied by the national industry, results in the absence of such more stringent national climate policies.

Interestingly, European cooperation may also influence climate governance within a Member State. A good example of this is the fate that befell the Netherlands Energy Accord. This agreement between major energy producers and consumers as well as environmental NGOs envisaged, inter alia, the coordinated closure of a number of ageing and thus relatively dirty coal-fired power plants. Such a coordinated closure resulted in concerns raised on the basis of competition law and the agreement was thus notified to the Netherlands competition authority, which issued an opinion on the agreement's compatibility with the cartel prohibition.[69] In assessing the Accord the competition authority balanced the negative welfare effects of the reduction of generation capacity with the positive effects of reduced emissions. In that regard, it stated that the Accord would not reduce emissions of carbon dioxide because the EU ETS would result in an increase of these emissions elsewhere.[70] The multi-level nature of EU climate law thus results in effects for the appraisal of national initiatives under national law. These effects follow from the (in)famous waterbed effect whereby allowances not used by one participant in the scheme can be used by another participant. One solution would be

---

[66] For an analysis in general and applied to the EU-Chinese context see Y. Zeng, *Obstacles to Linking Emissions Trading Systems in the EU and China – A Comparative Law and Economics Perspective*, PhD Groningen 2018.

[67] Directive 2010/75 (IED), OJ 2010 L 334/17. Recital 9 of the preamble in connection with Article 9(1) of the IED.

[68] See further recital 10 of the preamble to the IED.

[69] Available at www.acm.nl/en/publications/publication/12046/ACM-deal-over-closing-down-coal-power-plants-harms-consumers/ last accessed 16 June 2021.

[70] ACM Report on the Energy Accord, available at: www.acm.nl/en/publications/publication/12082/ACM-analysis-of-closing-down-5-coal-power-plants-as-part-of-SER-Energieakkoord/, p.4 last accessed 16 June 2021.

to cancel the allowances held by the operators of the coal-fired power plants to the amount that corresponds to the avoided emissions, to counteract this waterbed effect. This, however, would simultaneously result in a reduction of income for these operators, and thus affect their competitive position. Obviously, these companies are likely to show self-restraint in view of the need to preserve their competitive position. At the same time, the ETS Directive now allows the Member State to cancel allowances that are freed up due to more stringent national measures.[71] This allows for more Member State involvement in EU climate governance, insofar as this governance is beneficial to climate protection.

The above refers overwhelmingly to the central state. In terms of governance, it is also relevant to mention that EU climate governance also increasingly involves sub-national levels of governance. The Covenant of Mayors for Climate & Energy is a prominent example of how the European Commission has encouraged sub-national governments to become involved in climate governance.[72] This also makes sense given that the bulk of the impact of climate change will have most impact in the urban environment and thus affect cities primarily. At the same time, the bottom-up policies that emanate from cooperation such as the Covenant of Mayors are likely to shape the national policies of the Member States,[73] potentially reducing the role of central governments in climate governance.

In short, a Member State's core position in the governance of phase one has been replaced largely by a central role for the Commission. Where there is room for national governance, this is strictly circumscribed. Further evidence for this can be seen in the provisions on the Modernisation Fund set up pursuant to Article 10d of the ETS Directive. Again, we see Member State involvement in the governance of this fund, yet we also see that in the absence of consensus, decisions can be taken by simple majority.[74] It is clear that for climate governance the EU no longer wants to be hostage to the lowest common denominator between the Member States.

When we cast the climate law net a bit wider, one final example of the effect of a central role for the market in shaping the rules can be seen in the *Hinkley Point C* saga.[75] This case concerns a Commission decision that authorised UK state aid for the construction and operation of the Hinkley Point C nuclear power plant, required to ensure generation adequacy in light of expected RES uptake and electrification. In a nutshell, Austria objected against this decision because it considered that the environmental integration principle enshrined in Article 11 TFEU had not been complied with. In the framework of EU state aid supervision, the integration principle is, according to the Court, binding in the sense that the Commission cannot

---

[71]   This is made possible by Directive 2018/410, OJ 2018 L 76/3, that changed Article 12(4) of the ETS Directive, allowing for such cancellation of allowances.

[72]   It was created in 2008 by means of a Commission initiative: https://www.eumayors.eu/about/covenant-initiative/origins-and-development.html last accessed 16 June 2021.

[73]   For an analysis of this effect see M. Fraundorfer, 'The Role of Cities in Shaping Transnational Law in Climate Governance' 2017 *Global Policy*, pp.23–31, doi:10.1111/1758-5899.12365.

[74]   ETS directive, Article 10d(7).

[75]   Case C-594/18 P *Austria v Commission* ECLI:EU:C:2020:742.

allow state aid for an activity that violates EU environmental law.[76] Again, it is only through the market-based state aid rules that the EU plays its decisive role in steering the Member States.

10.4 EU CLIMATE LAW AND INTERNATIONAL MULTI-LEVEL GOVERNANCE

Companies' self-restraint in response to the need to remain competitive is an obvious trait of EU climate law when it is analysed from the perspective of international multi-level governance. The part of the climate acquis that deals with carbon leakage is an example of precisely that: the EU and the Member States showing self-restraint in the light of the international competition that faces certain industries located in the European Union. In terms of multi-level governance, the method of adopting the carbon leakage list under the EU ETS clearly involves the national and industry levels, with the latter submitting the economic data and the former drafting the list that is to be adopted by the Commission. The involvement of all three levels is eminently sensible in view of the fact that industry has a unique insight into its competitive position, the Member States can relate the data and findings of their industry to those of the other Member States and the Commission has oversight of the entire internal market and can function as an arbiter. More fundamentally, the Commission also plays a central role in the international relations between the EU and third countries and international organisations in the field of climate change. In this regard, the close connection between the EU's internal climate change goals and the international negotiations has always featured prominently on the Commission's agenda.[77] Interestingly, the Commission not only sees an ambitious climate change policy as a political goal, but increasingly presents it as a competitiveness issue, directly involving the industry involved.[78] This pivotal position for the Commission is even clearer in relation to the EU's external trade policy.

10.4.1 EU Climate Law and International Trade Law

We see this in the myriad cases concerning the EU common commercial policy. This policy is set out as an exclusive competence for the EU, which rules out Member State involvement unless this is envisaged in the EU acts in place.[79] The Commission's central position has meant

---

[76]   Case C-594/18 P *Austria v Commission* ECLI:EU:C:2020:742, para. 100.

[77]   K. Kulovesi, 'Climate Change in EU External Relations: Please Follow My Example (Or I Might Force You To)', in E. Morgera (ed), *The External Environmental Policy of the European Union*, Cambridge University Press 2012, at pp.133–8; M.A. Schreurs and Y. Tiberghien, 'Multi-Level Reinforcement: Explaining European Union Leadership in Climate Change Mitigation', 2007 *Global Environmental Politics* 7/4, at p 22.

[78]   Commission, 'Policy Framework for climate and energy policy in the period from 2020 to 2030', COM (2014) 15, pp.17, 18, where the EU presents itself as a 'global leader for low carbon technologies' that should seek 'to maintain its first mover advantage'. See further and more recently, the Commission's communication 'A Clean Planet for All', COM(2018) 773 final, pp. 3, 4.

[79]   Joined cases C-70/94 and C-83/94 *Criminal Proceedings against Leifer and Werner* ECLI:EU:C:1995:328.

that it has sought to protect its mandate to negotiate exclusively on behalf of the European Union in relation to, for example, energy efficiency labelling of office equipment in accordance with the US Energy Star scheme.[80]

It has also put the Union institutions in the driving seat in terms of the possible illegality of the EU's external climate change policies in the light of international trade law. As regards the concept of international trade law, this certainly encompasses the WTO/GATT-*acquis*.[81] For the purposes of this chapter, however, it is also assumed to involve the rules on the provision of international transport services. It is this body of law that has created most tensions with EU climate law, resulting in a judgment by the Court of Justice of the European Union.

The most prominent example of such tensions arises from the inclusion of international aviation in the EU ETS.[82] This inclusion has been controversial from the outset as a potentially extraterritorial measure that would be at odds with international aviation law. Viewed from the perspective of non-EU states, the EU appears to be using its market power as an economic powerhouse and air travel destination to force compliance with the EU climate rules. From the EU's perspective, the inclusion of aviation in the EU ETS is second-best to adopting an international multilateral agreement on this matter.[83] The illegal extraterritoriality is what essentially underlies the *Air Transport Association of America* case.[84] Ultimately, the Court found the application of the EU ETS to international air traffic departing from and landing at European airports compatible with international law, without ruling explicitly on the extraterritoriality of the regulation at hand.[85]

Whatever may be made of the claim of extraterritoriality or the qualification of the EU's approach as multilateral or unilateral, in terms of multi-level governance the EU's inclusion of aviation has been – for the moment at least – successful in putting climate change firmly on the international political agenda. The practical implementation of this is that the International Civil Aviation Organization (ICAO) has committed to negotiations with a view to coming to a global market-based mechanism that should be finalised in 2016 and enter into force in 2020. The contrast between extraterritoriality and the result of the EU's actions is interesting:

---

[80]   Case C-281/01 *Commission v Council* (Energy Star) ECLI:EU:C:2002:761.

[81]   To this date no case has arisen in this area. For an overview of possible legal issues see: H. van Asselt and F. Biermann, 'European Emissions Trading and the International Competitiveness of Energy-Intensive Industries: A Legal and Political Evaluation of Possible Supporting Measures', 2007 *Energy Policy* 35, pp.497–506.

[82]   This is the result of Directive 2008/101, OJ 2009 L 8/3.

[83]   Directive 2008/101, notably recitals 8–10 of the preamble and Regulation 2017/2392, OJ 2017 L 350/7, recitals 5–8 of the preamble.

[84]   Case C-366/10 *Air Transport Association of America and Others v Secretary of State for Energy and Climate Change* ECLI:EU:C:2011:864.

[85]   Ibid, para. 129. For an analysis see J. Scott and L. Rajamani, 'EU Climate Change Unilateralism', 2012 *European Journal of International Law* 23/2, pp.469–94. Similar debates can be expected in the light of the contemplated inclusion of international maritime shipping in the EU ETS: see COM (2013) 480, holding a proposal for the monitoring, reporting and verification of emissions from shipping Regulation that is seen as the first of three steps leading to the inclusion of shipping in the EU ETS.

extraterritoriality is traditionally seen as a threat to the international legal order, whereas the EU's actions have been framed in the light of the absence of an international legal order and geared towards the creation of such a legal order. This is evident from both the 'Stop the Clock' Decision and the Amending Regulation. In the 'Stop the Clock' Decision the European Union temporarily suspended the application of the EU ETS in order to facilitate the progress made at ICAO level.[86] On a similar note, the Amending Regulation that amends the ETS Directive so that – in a nutshell – international flights are assumed to comply with the obligations under the EU ETS is set firmly within an agenda that intends to reward and encourage international progress at ICAO level.[87] Seen from this perspective, the greening of the international trade *acquis* can be attributed in part to the EU.

10.4.2 EU Climate Law and International Climate Law

A similar phenomenon occurs, though with more limited effects, in relation to international climate law. There, the essential drivers within the EU are industry and Member States in a combined effort in order to protect the competitiveness of the European industry. It is well known that such competitiveness concerns have triggered the inclusion of carbon leakage provisions. In this regard it is worthy of mention that the carbon leakage mechanisms envisage a central role for the European Commission. Whether in relation to the adoption of the carbon leakage list or the approval of state aid to offset indirect carbon leakage risks, EU climate law puts the Commission in the driving seat.[88]

This centrality of the Commission can also be seen more generally in the Commission's position internally, that is, in relation to the negotiations leading to the 2009 and later amendment of the EU ETS, as well as externally in representing the EU.[89] To a certain extent this more central role for the Commission is a logical consequence of the changed constitutional arrangements after the entry into force of the Lisbon Treaty.[90] It is also, however, inextricably

---

[86] Decision 377/2013, OJ 2013 L 113/1, notably recitals 5 and 6 of the preamble.

[87] Regulation 421/2014, OJ 2014 L 129/1, notably recitals 2 and 3 of the preamble. This Regulation introduces Article 28a in the ETS Directive. This can also be seen in Regulation 2017/2392, OJ 2017 L 350/7, notably recitals 8–11, and Article 28b that is inserted into the ETS Directive, in particular paragraph 1 insofar as this requires the Commission to take into account the implementation of an ICAO market-based scheme in third countries. See further Article 25a of the ETS Directive.

[88] See Chapter 9.

[89] For an analysis see H. Vedder, 'The Formalities and Substance of EU External Environmental Law: Stuck between Climate Change and Competitiveness' in E. Morgera (ed), *The External Environmental Policy of the European Union*, Cambridge University Press 2012, pp.22–4.

[90] Articles 17(1) and 27(2) TEU for the first time explicitly mention this task for the Commission and the High Representative. It may be noted that the EU's external policy envisages a major role for the rule of law (Article 21(1) TEU) which entails a policy preference for multilateral law based on EU legalism, as opposed to other states that do not postulate legalism quite so much, see K. Kulovesi, 'Climate Change in EU External Relations: Please Follow My Example (Or I Might Force You To)', in E. Morgera (ed), *The External Environmental Policy of the European Union*, Cambridge University Press 2012, at p. 116.

linked to the substance of the matter: the centrality of competitiveness concerns. Apart from the central position of the Commission, the explicit connection between the EU's internal climate change policies and the international climate change mitigation effort must be noted here.

We see this in the ETS Directive's inclusion of what can be called sticks and carrots for international multilateral cooperation. The carrot exists in the form of the EU's pledge to reduce greenhouse gas emissions by 20 per cent and top that up (by an additional 10 per cent) to 30 per cent when there is an international accord on this matter.[91] The clearest stick for both developing and developed countries can be found in Article 11a(7) of the EU ETS Directive. This holds that, following the conclusion of an international agreement, only credits for project activities from countries that have ratified that agreement can be taken into account for the purpose of the EU ETS. In view of the EU's position as the world's biggest carbon market, this constitutes a significant incentive and integrates the international layer into the multi-level governance scheme laid down in the ETS Directive. The same holds true for Article 28 of the ETS Directive.[92] This provision requires the Commission to forward a report to the Council and European Parliament on the possibly more ambitious EU greenhouse gas abatement target. This report should essentially contain an appraisal of the international agreement as well as an assessment of the impact on national and EU policies and competitiveness.

As regards international governance as well as governance in third countries, the RES Directive also acknowledges a multi-level governance scheme. In particular where such renewables involve raw materials coming from primary forests, nature reserves, highly biodiverse grassland and land with high carbon stock, the EU keenly eyes the way in which these raw materials are grown and harvested in in third countries as well as the way in which this regulated in those countries.[93] Similarly, raw materials grown on peatlands will only be taken into account if there is evidence that their production does not involve drainage of land that was undrained before January 2008.[94] As a result, the biofuels resulting from these raw materials will not count towards the renewables targets of the Members State and will not command a price premium. The 2018 RES Directive continues along these lines and includes an incentive for third countries that seek to export biomass to the EU to be party to the Paris Agreement.[95]

In addition to incentivising biomass production through the Member States, the RES Directive envisages reporting by the EU itself of compliance by Member States,[96] as well as

---

[91] See e.g., Article 30 of the ETS directive.

[92] Another carrot can be found in Article 10(3)(c) ETS Directive, according to which the proceeds of the allowance auctions may be used to fund forestry projects, but only in third countries that have ratified the international agreement.

[93] Article 17 (3)(a), (b), (c) and (4) RES Directive.

[94] Article 17(5) RES Directive.

[95] Article 29(7) of Directive 2018/2001 OJ 2018 L 328/82.

[96] This applies to third countries as well as Member States. Concerning the latter this has the awkward consequence that the Commission will have to report on the application of the CITES Regulation in order to control

third countries with certain international conventions[97] and development standards.[98] This is a weaker instrument that appears to rely primarily on political declarations. It requires the Commission to report to the European Parliament and Council on the impact of the Union's biofuel policy upon food availability and affordability in developing countries. Insufficient environmental governance in third countries may thus also trigger the applicability of this reporting procedure. In view of the requirements imposed on the EU by the GATT/WTO rules, these corrective measures would most probably only involve political action or a further extension or reformulation of the list of items taken into account for the incentivisation effect.

All in all, EU climate law clearly has an international governance element to it. This international governance element is, however, less successful. For one, the Paris Agreement, though hailed as the first legally binding and universal climate agreement, still does not include several significant greenhouse gas emitters. In addition, the EU has been less than effective in setting the agenda for the recent round of international climate change negotiations. Despite these setbacks, the EU still sets much of its energy and climate agenda in accordance with international negotiations.[99]

10.5 CONCLUSION

Cost-effectiveness and solidarity between the Member States are major driving forces in shaping the multi-level framework for EU climate governance. In particular, cost-effectiveness, using market mechanisms to bring about greenhouse gas abatement at the lowest possible costs, can be seen as an important driver. The incentive schemes trigger a reduced mutual interdependence or solidarity between the Member States, which in turn reduces the role of the EU in multi-level governance. As soon as public funding is involved (such as subsidies to reduce the costs of renewable energy), the Member States take out the cross-border element and thus remove the EU from the governance framework that drives decisions on how this funding is distributed. We see this tension when reviewing the process resulting in the 2030 Climate and Energy Framework, where the European Council conclusions are succinct and gnomic on the governance of this framework, whereas the Commission envisaged a true review process for the national policies involved.

---

whether or not the Convention on International Trade in Endangered Species of Wild Fauna and Flora has been implemented: cf. Article 17(7), tenth indent, RES Directive.

[97]   The list of ILO Conventions and environmental treaties (the Cartagena Protocol on Biosafety and the Convention on International Trade in Endangered Species of Wild Fauna and Flora) appears to be exhaustive. It is not clear why other international treaties, such as the Ramsar Convention (on the protection of wetlands), were not included.

[98]   Article 17(7) Renewables Directive.

[99]   See European Council Conclusions 20/21 March 2014, p.7 and, more recently, Commission Communication: The Road from Paris: assessing the implications of the Paris Agreement and accompanying the proposal for a Council decision on the signing, on behalf of the European Union, of the Paris agreement adopted under the United Nations Framework Convention on Climate Change, COM(2016) 110 final.

The market mechanism that drives (market) integration in general in the EU also underlies the creation of an Energy Union. In this Energy Union ever more market integration is also seen as creating more solidarity between the Member States, ultimately resulting in opening up incentive schemes to companies across the EU instead of limiting the efficiencies that markets can bring to the national territory. This potential for cost-effectiveness has already resulted in (the evolution of) the EU ETS. Interestingly, it is exactly this involvement of the industry level, as well as the need to take into account the international level, that reinforces the role of the EU in the multi-level governance scheme.

The need for efficiency in EU climate law is expected to result in more solidarity between the Member States. This is also why, notably, the Commission presents an ambitious climate policy as an instrument to maintain or even increase competitiveness. This will increase the importance of the EU as the locus for decision-making that not only concerns myriad companies within the EU, but also reflects the fact that the EU is far from the only emitter of greenhouse gases. Ultimately, climate change is a truly global problem that can only be tackled collectively, which needs to go hand in hand with solidarity.

CLASSROOM QUESTIONS

1. Why is EU climate governance so complicated? Explain in your own words.
2. In which areas of EU climate law does governance take place mostly at Member State level and in which areas does this occur more at EU level?
3. Design and defend an easier scheme for climate governance in the EU.

SUGGESTED READING

Books

Bendlin L, *Orchestrating Local Climate Policy in the European Union (Springer Nature 2020)*.
Winter G, *Multilevel Governance of Global Environmental Change: Perspectives from Science, Sociology and the Law (Cambridge University Press 2006)*.
van Zeben J, *The Allocation of Regulatory Competence in the EU Emissions Trading Scheme (Cambridge University Press 2014)*.

Articles and chapters

Kulovesi K, *'Climate Change in EU External Relations: Please Follow My Example (Or I Might Force You To)' in E Morgera (ed), The External Environmental Policy of the European Union (Cambridge University Press 2012)*.
Roeben V, *Towards a European Energy Union: European Energy Strategy in International Law, chapters 3 and 4 (Cambridge University Press 2018)*.
Scott J and Rajamani L, *'EU Climate Change Unilateralism' (2012) 23(2) European Journal of International Law 469–94*.

Policy documents

No specific policy documents available (to our knowledge).

11
Human rights and EU climate law

ABSTRACT

- Climate change will negatively affect nearly *all* human rights, including the right to life, the right to health and the right to a healthy environment;
- Climate change affects the rights of people in vulnerable situations more severely and rapidly, like those in low-lying or arid areas, the poor, the elderly or children;
- All EU Member States are parties to binding treaties for the protection of human rights, including the EU Charter on Fundamental Rights (CFREU), the European Convention on Human Rights or treaties in UN human rights law;
- EU Member States and EU institutions must take into account their existing human rights obligations when designing and implementing climate policies, which includes particularly the CFREU, but also the wider international and European legal framework;
- Despite widespread recognition that human rights apply to climate policy-making, EU law-makers still often fail to adequately assess how new EU climate laws respect human rights, also in their wider international law context;
- Failures to adequately account for human rights may lead to rights-based climate litigation. Such litigation has been on the rise across Europe, with mixed results so far:
  - Litigation in the Dutch case *Urgenda v the Netherlands* successfully forced the Netherlands to reduce GHG emissions by at least 25 per cent by 2020 for the protection of the right to life and private life in Articles 2 and 8 of the European Convention on Human Rights, offering a unique interpretation of these rights;
  - In other jurisdictions, litigation meets procedural hurdles, especially 'standing' requirements that only grant access to courts when people can prove that climate action harms them in a way that others are not;
  - The first two EU rights-based climate cases, *Carvalho and Others* and *Sabo and others* at the CJEU, were both rejected based on such procedural grounds.

11.1 INTRODUCTION

The attention paid to human rights law's applicability to climate action is increasing globally. This is not surprising considering the wide-ranging and severe consequences predicted to result from unabated global warming. International organisations such as the Intergovernmental Panel on Climate Change (IPCC), the World Health Organization (WHO) and the European

Commission have for some time acknowledged the impact of climate change on human health and human rights.[1]

In 2015, the Paris Agreement affirmed that all its Parties should 'respect, promote and consider their respective obligations on human rights when taking climate action'. This statement did not create new human rights obligations, but serves as an important reminder that all Parties to the Agreement must comply with their respective existing human rights obligations when designing and implementing climate change policies.[2] The Paris Agreement also emphasized that especially relevant human rights would include 'the right to health, the rights of indigenous peoples, local communities, migrants, children, persons with disabilities and people in vulnerable situations, and the right to development, as well as gender equality, empowerment of women and intergenerational equity'.[3]

In this chapter we will explain how international and European human rights law are relevant to EU climate policy-making and implementation. We also show how, in the case of failure to protect human rights, individuals and non-governmental organizations may try to use human rights to demand better climate policies in the courts. Section 11.2 starts by outlining the relatively advanced recognition of rights and obligations in relation to climate change in UN human rights law. Section 11.3 turns to European human rights law, with special attention paid to the Charter of Fundamental Rights of the European Union (CFREU) and the European Convention on Human Rights and Fundamental Freedoms (ECHR). Section 11.4 investigates how human rights are already playing a role in EU climate law-making. It evaluates whether EU climate law pays sufficient attention to fundamental rights protection. Section 11.5 discusses some examples of emerging rights-based climate litigation in Europe, including especially the *Carvalho* and *Sabo* cases at the Court of Justice of the European Union (CJEU) and the *Urgenda* case in the Dutch legal system. Both challenged the adequacy of GHG emissions targets based on human rights. Section 11.6 concludes.

---

[1]   See e.g. IPCC, 'Climate Change: Impacts, Adaptation and Vulnerability – Chapter 11 on Human Health, Well-Being and Security' (IPCC Assessment Report 5, 2014) 709–54 via: www.ipcc.ch/pdf/assessment-report/ ar5/wg2/WGIIAR5-Chap11_FINAL.pdf last accessed 13 January 2021; WHO, 'Climate Change and Health' via; www.who.int/mediacentre/factsheets/fs266/en/ last accessed 13 January 2021; European Union, 'Seventh National Communication & Third Biennial Report from the European Union under the UN Framework Convention on Climate Change (UNFCCC): United for Climate' (December 2017) 111–12.

[2]   The reference was included in the non-operative 'preambular' paragraphs of the Paris Agreement, and the statement largely refers to States' existing obligations.

[3]   Preambles of Paris Agreement.

11.2 INTERNATIONAL HUMAN RIGHTS LAW AND CLIMATE CHANGE

The protection of international human rights has its origins in the adoption of the UN Charter in 1945.[4] The UN Charter established the United Nations (UN) organizations after the Second World War, aiming to promote international peace and security, global cooperation and respect for human rights.[5] The UN Charter refers explicitly to human rights in Articles 1, 55 and 56, but only in a general sense. Therefore, the UN General Assembly adopted the Universal Declaration of Human Rights (UDHR) in 1948, as the first ever detailed global statement on the human rights of every human being.[6] Despite its non-binding status, the UDHR is still a focal point for human rights protection within the UN today, and in international law generally.

After 1948, several binding international human rights law treaties were also adopted. There are currently nine of such treaties in force, including as particularly relevant to climate change:

- the International Covenant on Civil and Political Rights (ICCPR), adopted in 1966, entry into force 23 March 1976 (with 173 States parties);
- the International Covenant on Economic, Social and Cultural Rights (ICESCR), adopted in 1966, entry into force 3 January 1976 (171 States parties);
- the Convention on the Elimination of all Forms of Racial Discrimination (CERD), adopted in 1966, entry into force 12 March 1969 (182 States parties);
- the Convention on the Elimination of All Forms of Discrimination Against Women (CEDAW), adopted in 1979, entry into force 3 September 1981 (189 States parties);
- the Convention on the Rights of the Child (CRC), adopted in 1989, entry into force 2 September 1990 (196 States parties);
- the Convention on the Rights of Persons with Disabilities (CRPD), adopted in 2006, entry into force 3 May 2008 (182 States Parties).

As can be seen, most of these treaties enjoy near universal ratification. All EU Member States are parties to them, and the EU is a party to the CRPD directly.[7]

In terms of supervision, the UN Human Rights Council (UN HRC) is the most important body for monitoring human rights under the UN Charter and UDHR.[8] It is the main political organ for human rights within the UN, consisting of a rotating membership of 47 UN Member States. It can review all UN Member States' human rights records, adopt non-binding resolu-

---

[4]   See for further background: Surya P. Subedi, *The Effectiveness of the UN Human Rights System: Reform and the Judicialisation of Human Rights* (Routledge 2018).

[5]   See articles 1–2 of the UN Charter, adopted in San Francisco on 26 June 1945, entry into force 24 October 1945, via: www.un.org/en/sections/un-charter/un-charter-full-text/index.html last accessed 13 January 2021.

[6]   UN General Assembly Resolution 271 A (III) International Bill of Rights: Universal Declaration of Human Rights (10 December 1948) UN Doc. A/RES/217(III). (UDHR).

[7]   The texts and ratification status of each treaty can be consulted through: indicators.ohchr.org.

[8]   UN General Assembly Resolution 60/251, 'Human Rights Council' (15 March 2006) UN Doc. A/RES/60/251.

tions on rights violations and solicit studies from other UN bodies to help clarify States' human rights obligations on contemporary challenges.[9] The UN HRC is supported it in its work by the UN Office of the High Commissioner of Human Rights (OHCHR), and by its 'UN Special Procedures'. The latter are independently appointed experts that receive a specific thematic or country supervision mandate.[10] Although there is currently no Special Procedure on 'Human Rights and Climate Change', several of the current Special Procedures have worked on climate change in the past years, including in particular the UN Special Procedure on Human Rights and the Environment (see, for an overview of their work, Table 11.1).[11]

The individual human rights treaties are each supervised by their own 'treaty committees'. These committees, typically consisting of 18–23 independent experts, *inter alia*, are responsible for drafting so-called General Comments that interpret the meaning of specific treaty provisions, reviewing periodic implementation reports by States and deciding on individual complaints brought by civil society against States parties.[12]

Due to the multitude of international human rights law instruments and related supervisory bodies, it is difficult to give a succinct overview of *all* their contributions to the clarification of States' human rights obligations related to climate change made over the years.[13] The following sections will therefore chiefly highlight the most important past and recent developments, whereas Table 11.1 provides the reader with an overview of where to find further guidance on the protection of specific rights. It is important to briefly note upfront that even though none of these legal opinions are formally legally binding, they are typically viewed as offering valuable legal guidance on States' binding obligations under relevant legal instruments. As explained in Section 11.4.1, EU law-makers are expected to take relevant legal opinions of UN bodies into account when designing EU climate policies.[14]

---

[9]   Subedi (n. 4) 100–39.

[10]   See e.g. Subedi (n. 4) 140–95, or OHCHR, *Manual of Operations of the Special Procedures of the Human Rights Council* (adopted by the 15th Annual Meeting of the Special Procedures in June 2008) via: www.ohchr.org/Documents/HRBodies/SP/Manual_Operations2008.pdf last accessed 13 January 2021.

[11]   See overview of work and mandates of the Special Procedures: www.ohchr.org/EN/HRBodies/SP/Pages/Welcomepage.aspx last accessed 13 January 2021.

[12]   See e.g. Subedi (n. 4) 71–99; Helen Keller, Geir Ulfstein and Leena Grover (eds), *UN Human Rights Treaty Bodies: Law and Legitimacy* (CUP 2012).

[13]   See e.g. Margaretha Wewerinke-Singh, *State Responsibility, Climate Change and Human Rights under International Law* (Hart Publishing 2018); Erika Lennon et al., 'Rights in a Changing Climate: Human Rights under the UN Framework Convention on Climate Change' (CIEL, December 2019) via: www.ciel.org/wp-content/uploads/2019/12/Rights-in-a-Changing-Climate_SinglePage.pdf last accessed 13 January 2021.

[14]   For discussion e.g: Michael Addo, *The Legal Nature of International Human Rights* (Brill Nijhoff 2010); Subedi (n. 4) 81, 90, 100, 222, 225, 237; Keller, Ulfstein and Grover (n. 12).

11.2.1 Development of Climate Change as a Human Rights Concern

International attention to the topic of human rights and climate change commenced with the adoption of HRC Resolution 10/4 on 'Human Rights and Climate Change' in 2009.[15] HRC Resolution 10/4 was significant because it was adopted by the main inter-governmental human rights body of the UN. It concluded for the first time in bold terms that global warming leads to *both direct and indirect human rights impacts*. Moreover, climate change was expected to ultimately affect all individuals and communities, but certain segments of society more acutely. This is especially the case for those living in already vulnerable conditions, owing to factors of geography, poverty, gender, young age, old age, indigenous status, minority status or disability. The resolution further underscored that the human rights most likely to be affected include the right to life, the right to adequate food, the right to health, the right to adequate housing, the right to self-determination and the right to water and sanitation.[16] HRC Resolution 10/4 was cited in the UNFCCC's Cancún Agreements a year later, in support of the assertion that all UNFCCC Parties 'should, in all climate change related actions, fully respect human rights'.[17]

After 2009, States' human rights obligations under UN human rights law rapidly gained more attention in the work of other UN monitoring bodies too, and an overview of their work on different issues and rights, or the rights of children, women, migrants, the elderly and persons with disabilities, is found in Table 11.1.

11.2.2 The Right to Life

The right to life is arguably one of the most fundamental human rights (ultimately) at threat from climate change. As we see later on, it is also often invoked in European climate litigation. In international human rights law, the right to life is recognized in Article 6 of the ICCPR, and in various provisions of the CRC, CEDAW and CRPD.

In 2018, the UN Human Rights Committee, the treaty committee responsible for supervising the implementation of the ICCPR, issued its long-awaited 'General Comment No. 36' on the right to life, thereby replacing an outdated 'General Comment' on Article 6 ICCPR dating back to 1984.[18] General Comment No. 36 affirmed that climate change is by now 'one of the most pressing and serious threats' to present and future generations' ability to enjoy their

---

[15] UN Human Rights Council Resolution 10/4. 'Human Rights and Climate Change' (25 March 2009) UN Doc. A/RES/10/4; HRC, 'Report of the Office of the High Commissioner on Human Rights on the Relationship between Climate Change and Human Rights' (15 January 2009) UN Doc. A/HRC/10/6.

[16] Ibid.

[17] Lennon et al. (n. 13); UNFCCC Decision 1/CP.16, 'The Cancún Agreements' (15 March 2010) UN Doc. FCCC/CP/2010/7/Add.1, preambular paragraphs, p.2, and paras 8, 86–87, Appendix I.

[18] Human Rights Committee, 'General Comment No. 36 on Article 6 ICCPR (Right to Life)' (31 October 2018) UN Doc. CCPR/C/GC/36, para. 1.

Table 11.1 Key interpretative documents on human rights and climate change in UN law

UN Human Rights Council, with support of OHCHR	Special Procedures of the UN Human Rights Council	UN Treaty Bodies
Report on the **relationship between climate change and human rights** (2009) UN Doc. A/HRC/10/6.	Report of the Special Rapporteur on the right to housing on **the right to adequate housing and climate change** (6 August 2009) UN Doc. A/64/255.	Committee on the Rights of the Child (CRC), 'General Comment No. 15 (2013) on the **right of the child to the enjoyment of the highest attainable standard of health** (art. 24)' (2013) UN Doc. CRC/C/GC/15, paras. 5, 50.
Analytical study on the relationship between climate change and the **human right of everyone to the enjoyment of the highest attainable standard of physical and mental health** (2016) UN Doc. A/HRC/32/23.	Focus report: **Mapping human rights and climate change**, prepared for the independent expert on the issue of human rights obligations relating to the enjoyment of a safe, clean, healthy, and sustainable environment' (June 2014) via: www. ohchr.org.	Committee on the Elimination of Discrimination against Women (CEDAW). 'General Recommendation No. 37 on **gender-related dimensions of disaster risk reduction in the context of climate change**' (2018) UN Doc. CEDAW/C/GC/37.
Analytical study on the relationship between climate change and the full enjoyment of the **rights of the child** (2017) UN Doc. A/HRC/35/13.	Report of the Special Rapporteur on the right to food on the **impact of climate change on the right to food** (2015) UN Doc. A/70/287.	Human Rights Committee, 'General Comment No. 36 on Article 6 ICCPR on **right to life**' (2018) UN Doc. CCPR/C/GC/36.
Report on the slow onset effects of climate change and **human rights protection for cross-border migrants** (2018) UN Doc. A/HRC/37/CRP.4.	Report of the Special Rapporteur on human rights obligations relating to the enjoyment of **a safe, clean, healthy and sustainable environment on climate change** (2016) UN Doc. A/HRC/31/52.	Human Rights Committee, *Ioane Teitiota v New Zealand*, Communication No. 2728/2016 (Views adopted on 4 October 2019) UN Doc. CCPR/C/127/D/2728/2016.
Analytical study on gender-responsive climate action for the full and effective enjoyment of **the rights of women** (2019) UN Doc. A/HRC/41/26.	Report of the Special Rapporteur on **the rights of indigenous peoples on the impacts of climate change and climate finance** (2017) UN Doc. A/HRC/36/46.	
Analytical study on the promotion and protection of **the rights of persons with disabilities** in the context of climate change (2020) UN Doc. A/HRC/44/30.	Report of the Special Rapporteur on human rights obligations relating to the enjoyment of **a safe, clean, healthy and sustainable environment on a safe climate** (2019) UN Doc. A/74/161.	
Analytical study on **the rights of older persons** in the context of climate change (2021) as requested by A/HRC/RES/44/7.	Report of the Special Rapporteur on extreme poverty and human rights on **climate change and poverty** (2019) UN Doc. A/HRC/41/39.	

rights to life.[19] States have duties to protect people's lives against any harmful general conditions in society that could give rise to direct threats to life, but also against conditions that may 'prevent individuals from enjoying their right to life with dignity'.[20] Such harmful conditions expressly include environmental threats, like climate change. Because people's ability to enjoy their rights under Article 6 ICCPR depends 'on the measures that States implement to preserve the environment and protect it against harm', including climate change as caused by public and private actors, States parties to the ICCPR are obliged to: (a) develop and implement relevant substantive environmental standards; (b) ensure the sustainable use of natural resources; (c) conduct environmental impact assessments; and (d) pay due regard to the precautionary approach.[21]

The right to life was recently also invoked in the first climate cases at UN treaty committees. One pending complaint was lodged by Greta Thunberg and 15 other youth climate activists against several Parties to the CRC, another application was filed by inhabitants of the Australian Torres Strait Islands against Australia under the ICCPR.[22] Each of these cases argue that States are not acting sufficiently ambitiously to protect them from dangerous climate change. A third case, *Teitiota v New Zealand*, on the controversial topic of cross-border human migration caused by climate change, was recently decided by the Human Rights Committee.[23]

In a landmark decision, this committee reaffirmed that climate change may indeed lead to life-threatening risks for, for example, the nationals of small island developing states, such as Kiribati national Teitiota. Relevant threats to life could stem from sudden-onset events associated with climate change (intense storms or flooding) or slow-onset events (sea-level rise, salinization or land degradation). Moreover, in this context the right to life might (ultimately) also give rise to cross-border protection through migration, meaning that a *third State*, such as New Zealand, could be compelled to protect a *foreign national*, such as Teitiota, against threats to life they may experience due to climate change in their home country.[24]

Teitiota's claims were dismissed, however, because the Committee found no evidence of 'clear arbitrariness, error or injustice' in New Zealand's evaluation of any existing 'real, personal and foreseeable risks' to Teitiota's life, at that point in time, should he be sent back to Kiribati. For example, Teitiota could not prove he might face violent disputes over scarce land, inability to grow food, a lack of access to potable water or otherwise life-threatening con-

---

[19] Ibid. paras. 26, 62.

[20] Ibid. para. 26.

[21] Ibid. para. 62.

[22] The petitions can be found via the Sabin Center for Climate Law's Climate Case Chart: http://climatecasechart .com/non-us-case/sacchi-et-al-v-argentina-et-al/ and http://climatecasechart.com/non-us-case/petition-of-torres -strait-islanders-to-the-united-nations-human-rights-committee-alleging-violations-stemming-from-australias -inaction-on-climate-change/ last accessed 13 January 2021.

[23] Human Rights Committee, *Ioane Teitiota v New Zealand*, Communication No. 2728/2016 (Views adopted on 4 October 2019) UN Doc. CCPR/C/127/D/2728/2016, paras 9.1–9.5.

[24] Ibid, paras 1.1–2.10, 9.11.

ditions.[25] More specifically, his situation was also not held to be sufficiently *personal* to him, as other Kiribati nationals are facing similar threats. He would not face personal persecution, nor any personal rights violations upon his return. On the contrary, the Kiribati government seemed sufficiently able and willing to take protective measures itself in the coming 10–15 years, providing for Teitiota's basic necessities of life or ensuring his relocation otherwise – if necessary with the help of the international community.[26] Therefore New Zealand and the Committee argued that Teitiota's home State would (for now) be able to sufficiently protect Teitiota's right to life with dignity within Kiribati itself.

Due to the unique aspect in this case of cross-border or 'extra-territorial' human rights protection against the effects of climate change, it will be of great interest to see how the Human Rights Committee and CRC Committee will deal with the other two pending cases, which deal more directly with protection of a State's own nationals against climate change and obligations to reduce GHG emissions faster. In any case, the *Teitiota* decision has attracted some critique for having set an 'unreasonable burden of proof' for proving risks and dangers to the right to life (with dignity). The irreversible and severe nature of the threat of climate change, and the short-term possibilities of loss of life, should have been given greater weight and more serious attention.[27]

11.2.3 The Right to a Healthy Environment and a Stable Climate

Since many rights have been linked to climate change, there have been increasingly strong calls for the recognition and protection of peoples' rights to a healthy environment or safe climate.[28] In particular, the UN Special Procedure on Human Rights and the Environment recently made several important contributions to the clarification of States' human rights obligations on climate change.[29] In doing so, it built on and synthesized the work of other UN bodies, and produced some mapping reports, as well as supported the recognition of a new right to a healthy environment, including a stable climate.[30]

The Special Procedure's synthesis report of 2019 particularly clarified a range of obligations for States, including the fact that States '*must not violate the right to a safe climate* through their own actions; must protect that right from being violated by third parties, especially businesses;

---

[25]   Ibid. paras 9.6–9.7.

[26]   Ibid. paras 9.6, 9.13–9.14.

[27]   Ibid. Annex I, for dissenting opinion of Committee Member Duncan Laki Muhumuza; see also Jane McAdam, 'Protecting People Displaced by the Impacts of Climate Change: The UN Human Rights Committee and the Principle of Non-refoulement' (2020) 114(4) *American Journal of International Law* 708–25.

[28]   Ademola Oluborode Jegede, 'Arguing the Right to a Safe Climate under the UN Human Rights System' (2020) 9 *International Human Rights Law Review* 184–212.

[29]   HRC, Report of the Special Rapporteur on human rights obligations relating to the enjoyment of a safe, clean, healthy and sustainable environment on a safe climate (15 July 2019) UN Doc. A/74/161, paras 43–44, 65, 96.

[30]   See these reports listed in Table 11.1.

and must establish, implement and enforce laws, policies and programmes to fulfil that right. States also must avoid discrimination and retrogressive measures.'[31]

The report further considered that international environmental law reinforces States' human rights obligations and vice versa. The 'no-harm rule' in customary international environmental law, for example, implies that States must not contribute, cumulatively, to the creation of foreseeable serious harmful effects to the environment and territories of other States, including any harmful effects to human rights.[32] More specifically, States' failures to fulfil international climate commitments under the UNFCCC and Paris Agreement could essentially be viewed as a '*prima facie violation*' of States' obligations to fulfil human rights, due to the severely catastrophic effects that are expected to result from such failures.[33]

The report drew inspiration in this respect from a set of recommendations published a year earlier by the treaty committee for the ICESCR, in response to the IPCC's Special Report on Global Warming of 1.5°C.[34] This committee held that all ICESCR States parties owe human rights duties 'to prevent foreseeable human rights harm caused by climate change'.[35] In discharging such duties, States must mobilize their 'maximum available resources', and act in accordance with the best available scientific evidence. Additionally, when designing and implementing voluntary 'nationally determined contributions' (NDCs) under Article 4(2) of the Paris Agreement, States must take guidance from human rights protected by the ICESCR. These rights include the rights to food, water, health and adequate housing. The ICESCR committee further noted with concern that the currently communicated NDCs so far are 'insufficient to meet what scientists tell us is required to avoid the most severe impacts of climate change'. If States 'want to act consistently with their human rights obligations' they should revise their NDCs communicated under the PA to better reflect their 'highest possible ambition', as referred to in Article 4(3) of the PA.[36]

Finally, aside from these more general obligations on appropriate GHG emissions reduction target setting, the 2019 synthesis report included several highly specific recommendations. These included, among others, the immediate termination of fossil fuel subsidies; the prohibition of further exploration of fossil fuel reserves (because using up all currently tapped reserves will already result in breaching available global carbon budgets); and rejecting or prohibiting the expansion of new fossil fuel infrastructures, especially the most destructive ones, such

---

[31]   HRC (n. 29) paras 64–65.

[32]   Ibid. para. 66. The no-harm rule typically prohibits 'causing serious harm to the environment or the territories or peoples of other States, or to areas beyond the limits of national jurisdiction'.

[33]   Ibid. paras 70, 74–75.

[34]   IPCC, 'Special Report on the Impacts of Global Warming of 1.5°C' (IPCC/WMO: Geneva, 2018) via: www.ipcc .ch/sr15/.

[35]   OHCHR, 'Statement of the Committee on Economic, Social and Cultural Rights: Climate Change and the International Covenant on Economic, Social and Cultural Rights (8 October 2018) paras 3, 5–6.

[36]   Ibid. para. 6.

as fracking.[37] This would suggest that States' discretion to pursue certain climate policies is increasingly curtailed.

All in all, UN human rights bodies thus play a considerable role in affirming the applicability of human rights to climate action undertaken under the UNFCCC and the Paris Agreement, as well as in providing interpretations of States' obligations under specific rights and instruments. The following sections show how international rights are to a large extent mirrored in European law, and will reiterate that European law-makers are expected to take note of relevant UN interpretations when interpreting similar EU rights.

11.3 EUROPEAN HUMAN RIGHTS LAW AND CLIMATE CHANGE

In parallel to the establishment of the United Nations as a global initiative, several European initiatives were also undertaken after the Second World War, to promote democracy, rule of law, human rights and permanent peace on the European continent. The following sections explain the development of European human rights law within the Council of Europe and within the European Union, and their respective relevance to climate action.

11.3.1 Council of Europe

One of the first initiatives to promote peace, prosperity and European cooperation, as well as rule of law, democracy and human rights, was the establishment of the Council of Europe in 1949, and this organization's immediate adoption in 1950 of the European Convention on Human Rights and Fundamental Freedoms (ECHR), which entered into force in 1953.[38]

Today, the Council of Europe consists of 47 Member States, including all 26 EU Member States. All Council of Europe members must ratify the ECHR when they join the organization. They do not have to ratify all later Protocols to the ECHR, nor the Council of Europe's other main human rights treaty: the European Social Charter (ESC).[39] The latter was adopted to complement the ECHR for the first time in 1961, and revised in 1995. Both versions of the ESC are in force, and all EU Members are party to either the original version of 1961 or the European Social Charter (Revised) of 1995, which entered separately into force in 1999.[40]

---

[37] HRC (n. 29) paras 77–78.

[38] Statute of the Council of Europe (adopted in London on 5 May 1949, entry into force 3 August 1949) ETS No. 001; Convention for the Protection of Human Rights and Fundamental Freedoms (adopted by Council of Europe in Rome on 4 November 1950, entry into force 3 September 1953) ETS No. 005.

[39] European Social Charter (adopted by Council of Europe in Turin on 19 October 1961, entry into force 26 February 1965) ETS No. 35; European Social Charter (Revised) (adopted by the Council of Europe in Strasbourg on 3 May 1995, entry into force 1 July 1999) ETS No. 163.

[40] See further explanations: Carole Benelhocine, *The European Social Charter* (Council of Europe Publishing 2012).

The ECHR is supervised by the European Court of Human Rights in Strasbourg, which has the competence to adopt binding decisions in complaints against States brought by individuals or non-governmental organizations claiming to be the victim of a violation of their rights under the ECHR.[41] The ECHR mostly protects civil and political rights (e.g. right to life, right to private life, right to freedom of opinion). The ESC(R) is supervised by the European Committee of Social Rights, and mostly protects social and economic rights (such as the right to health, the right to adequate housing, the right to protection of the family). It has the competence to take non-binding decisions on collective complaints against States parties, brought by non-governmental organizations on behalf of individuals.

Both the ECHR and the ESC(R) have been applied and interpreted in the context of environmental threats, if not necessarily directly in relation to climate change so far.[42] The European Court of Human Rights in particular developed a well-known and vast body of 'environmental' case-law under the rights to life and private life in Articles 2 and 8 ECHR.[43] It stated that even if the ECHR lacks an explicit right to a healthy environment, Articles 2 or 8 ECHR can still be violated when environmental pollution 'significantly impairs' people's life, health or quality of life. In such cases, the pollution must reach a minimum level of severity and be assessed in light of its context, duration and intensity, impacts on mental and physical health, and any relevant personal factors such as age, profession or personal lifestyle.[44]

Significantly, Articles 2 and 8 ECHR also apply to the mitigation of 'natural' hazard risk, such as floods, mud slides or earthquakes.[45] In such cases, States' positive obligations for the protection of life and private life will depend 'on the origin of the threat' and the extent to which the risk is 'susceptible to mitigation'. Protective obligations also apply only insofar as hazards are *clearly identifiable* and *imminent*. In practice, this means that States may have both obligations to prevent and mitigate a known existing natural hazard risk itself (e.g. reduce a flood risk), as well as to take steps to minimize the negative impacts of such hazardous risk on life or private life, in the case that the hazard itself is beyond their (reasonable) control (e.g. in the case of an earthquake).[46]

---

[41] See Articles 34–35 ECHR.

[42] Council of Europe, 'Factsheet on Environment and the European Convention on Human Rights' (version of December 2020) via: www.echr.coe.int/documents/fs_environment_eng.pdf last accessed 13 January 2021; European Committee of Social Rights, Complaint No. 30/2005, *MFHR v Greece* [Decision on Merits of 6 December 2006] paras 195–196, deciding the 'right to health' in Article 11 ESC also includes the 'right to a healthy environment'.

[43] Council of Europe (n. 42); Sanja Bogojevic and Rosemary Rayfuse (eds), *Environmental Rights in Europe and Beyond* (Hart Publishing 2018).

[44] See e.g. ECHR, *Kyrtatos v Greece* (22 May 2003) app. no. 41666/98, ECHR 2003-VI, para. 52; ECHR, *Lopez Ostra v Spain* (9 December 1994) Series A no. 303-C, para. 51; ECHR, *Dubetska v Ukraine*, app. no. 30499/03 (10 February 2011) paras 105–106; ECHR, *Fadeyeva v Russia* (9 June 2005) app. no. 55723/00, paras 68–69.

[45] See especially cases of *Budayeva, Öneryildiz, Koldyadenko, Hadzhiyska* and *Özel* listed in Council of Europe (n. 42).

[46] E.g. ECHR, *Özel and others v Turkey* (17 May 2015) app. no. 14350/05, paras 170–174.

In the context of climate change, the (violent) natural hazards predicted to result from unabated global warming can arguably largely still be prevented, by successfully implementing the Paris Agreement. Section 11.5 further discusses the ECHR and climate change in light of recent climate litigation, and in fact, the European Court of Human Rights may soon be asked to give its own opinion: the first couple of climate cases are pending at the time of writing.[47]

11.3.2 European Union

The establishment of (what is now) the European Union took place in the same period, through the establishment of the European Economic Community (EEC) by the Treaty of Rome in 1957.[48] The primary aim of the EEC was viewed as securing peace, economic and social progress, and realizing closer harmonious relations between European nations through economic integration.[49] As such, the initial founding treaties did not (yet) include any express references to fundamental rights.

Over the years, the number of Member States gradually increased, and several revisions of the Treaty of Rome integrated other policy areas, such as social cohesion, environmental protection and consumer protection. Although 'fundamental rights' were not mentioned in the early Community treaties, the European Court of Justice (ECJ) affirmed by the start of the 1970s that fundamental rights were a part of the general principles of Community law. They could be derived from the common traditions of its Members – including the ECHR and international human rights law – and form the basis of a successful challenge to a European law.[50]

In 1992, the Treaty of Maastricht reflected this jurisprudential development through a first express reference to fundamental rights in the founding treaties, and since the Lisbon Treaty of 2007, which entered into force in 2009, Article 2 TEU reaffirmed as the European 'values' upon which the EU is founded the 'respect for human dignity, freedom, democracy, equality, the rule of law and respect for human rights, including the rights of persons belonging to minorities'.[51]

---

[47] See e.g. ECHR, *Cláudia Duarte Agostinho and others v Portugal and 32 other States* (Communication of 7 September 2020) app. no. 39371/20; and the *Swiss Grannies* application of November 2020 discussed in Section 5.

[48] Treaty establishing the European Economic Community (signed by Belgium, Federal Republic of Germany, Netherlands, Italy and Luxembourg in Rome on 25 March 1957) https://eur-lex.europa.eu/legal-content/EN/TXT/?uri=CELEX:11957E/TXT.

[49] Ibid. preambles and Article 2.

[50] See e.g. Allan Rosas, 'The Charter and Universal Human Rights Instruments' in Steve Peers et al. (eds), *The EU Charter of Fundamental Rights: A Commentary* (CB Beck/Hart 2014) pp.1687–1700, referring to the early cases of *Stauder, Internationale Handelsgesellschaft*, and *Nold*.

[51] Treaty on European Union, OJ C 191/1, 29.7.1992; Consolidated version of the Treaty on European Union, OJ C 236/17, 7.8.2012; see also for further discussion Manuel Kellerbauer, Marcus Klamert and Jonathan Tomkin (eds), *The EU Treaties and the Charter of Fundamental Rights: A Commentary* (OUP 2019); Stephen Weatherill, *Law and Values in the European Union* (OUP 2016) 139–52.

Significantly, the EU also proceeded to develop its own charter of rights in the form of the Charter of Fundamental Rights of the EU (CFREU) adopted in 2000.[52] The CFREU protects a comprehensive range of rights and principles of both a civil, political, social, economic and cultural nature, and gained legally binding force through the Lisbon Treaty in 2009. The rights most relevant to climate change include the right to human dignity; the right to life; the right to protection of private life, family life and the home; the right to property; the protection of health; a high level of environmental protection and improvement of the quality of the environment; and rights to cultural diversity, the rights of children, the rights of elderly people and the rights of persons with disabilities.[53] As shown in Section 11.4, these rights have so far received limited attention in the context of EU climate law-making. Section 11.5 will show how litigants in the *Armando Carvalho* case invoked several of these rights in seeking the annulment of several legislative climate acts at the CJEU.

Finally, in support of the new CFREU, the Council established the EU Fundamental Rights Agency (FRA) through Regulation 168/2007.[54] This FRA essentially has a promotive rather than a supervisory mandate.[55] It provides information, assistance and expertise on fundamental rights, and may collect information, carry out studies, write reports, opinions and conclusions and raise public awareness of human rights. It has not addressed climate change so far. When the FRA was established, the scope of the CFREU was also re-affirmed, in that the CFREU

> bearing in mind its status and scope, and the accompanying explanations, reflects the rights as they result, in particular, from the constitutional traditions and international obligations common to the Member States, the Treaty on European Union, the Community Treaties, the European Convention for the Protection of Human Rights and Fundamental Freedoms, the social charters adopted by the Community and by the Council of Europe and the case law of the Court of Justice of the European Communities and of the European Court of Human Rights.[56]

The CFREU is now understood to be the main human rights instrument for the EU, even if its origins can still clearly be traced to the wider fundamental rights traditions of EU Member States, and international law and Council of Europe law.[57]

---

[52] Charter of Fundamental Rights of the European Union, OJ C 326/391, 26.10.2012.

[53] Articles 1, 2, 7, 17, 21–26, 35, 37 CFREU.

[54] Council Regulation (EC) 168/2007 of 15 February 2007 establishing a European Union Agency for Fundamental Rights, OJ L 53/1.

[55] See for commentaries on the FRA: Rosemary Byrne and Han Entzinger (eds) *Human Rights Law and Evidence-Based Policy: The Impact of the EU Fundamental Rights Agency* (Routledge 2018); Philip Alston and Olivier de Schutter (eds) *Monitoring Fundamental Rights in the EU: The Contribution of the Fundamental Rights Agency* (Hart 2005).

[56] Ibid. recital (2).

[57] Rosas (n. 50).

The EU human rights system is still evolving. This is visible from the Union's adoption of the (non-binding) EU Pillar on Social Rights in 2017.[58] Many of the Pillar's 20 principles on social rights focus heavily on worker' rights, but several principles refer to the rights of all. This includes the novel right of access to essential services of good quality, including access to services such as energy, water or digital communications.[59]

11.3.3 Multi-layered Nature of Human Rights Protection in the EU

According to Article 51 CFREU, the CFREU applies to the EU institutions and Member States only insofar as they are implementing EU law; thus, when they act within the scope of EU law. Nevertheless, it is important to keep in mind the wider international human rights law context for the protection of EU citizens' human rights, for several reasons. First, the CFREU itself acknowledges the 'multi-layered' nature of human rights protection within the EU, for example by stating in Article 53 that the CFREU cannot be 'interpreted as restricting or adversely affecting' any existing rights and freedoms recognized by Union law, the ECHR or international legal agreements to which the Union or all Member States are party. This includes all the international human rights treaties discussed in Section 11.2. Second, according to Article 52, rights in the CFREU are considered to be the same in 'meaning and scope' as corresponding rights protected by the ECHR.

Such interpretative principles must thus be borne in mind, for example, when assessing whether new EU climate laws comply with relevant rights under the CFREU, as will be explained in the following Section 11.4.

11.4 EFFECTUATING HUMAN RIGHTS WITHIN EU CLIMATE LAW

To improve the overall quality of EU law and policy-making, and to ensure better attention to the protection of fundamental rights in the day-to-day work of the EU, the EU institutions have been implementing a so-called Better Regulation Agenda since 2000. A concrete objective of this agenda is to ensure that *all* EU laws and policies, including climate laws and policies, comply fully with fundamental rights.[60]

---

[58] Interinstitutional Proclamation on the European Pillar of Social Rights C 428/10, 13.12.2017.

[59] Principles 19 and 20 of the EU Pillar of Social Rights.

[60] See Inter-institutional Agreement between the European Parliament, the Council of the European Union and the European Commission on Better Law-Making OJ L 123, 12.5.2016, pp.1–14; European Commission, 'Impact Assessment Guidelines' SEC(2009) 92, 5, 8, 16, 22, 31, 33, 35–6, 39; European Commission, 'Strategy for the Effective Implementation of the Charter of Fundamental Rights by the European Union' COM(2010) 573 final; Commission Staff Working Paper, 'Operational Guidance on Taking Account of Fundamental Rights in Commission Impact Assessments' SEC(2011) 567 final; Commission Staff Working Document, 'Better Regulation Guidelines' SWD(2017) 350, pp.18, 22–5, 27, 31; and the 'Better Regulation Tool #28 on Fundamental Rights and Human Rights' (2017) via: https://ec.europa.eu/info/sites/info/files/file_import/better-regulation-toolbox-28_en_0.pdf last accessed 13 January

11.4.1 The EU Better Regulation Framework as a Human Rights Tool

A key tool for 'Better Law Making' includes the Commission's so-called comprehensive Impact Assessments (IAs) that it must carry out before proposing any new policy initiatives that might have 'significant economic, environmental or social impacts'.[61] According to the EU's Inter-Institutional Agreement on Better Law-Making, IAs should cover the 'existence, scale and consequences' of any problem to be regulated by EU law, and explain 'whether or not Union action is needed' in that sphere, keeping in mind, for example, principles of subsidiarity and proportionality.[62] The IA should also map out possible alternative solutions and short-term and long-term costs and benefits of proposals, and assess economic, environmental and social impacts in an integrated and balanced way. More importantly, IAs play a significant role in assessing whether legislative proposals 'fully respect fundamental rights', as further explained below. Aside from the IAs, the Commission must also produce an 'explanatory memorandum' for new legislative initiatives. These must explain *how* proposed measures 'are compatible with fundamental rights'.[63]

Since 2009, the Commission has been developing various Impact Assessment Guidelines to guide EU policy-makers in the practice of IAs.[64] In relation to human rights, these clarify that the CFREU legally limits the EU's rights to act, and therefore IAs should *fully identify and assess qualitatively* whether proposals or initiatives raise any fundamental rights issues.[65] Limits to act may also result from prior political choices or other existing legal obligations, including those following from 'international law, or obligations related to fundamental rights'.[66] IAs should therefore explain, in light of relevant pre-existing legal obligations, why certain policy options are preferable, necessary or not feasible, and consider how different policy options might affect fundamental rights differently, both directly and indirectly. They must also pay attention to different groups in society and identify both *beneficial* effects (promote fundamental rights) and *negative* effects (limit fundamental rights) of new legislative proposals.[67]

2021. See also: Israel de Jesús Butler, 'Ensuring Compliance with the Charter of Fundamental Rights in Legislative Drafting: The Practice of the European Commission' (2012) 37 *European Law Review* 397; Olivier de Schutter, 'The Implementation of the Charter of Fundamental Rights in the EU Institutional Framework: Study for the AFCO Committee' (European Union, November 2016); Mark Dawson, *The Governance of EU Fundamental Rights* (CUP 2017) 84–97.

[61] Interinstitutional Agreement on Better Law-Making (n. 60), paras. 12-18.

[62] Ibid. para. 13.

[63] Ibid. para. 25.

[64] See n. 60 above, including especially: SEC(2009) 92, pp.13, 16, 22, 39; SWD(2017) 350, pp.18, 22–4, 27; SEC(2011) 567 and Better Regulation Tool#28.

[65] SEC(2009) 92, p. 39; SEC(2011) 567, p.17.

[66] SEC(2011) 567, p.17.

[67] SEC(2011) 567, pp.14–15, 17–22 (this document highlights differential impacts for *inter alia* persons with disabilities, children, women); SEC(2009) 92, pp.13, 16, 33, 35, 38–9.

Clearly, the above implies a very comprehensive and ambitious type of human rights assessment, involving considerable justification of why and how specific fundamental rights are (not or least) affected by a specific piece of proposed legislation. In carrying out their IAs, EU policy-makers can use the 'fundamental rights check-list', as shown in Box 11.1.

Box 11.1   Fundamental Rights Check-List in the IA Guidelines[68]

- What fundamental rights are affected?
- Are the rights in question absolute rights (which may not be subjected limitations, examples being human dignity)?
- What is the impact of the various policy options under consideration on fundamental rights? Is the impact beneficial (promotion of human rights) or negative (limitation of fundamental rights)?
- Do the options have both a beneficial and a negative impact, depending on the fundamental rights concerned?
- Would any limitation of fundamental rights be formulated in a clear and predictable manner?
- Would any limitation of fundamental rights:
  - Be necessary to achieve the objective of general interest or the protect the rights and freedoms of others (which)?
  - Be proportionate to the desired aim?
  - Preserve the essence of the fundamental rights concerned?

Finally, the IA guidelines explicitly clarify that EU policy-makers are required to consult the case-law of the CJEU and ECHR in trying to understand the scope of relevant fundamental rights, and, as appropriate, the various legal opinions and General Comments of UN human rights monitoring bodies as discussed in section 11.2.[69]

Despite these elaborate instructions to policy-makers, and the explicit wish of the Commission to foster a 'fundamental rights culture', it is questionable whether human rights really achieve the level of attention in EU law and policy-making demanded by the Better Regulation initiative. There is clear evidence that IA procedures still insufficiently address human rights, and that the Better Regulation initiative has not (yet) led to the mainstreaming of fundamental rights in EU decision-making processes.[70]

Various explanations could be offered for piecemeal approaches to fundamental rights protection in EU law-making so far. First, EU policy-makers may still lack sufficient awareness,

---

[68]   SEC(2011) 567, pp.6–7.

[69]   Better Regulation Tool #28 (2017) 212; SEC(2011) 567 final, pp.7–9, and Annex I.

[70]   Dawson (n. 60) 88–9, 92–4; Butler (n. 60) 398, 403, 407–10, 414–15, 417. SEC (2009) 92, p.39; SEC (2011) 567 final, p.10; de Schutter (n. 60) 19, 24, also noting that the opinions of the Legal Service responsible for checking final proposals are not published, and thus cannot be assessed by the public. Yet, there is evidence that also the Legal Service is unable to adequately identify relevant human rights issues or is sometimes bypassed for political reasons.

skill and familiarity to fully understand, apply and promote the broad body of European and international human rights law to a wide variety of topics. Second, at times, plain political expediency could explain why fundamental rights are only tested minimally.[71]

Below we will consider more closely whether and how human rights have featured in EU climate law-making.

11.4.2 Human Rights and EU Climate Target-setting and Regulation

As explained in Chapter 2 of this book, EU climate law currently consists of an aggregate set of policies that together enable the implementation of the GHG emissions reduction targets of the EU under the UNFCCC and Paris Agreement. These policies include, *inter alia*, the EU Emissions Trading Scheme Regulation, the Effort Sharing Decision and Regulation, the Land-Use and Land Use-Change and Forestry (LULUFC) Regulation and the Renewable Energy Directive.[72] This sub-section first considers whether and how human rights played a role in the actual design and implementation of the EU's GHG emissions reduction targets under the UNFCCC and Paris Agreement. The following sub-section takes a narrower focus, on human rights impacts that may arise from specific GHG mitigation activities that aim to achieve such targets, such as the production of renewable energy or the management of climate sinks.

As explained elsewhere in this volume, until recently the EU committed itself to a GHG reduction target of 20 per cent by 2020 compared to 1990 levels. It also committed itself to a 40 per cent reduction by 2030. Additionally, it agreed to a conditional target of 30 per cent by 2020 as part of its second Kyoto Protocol commitment period, if other countries listed in Annex I to the UNFCCC were prepared to commit to similar ambitious targets.[73] Since the adoption of the Paris Agreement, the EU has been debating the need for more ambitious targets, and by the end of 2020 ambitions were ratcheted up as part of its NDC process.[74] The EU has now agreed to a reduction target of 55 per cent by 2030, and to achieving climate neutrality by 2050. Even more ambitious targets proposed by members of the European Parliament, in the range of a 60–65 per cent reduction, did not materialize.[75]

---

[71] See Butler (n. 60); Dawson (n. 60); De Schutter (n. 60).

[72] See for further explanation of these instruments Chapter 2.

[73] See for a history of the EU's targets also Chapter 2.

[74] See e.g. European Commission, 'Communication on Stepping Up Europe's 2030 Climate Ambition', COM(2020) 562 final; European Commission, 'Communication on Clean Planet for All, COM(2018) 773 final; European Commission, 'Communication on the European Green Deal', COM(2019) 640 final; Proposal for a Regulation of the European Parliament and of the Council establishing the framework for achieving climate neutrality and amending Regulation (EU) 2018/1999 (European Climate Law) COM(2020) 80 final.

[75] European Council, 'Council Agrees on Full General Approach on European Climate Law Proposal' (17 December 2020) via: www.consilium.europa.eu/en/press/press-releases/2020/12/17/council-agrees-on-full-general -approach-on-european-climate-law-proposal/# last accessed 13 January 2021; Florence Schulz, 'EU Lawmakers Debate 65% Climate Target Proposal' (EurActiv, 29 May 2020) via: www.euractiv.com/section/energy-environment/ news/eu-parliaments-climate-committee-divided-over-65-climate-target/ last accessed 13 January 2021.

Of course, these more ambitious GHG reduction commitments will now require a set of new or revised robust laws and policies to help facilitate their implementation.[76] An important question is, then, whether and to what extent human rights should and have played a role in the setting of past and present GHG reduction targets for the EU, and in the legislation that supports the realization of such targets.

Taking a look at the 'first generation' of EU GHG reduction targets and laws, adopted up to 2009, it is clear that human rights have been addressed and even included in the final texts of some instruments. Yet, any actual assessments of *how* fundamental rights might apply fully are largely absent. Thus, while the EU ETS Directive 2003/87/EC for example states that it 'respects the fundamental rights and observes the principles recognised in particular by the Charter of Fundamental Rights of the European Union',[77] the legislative history for this directive does not really reveal any meaningful assessment of *how* fundamental rights constrained or shaped it.[78] There are just some hortatory statements *that* the directive respects the CFREU, supposedly in its totality. The Effort Sharing Decision 406/2009/EC did not refer to human rights at all, either in its text or in the drafting process.[79]

Arguably, this limited attention to human rights in early EU climate policies might be explained by the fact that during their negotiation stages, neither UN HRC Resolution 10/4 (2009) nor the UNFCCC's Cancún Agreements (2010) had yet been adopted. EU policy-makers may thus have lacked awareness, or the right (interpretative) tools to properly assess EU climate targets and policies from a human rights perspective.

The same, however, cannot be said for 'second and third generations' of EU climate and energy laws and policies. A closer look at some of the newer policies, including currently pending proposals for 2021–2, reveals that while there is more attention to human rights, this is only marginally better (so far). The explanatory memorandum for Effort Sharing Regulation 2018/842, the successor of Effort Sharing Decision 406/2009/EC, for example only vaguely notes that because it 'primarily addresses Member States as institutional actors', this instrument must be considered 'consistent with the CFREU'.[80] There is no reference to fundamental rights in its text. The latter seems somewhat curious, because Effort Sharing Regulation 2018/842 actually provides the main legal basis for imposing binding GHG reduction targets on all EU

---

[76]   See Chapter 2, COM(2020) 562 final (n. 74), p.2.

[77]   Directive 2003/87/EC establishing a scheme for greenhouse gas emission allowance trading amending Council Directive 96/61/EC, OJ L 275/32, 25.10.2003, recital (27). See a similar reference in its successor: Directive (EU) 2018/410 of 14 March 2018, OJ L 76/3, recital (32).

[78]   See e.g. Proposal for an EU ETS Directive amending Council Directive 96/61/EC, COM(2001) 581 final; Commission Staff Working Document, 'Annex I to the Impact Assessment', SEC(2008) 85, pp.127–8.

[79]   See e.g. Decision No. 406/2009/EC, OJ L 140/136, 5.6.2009; European Commission, Proposal for a Decision on the Effort of Member States to Reduce their GHG Emissions to meet the Community's GHG Emission Reduction Commitments up to 2020', COM(2008) 17 final.

[80]   European Commission, 'Proposal for a Regulation on binding annual greenhouse gas emission reductions by Member States from 2021 to 2030 for a resilient Energy Union and to meet commitments under the Paris Agreement' COM/2016/0482 final, p.7.

Member States for the period 2021–30 in line with the EU's collective targets – at least until these were amended for 2030 and 2050.[81] In this sense, it is noteworthy that the (short) inception impact assessments for currently pending revisions of the EU ETS Regulation, the Effort Sharing Regulation and the LULUCF Regulation, affirm their human rights relevance more firmly.[82] Each states that the proposed instrument aims to 'advance environmental protection' and to pursue the objective of a high level of environmental protection and the improvement of environmental quality, in pursuit of Article 37 CFREU.[83] Yet, again, *how* exactly this is done, or whether and how certain GHG reduction targets, or any other options, might be *more or less beneficial* or *comply* with the rights in the CFREU, has not received attention so far. Very similarly, the proposal for the first EU Climate Law, to be adopted over the period 2021–2 to underpin the EU's new GHG targets, also notes that it will respect

the fundamental rights and observe the principles recognised in particular by the Charter of Fundamental Rights of the European Union. In particular, it contributes to the objective of a high level of environmental protection in accordance with the principle of sustainable development as laid down in Article 37 of the Charter of Fundamental Rights of the European Union.[84]

Again, although this statement 'on paper' affirms the relevance of the CFREU to the new EU Climate Law, there is no further guidance or detail provided on *how* it aligns with Article 37 Charter or other rights commonly invoked in the context of climate change and GHG mitigation, such as the right to life, human dignity or private life. There is certainly no reference to the various relevant legal developments in UN law, or evolving understanding of the applicability of the ECHR.

All in all, there is thus limited evidence that human rights played any significant role in discussing the appropriateness of specific EU GHG emission reduction targets for 2020, 2030 or 2050. In light of the significant developments on human rights and climate change in the UNFCCC, in UN human rights law and in European climate litigation, this displays poor sensitivity of EU law-makers to ongoing debates on human rights and climate mitigation. More specifically, poor assessment of the impacts of EU climate policies on human rights may open up legislative acts to rights-based litigation, similar to climate litigation seen in national courts across Europe.

---

[81]   Regulation (EU) 2018/842 on binding annual greenhouse gas emission reductions by Member States from 2021 to 2030 contributing to climate action to meet commitments under the Paris Agreement and amending Regulation (EU) No 525/2013, OJ L 156/26, 19.6.2018.

[82]   See for an explanation of these instruments and their history, Chapter 2.

[83]   European Commission, 'Inception Impact Assessment for Amendment of the EU Emissions Trading Scheme', Ares (2020)6081850 – 29/10/2020; European Commission, Inception Impact Assessment for Amendment of the LULUFC Regulation (EU) 2018/84, Ares(2020)6081753 – 29/10/2020; European Commission, 'Inception Impact Assessment for Amendment of Regulation (EU) 2018/842', Ares(2020)6081605 – 29/10/2020.

[84]   European Commission, 'Proposal for a Regulation establishing the framework for achieving climate neutrality and amending Regulation (EU) 2018/1999 (European Climate Law)' COM(2020) 80 final, p.6.

11.4.3 Human Rights and Specific GHG Reduction Activities Supported by EU Climate Laws

There has been some recognition that negative human rights impacts may also result from EU Member States' implementation of specific GHG reduction activities, both within EU countries and in third countries. The following paragraphs briefly highlight several concerns that arose during negotiations for the LULUFC Regulation and Renewable Energy Directive(s), for example as related to the management of climate sinks, land-use change and deforestation, renewable electricity projects and the production of agrofuels with feedstocks.

Human rights, climate sinks, biomass and forestry projects

The legislative history for the LULUFC Regulation (EU) 2018/841 – which accounts for States' GHG emissions from (changes to) the land-use and forestry sectors – reveals that several concerns were raised about possible human rights violations that could result from the use of forests and lands for (forestry-related) biomass production, and from the preservation of forests and plant stocks as climate sinks.[85] The Committee on Development of the European Parliament, especially, noted that the new LULUFC Regulation would likely set an important global precedent regarding GHG emissions accounting from sinks and the land-use sector, and therefore the EU should aim for high standards of protection for local communities. In particular, the EU must lead by example by affirming the principles of equity, sustainable development and the promotion of international human rights. In particular, poor local communities and indigenous people that depend on forests and lands may not face displacement due to carbon sink-related projects (e.g. deforestation, afforestation or reforestation).[86]

The European Parliament concretely proposed that LULUFC Regulation (EU) 2018/841 should affirm that forests must be managed responsibility; that deforestation of sensitive ecosystems must be prohibited; and that land rights, the rights of indigenous communities, human rights and workers' rights must be respected.[87] These proposals were not accepted in the final texts, but the LULUFC discussion nevertheless shows that carbon sinks and (forestry-related) biomass projects, supported by the EU and its Member States, may negatively affect the human rights of communities where they are implemented.

Interestingly, such concerns have been raised both in the UNFCCC Cancún Agreements and in UN human rights law as well. The Cancún Agreements, for example, noted the relevance of sustainable forest management, deforestation, forest carbon stocks and related eco-system services, and linked this to the rights of indigenous peoples and the UN Declaration on the Rights

---

[85]   Regulation (EU) 2018/841 on the inclusion of greenhouse gas emissions and removals from land use, land use change and forestry in the 2030 climate and energy framework OJ L 156/1, 19.6.2018.

[86]   Opinion of the Committee on Development of the European Parliament of 3 May 2017, p.42 et seq. via: https://www.europarl.europa.eu/doceo/document/A-8-2017-0262_EN.pdf last accessed 13 January 2021.

[87]   Proposal for Amendment 16 by the European Parliament on 13 September 2017, via: www.europarl.europa.eu/doceo/document/TA-8-2017-0339_EN.pdf last accessed 13 January 2021.

of Indigenous Peoples as adopted by the UN General Assembly in 2007.[88] The CRC treaty committee has expressed its concerns that deforestation could severely affect the lives and subsistence possibilities of people living in deforested areas, including children, for example through 'forced displacement, degradation of biodiversity and erosion of riverbanks'.[89]

Human rights and the implementation of renewable electricity projects

The implementation of renewable electricity projects has been cited as raising just as many issues for the protection of rights of local communities, especially during the negotiations for the Renewable Energy Directive(s). Indeed, both the text and drafting history for the Renewable Energy Directive (EU) 2018/2001 refer to human rights of indigenous peoples and local communities. The most concrete legal obligation on human rights can be found in Article 22, which states that joint implementation of renewable electricity projects *with third countries* must take place 'in accordance with international law', and only in countries that are signatories to the ECHR or other international human rights treaties.[90]

This simple reference to international law might in fact imply rather comprehensive 'human rights impact assessments' for such projects, with international legal guidance available through documents such as the UN Guiding Principles on Human Rights and Project-Induced Displacement, the UN Guiding Principles on Business and Human Rights and the UN Declaration on the Rights of Indigenous Peoples.[91]

Finally, even though Article 22 only refers to projects *with and in third countries*, human rights obligations naturally also apply to projects implemented by EU countries within the EU. The UN Special Rapporteur for the Rights of Indigenous Peoples has in fact previously warned Finland, Norway and Sweden that they must ensure their decarbonization policies respect the rights of European indigenous peoples. They should consolidate climate policies in consultation with the representative organs of such peoples, and ensure that any 'measures to promote renewable energy sources, such as wind farms, do not themselves adversely affect Sami livelihoods'.[92]

Human rights and the production of agrofuels

As a third and final issue, the production of (liquid) agrofuels with food stocks and lands for the production of food arose as a specific issue in the negotiations for the Renewable Energy

---

[88] See UNFCCC Cancún Agreements (2010), paras 86–87, p.26, Appendix I; Ibid.

[89] CRC, 'Concluding Observations on Lao Democratic Peoples Republic' (2018) UN Doc. CRC/C/LAO/CO/3-6, para. 36.

[90] Directive (EU) 2018/2001 on the promotion of the use of energy from renewable sources, OJ L 328/82, 21.12.2018.

[91] UN Special Rapporteur on the Right to Housing, 'Basic Principles and Guidelines on Development-Based Evictions and Displacement' (2008) UN Doc. A/HRC/4/18; Lidewij van der Ploeg and Frank Vanclay, 'A Human Rights Based Approach to Project Induced Displacement and Resettlement' (2017) 35 *Impact Assessment and Project Appraisal* 34–52.

[92] Human Rights Council, Report of the Special Rapporteur on the Situation of Human Rights and Fundamental Freedoms of Indigenous People (6 June 2011) UN Doc. A/HRC/18/35/ADD.2, paras 55–61, 86.

Directive (EU) 2018/2001.[93] There is a particular concern that such fuels create problems for food security, land rights and sufficiently sustainable livelihoods.

Members of the European Parliament feared that agrofuel production could have both implicit and explicit negative effects on human rights enjoyment through land-grabs, biodiversity degradation and various other negative social and economic impacts on lives and livelihoods, and indeed the right to food.[94] They proposed that the Renewable Energy Directive obliges Member States to 'refrain from subsidizing or mandating feedstocks for energy where such use would have a negative impact on land rights, food rights, biodiversity, soil or overall greenhouse balance' and essentially called for the phasing out of all policy incentives for biofuels, bioliquids and biomass fuels produced from food and feed crops, or other crops grown on productive agricultural land.[95] Additionally, renewable energy policy should be consistent with relevant international instruments for the protection of indigenous and other local communities' rights, including the following: International Labour Organization (ILO) Convention No. 169 on Indigenous and Tribal Peoples; the UN Voluntary Guidelines on Responsible Governance of Tenure of Land, Fisheries and Forests in the Context of National Food Security; the UN Principles for Responsible Investment in Agriculture and Food Systems; and OECD-FAO Guidance for Responsible Agricultural Supply Chains.[96]

Again, there is limited explicit attention to human rights in the final text of Renewable Energy Directive 2018/2001, which is somewhat surprising, because the former Renewable Energy Directive 2009/28/EC actually included fairly elaborate references to human rights.[97] The same is the case for its biofuel-related amendments through Directive 2015/1513. The latter, for example, explicitly stated that:

Good governance and a rights-based approach, encompassing all human rights, in addressing food and nutrition security, at all levels, are essential, and coherence between different policies should be pursued in cases of negative effects on food and nutrition security. In this context, the governance and security of land tenure and land-use rights are of particular importance.[98]

---

[93] See Max Jansson and Seita Romppanen, 'Biofuels' in Elisa Morgera and Kati Kuloves, (eds) *Research Handbook on International Law and Natural Resources* (Edward Elgar Publishing 2016) 281–304; Maria León-Moreta, 'A Threat to the Environment and Human Rights? An Analysis of the Production of Feedstock for Agrofuels on the Rights to Water, Land and Food' (2011) 1 *European Journal of Legal Studies* 102; Asbjørn Eide, *The Right to Food and the Impact of Liquid Biofuels (Agrofuels)* (FAO: Rome, 2008).

[94] Opinion of the Committee on Development of the European Parliament of 24 October 2017, p.161 et seq. via: www.europarl.europa.eu/doceo/document/A-8-2017-0392_EN.pdf last accessed 13 January 2021.

[95] Ibid.

[96] Ibid.

[97] See Article 17(7) of Directive 2009/28/EC of the European Parliament and of the Council of 23 April 2009 on the promotion of the use of energy from renewable sources, OJ L 140/16, 5.6.2009 for various reporting commitments for the Commission regarding human rights and EU biofuel policy and activities.

[98] Recital 26 of Directive (EU) 2015/1513, OJ L 239/1, 15.9.2015 [italics added by author].

Directive 2015/1513 further affirmed that when implementing biofuel-related activities, Member States should respect the UN's Principles for Responsible Investment in Agriculture and Food Systems, and take into account the UN's Voluntary Guidelines on the Responsible Governance of Tenure of Land, Fisheries and Forests in the Context of National Food Security.[99] An important feature of Directive 2015/1513 is that it set a 7 per cent cap on the extent to which food-based biofuels could be counted towards renewable energy objectives.[100]

Again, these agrofuels debates can be understood in light of wider international human rights law.[101] Thus, aside from critical comments by the UN Special Rapporteur on the Right to Food on the production of first generation agrofuels in its report on climate change,[102] the ICESCR committee recommended to Belgium that it should 'systematically conduct human rights impact assessments of Belgian firms' agrofuel activities in third countries', so as to ensure that 'projects promoting agrofuels' do not lead to 'negative impact on the economic, social and cultural rights of local communities' where they are implemented.[103] It is unclear why Renewable Energy Directive 2018/2001 omitted more explicit references to human rights, as in fact the directive still allows (some) foodstock-related biofuels, bioliquids and biomass fuels.[104] The *Sabo* case to be discussed in Section 5 will also further show how the directive later gave rise to rights-based litigation related to wood-based biomass production.

Limited attention to human rights in EU climate law-making

All in all, it can be concluded that while the 'Better Regulation Agenda' requires the Commission to carry out rigorous IAs, with firm attention to human rights and elaborate evaluations of how different policy options affect people's human rights differently, it is clear that IA practice, at least for EU climate law, is disappointing so far.

In fact, despite broad recognition of the applicability of human rights law to climate mitigation action, including the design of ambitious GHG mitigation targets, EU law-makers have paid scant meaningful attention to fundamental rights so far. Nevertheless, the absence of any direct references to human rights in legislative acts or procedures does not automatically mean that no relevant human rights apply to their implementation.

What follows shows how citizens are in fact trying to use fundamental rights in court to challenge various EU and EU Member States' climate policies.

---

[99] Ibid. recital 26 and Article 2.

[100] See also Chapter 5 in this book.

[101] Human Rights Council, 'Report of the UN Special Rapporteur on the Right to Food: Climate Change' (5 August 2015) UN Doc. A/70/287, paras 6062.

[102] Ibid.

[103] UN Committee on Economic Social and Cultural Rights, 'Concluding Observations on Belgium' (2013) UN Doc. E/C.12/BEL/CO/4, paras 22–23.

[104] See Chapter 5 in this book.

11.5 HUMAN RIGHTS AND EUROPEAN CLIMATE LITIGATION

Rights-based litigation on climate change is on the rise across Europe at the moment, although such litigation typically takes place nationally based on constitutional rights or ECHR rights, rather than the CFREU and EU climate law.[105] This section briefly discusses the two fundamental rights cases that have been directly lodged at the CJEU so far, as well as the successful *Urgenda* case in the Netherlands, which gave rise to a novel interpretation of Articles 2 and 8 ECHR in relation to States' individual obligations to mitigate GHGs according to particular pathways. A few comparisons to other interesting cases will also be made where relevant.

11.5.1 The *Carvalho* and *Sabo* cases

The *Carvalho* and *Sabo* cases were filed and decided quickly one after another between 2018 and 2020. Each application was filed on behalf of citizens and non-governmental organizations in the EU as well as some outside, for example in Kenya and Fiji (in *Carvalho*) and the US (in *Sabo*). The cases also each asked for the (partial) annulment of specific EU climate laws, which applicants held to be in violation of their fundamental rights and other higher ranking (EU) laws. The facts and claims presented by applicants are, however, quite different.

First, plaintiffs in the *Carvalho* case, also known as the *People's Climate Case*, complained that the EU's former GHG reduction target of 40 per cent by 2030 compared to 1990 emissions levels will be inadequate to sufficiently safely curb dangerous global warming. They therefore demanded a partial annulment of the three relevant implementing acts for such targets: the EU ETS Directive of 2018, the Effort Sharing Regulation of 2018 and the LULUFC Regulation of 2018.[106] They additionally asked the court to order the EU to adopt measures which can guarantee at least a reduction of 50–60 per cent by 2030 (compared to 1990).[107] As explained in Section 11.4.2, the EU recently committed itself to such targets, that is, a target of 55 per cent by 2030.

The children, adults and Sami Youth Organization supporting the *Carvalho* application particularly invoked their rights to life (in Article 2 CFREU), physical integrity (Article 3), to engage in work and to pursue a freely chosen or accepted occupation (Article 15), to conduct a business (Article 16) and to property (Article 17), as well as the rights of children (Article 24), and obligations of environmental protection (Article 37).[108] They did so by arguing that climate change is already affecting them, and will further affect them, as a result of droughts,

---

[105] See for an overview of such cases: http://climatecasechart.com/non-us-case-category/human-rights/ last accessed 13 January 2021.

[106] Case T-330/18, *Armando Carvalho and others v EP and Council* (Order of the General Court of 9 May 2019) ECLI:EU:T:2019:324; See also: https://peoplesclimatecase.caneurope.org/documents/ last accessed 13 January 2021.

[107] Ibid. paras 18, 22–24.

[108] See Case-T-330/18, *Armando Carvalho and others v EP and Council* (Application of 24 May 2018) paras 158, 161–197, 253, 381, 389, via https://peoplesclimatecase.caneurope.org/documents/ last accessed 13 January 2021. Also

flooding, heat waves, sea-level rises and disappearing cold seasons. This in turn affects their physical well-being and various livelihoods in the farming and tourism sectors. The Sami Youth Association further claimed that climate change negatively impacts on their cultural practices and indigenous ways of living, such as reindeer herding.[109]

Curiously, the *Carvalho* application did not directly rely on the right to private and family life, as protected in Articles 7 CFREU and 8 ECHR, and commonly invoked in domestic litigation across Europe, despite referring to the environmental case-law of the European Court of Human Rights on this right in their application.[110] They did not refer to the protection of health in Article 35 CFREU either, nor drew inspiration from the European Social Charter, the EU Pillar of Social Rights or UN human rights treaties.

The *Sabo* case stands in contrast to this. In this case, also known as the *EU Biomass Case*, again several individuals and organizations from both EU and non-EU countries argued that the Renewable Energy Directive (EU) 2018/2001 must be annulled because its support for wood-based biomass fuels will likely accelerate widespread deforestation. Moreover, the EU inadequately accounts for the GHG emissions resulting from wood-fuel production and burning, meaning that the Directive cannot adequately contribute to mitigating global warming, as intended.[111]

The CFREU rights claimed to be violated by the Directive included respect for private and family life (Article 7), the right to property (Article 17), a high level of human health protection (Article 35), a high level of environmental protection (Article 37) and the freedom to manifest religion (Article 10). These rights are said to be impaired, *inter alia*, by the ongoing (and future) destruction of ancient cultural and religious heritage sites in Estonia, Romania and Slovakia, such as forest groves; the excessive noise, wood dust and air pollution associated with both the production of wood-pellets and the operation of wood-fuelled power plants; and to damage to property and loss of traditional (family) hunting values, wildlife habitat and flood protection due to the clearing of trees in personal surroundings for wood-pellet production.[112]

Procedural obstacles at the CJEU: locus standi criteria

Both *Carvalho* and *Sabo* are extremely rich cases, giving rise to various pertinent EU fundamental rights law questions. These include whether and which CFREU rights are relevant to EU climate policies, or if CFREU rights are 'universal and must be respected for both EU and non-EU applicants'.[113] Yet, neither case was likely to receive a proper substantive review.

---

see Gerd Winter, '*Armando Carvalho and Others v EU*: Invoking Human Rights and the Paris Agreement for Better Climate Protection Legislation' (2020) 9 *Transnational Environmental Law* 137–64.

[109] Case-T-330/18 (n. 108) paras 1, 24–49.

[110] Case-T-330/18 (n. 108) paras 161–173.

[111] See for the history of this case, including application submitted to the CJEU on 4 March 2019, http://eubiomasscase.org/the-case/.

[112] See ibid. and Case T-141/19, *Sabo and Others v Parliament and Council* [Action brought on 4 March 2019] OJ C 148/60 29.4.2019.

[113] Ibid.

This is due to the rather strict 'standing' (*locus standi*) requirements for individuals and non-governmental organizations under Article 263(4) TFEU. Article 263(4) provides:

> Any natural or legal person may, under the conditions laid down in the first and second paragraphs, institute proceedings against an act addressed to that person or which is of *direct and individual concern to them*, and against a regulatory act which is of direct concern to them and does not entail implementing measures.[114]

In particular, the criteria for evidencing a 'direct and individual concern' as a result of legislative acts such as directives and regulations have been fairly strictly interpreted by the CJEU ever since its *Plaumann* judgment.[115] The 'individual concern' criterion has been particularly criticized for severely limiting people's access to justice, as it essentially requires applicants to prove that there are 'certain attributes peculiar' to them, or that they are in a particular 'factual situation which differentiates them from all other persons, and distinguishes them individually'.[116] Applicants thus had to prove – fairly similar to the Kiribati resident discussed in Section 11.2.2 – that they were somehow personally affected by EU climate laws, in a way that others are not.[117]

In doing so, applicants in both EU cases tried to argue that the *Plaumann* criteria are outdated and should not be interpreted too strictly, especially where human rights infringements are concerned. As stated by the *Sabo* applicants, the 'best interpretation is that individual concern is established by harm to individual rights'.[118] Also, the *Carvalho* plaintiffs argued that that each of them would be affected in terms of their fundamental rights in their own unique ways – such as by virtue of specific personal interests, circumstances, locations, ages or prior legal positions. In short, all of them are affected in a 'distinct and idiosyncratic manner', incomparable to the experiences of another person.[119] The European Parliament responded to these arguments in *Carvalho* that it was 'obviously absurd' to consider that a farmer affected by floods would be differently affected than a farmer affected by a drought.[120] It also called it 'fallacious from a logical perspective' to argue that climate change affects individual people differently: this would imply that 'each and every person around the world is individually

---

[114] Article 263(4) TFEU; See also Rudolf Geiger, Daniel Erasmus-Khan and Marcus Kotzur, *European Union Treaties: A Commentary* (C.H. Beck/Hart 2015) 874–88, paras 11, 28–33, with references to C-583/11 P, *Inuit Tapiriit Kanatami and Others v Parliament and Council* (3 October 2013) ECLI:EU:C:2013:625; and Case T-330/18 (n. 108) paras 33–41.

[115] Case 23-62, *Plaumann* (15 July 1963) ECLI:EU:C:1963:17.

[116] Ibid. para. 15; See for discussion Geiger, Khan and Kotzur (n. 114) 875–6, 880–3, and compare with C-583/11 P (n. 114) and Case T-600/15, *Pesticide Action Network Europe (PAN Europe) and others v Commission* (28 September 2016) ECLI:EU:T:2016:601.

[117] Case T-330/18 (n. 108) paras 44–46 *et seq*.

[118] See for example the application in the *Sabo* case (n. 112), para. 121.

[119] See application in Case T-330/18 (n. 108) paras 150–151.

[120] This is suggested in the C-T330/18 – Applicant's Reply on Admissibility (2018) 5, 15.

concerned' by the contested climate laws in that case.[121] In turn, the *Carvalho* applicants responded by stating that strict adherence to the *Plaumann* criteria would lead to the paradoxical outcome that the more serious and widespread the damage is, or the higher the number of people affected, the less judicial protection is available.[122]

As generally expected, the EU General Court sided with respondents, by stating that a simple invocation of fundamental rights is not enough to meet the criterium of individual concern in Article 263 TFEU.[123] In *Sabo*, it did also not accept that these applicants belonged to a more 'limited category of persons' affected by the EU's biomass policy.[124] Thus, even though the Court admits 'that every individual is likely to be affected one way or another by climate change', this was not sufficient to meet or abandon the *Plaumann* criteria. In fact, in *Sabo*, this argument was taken as evidence of *not* meeting the criteria.[125] Both the *Carvalho* applicants and the *Sabo* applicants also tried to force a different interpretation of the *Plaumann* criteria by appealing to notions of 'access to justice' and 'rights to remedy' protected by Article 47 of the CFREU, the ECHR and the Aarhus Convention on Access to Information, Public Participation and Access to Justice in Environmental Matters.[126] Yet, they did so in vain. It was pointed out that Union law does not preclude that individuals lodge cases nationally, and through such proceedings pursue a reference for a preliminary ruling from the CJEU.[127] The argument that this would be very cumbersome, and likely unsuccessful, because each system will have its own judicial procedures and remedies did not change the outcome.[128]

Finally, it is worthwhile to note that in the *Sabo* judgment the Court actually went a step further in proposing a solution, by noting that it may in fact be 'desirable' to widen people's opportunities for direct access to the CJEU in matters of environmental rights, but that only the EU Member States can reform the current system.[129] At the time of writing, both cases are pending for appeal with the CJEU.

Comparing locus standi arguments in other jurisdictions

The argument in domestic courts around Europe that EU citizens might face difficulties in effectively bringing rights-based climate cases was not without merit, because in fact *locus standi* hurdles have been known to plague climate cases generally.

In Switzerland, for example, a group of elderly women failed to convince Swiss courts that their rights under Articles 2 and 8 ECHR were sufficiently differently or 'intensely' affected as

---

[121] Case T-330/18 (n. 108) para. 28.

[122] Ibid. paras 31–32.

[123] Ibid. paras 46–48.

[124] Case T-141/19, *Peter Sabo and others v European Parliament and Council* (Order of the General Court of 6 May 2020) ECLI:EU:T:2020:179, paras 27–30.

[125] Ibid. paras 30–31; Case T-330/18 (n. 108) paras 48–50.

[126] Ibid. paras 39–45, 32, 52.

[127] Case T-330/18 (n. 108) paras 28, 53.

[128] Ibid. paras 32, 52–54.

[129] Case T-141/19 (n. 124) para. 45.

compared to the *general public*, owing to their exceptional vulnerability to heat stress and heat-waves. To effectively dispute Swiss climate policy in Swiss courts the women had to prove that their *own* special interest or rights were affected, which was difficult due to the wide range of impacts of climate change for many Swiss people. Heatwaves and heat stress are just one such phenomenon.[130] The Swiss Federal Supreme Court noted on the issue of standing in Swiss law that while the bar for accessing courts 'must not be set too high' in the interest of legal protection, *locus standi* requirements generally exist to prevent 'floods of appeals' or so-called *actio populars* claims on behalf of the entire population.[131] Therefore, there was a good reason to require that the women somehow showed their personal rights were affected, with a measure of sufficient intensity. After the Supreme Court dismissed the claims, the women appealed to the European Court of Human Rights on 26 November 2020.

The highly unsatisfactory 'injury to all is injury to none' approach was also an issue in other cases, but was explicitly rejected by Judge Ann Aiken of the Oregon District Court in the landmark US case of *Juliana v the United States*. Aiken opined that actually all 21 young applicants in this case had somehow proven that they experienced or faced personal 'concrete, particularized, and actual or imminent' harms to their homes, health, families and lives from climate change, *inter alia* owing to episodes of drought, floods or forest fires. Importantly, she decided that it does not matter if 'the experience at the root of [the] complaint was shared by virtually every American'; what matters is that the 'shared experience caused an injury that is concrete and particular to the plaintiff'.[132]

This seems a more satisfactory and appropriate approach to understanding the direct, personal harms that will likely affect many, if not all, human beings in very real, profound and personal ways over the coming years and decades. As acknowledged by the EU itself, weather-related disasters are anticipated to affect approximately two-thirds of the EU population annually by the year 2100 (with 351 million persons exposed per year during the period 2071–2100).[133] In this sense, *locus standi* requirements may reveal the possible limitations of trying to enforce effective human rights protection *through court proceedings*. It simultaneously reveals the importance of ensuring that human rights are adequately applied and evaluated from the start in climate law-making procedures.

---

[130] Cordelia Christiane Bähr et al., '*KlimaSeniorinnen*: Lessons from the Swiss Senior Women's Case for Future Climate Litigation' (2018) 9 *Journal of Human Rights and the Environment* 194–221. See for history and legal documents: http://climatecasechart.com/non-us-case/union-of-swiss-senior-women-for-climate-protection-v-swiss -federal-parliament/ last accessed 13 January 2021.

[131] Federal Supreme Court, *Verein KlimaSeniorinnen Schweiz et al. v DETEC*, Judgment 1C_37/2019 (5 May 2020) paras 4.1–4.4, 5.4–5.5; Farid Ahmadov, *The Right of 'Actio Populars' before International Courts and Tribunals* (Brill Nijhoff 2018) 156–61.

[132] District Court Oregon, *Juliana and others v United States* (11 October 2016) 217 F. Supp. 3d 1224 (D. Or. 2016) paras 50–51.

[133] European Union (n. 1) 107–9.

Urgenda v the Netherlands

The *Urgenda* case is currently viewed as one of the most important European and global precedents for successful rights-based climate litigation. In this case, the Dutch non-governmental organization Urgenda, initially together with nearly 900 Dutch citizens, asked the civil courts to order the Netherlands to reduce its GHG emissions by at least 25–40 per cent by 2020, as compared to 1990 levels. More precisely, *Urgenda* claimed that the Dutch government had wrongly lowered a previous reduction target for 2020 from 30 per cent to 17 per cent – which was done partially in fulfilment of EU effort-sharing agreements. All three courts which heard the case granted the reduction order.[134]

To support their claims, the applicants invoked an aggregate of legal bases in national, European and international law, including Dutch tort law concepts of 'wrongful act' and 'endangerment', constitutional environmental duties of care, Articles 2 and 8 ECHR, the no-harm rule in customary international law, the precautionary principle (as mentioned in the UNFCCC and in environmental case-law of the ECHR) and the international law of state responsibility.[135] In the end, however, the case was nearly exclusively decided based on the ECHR, as interpreted in light of its wider international law background.

Procedural hurdles? Locus standi in Urgenda

In terms of 'standing', it must be noted that ultimately the Appeals Court and Supreme Court allowed Urgenda to proceed with its claims for the protection of human rights in Articles 2 and 8 ECHR on behalf of Dutch citizens generally, due to fairly liberal standing rights granted to non-governmental organizations in the Dutch Civil Code – essentially akin to an *actio popularis*. As a result, Urgenda was able to argue in a more abstract sense that Dutch citizens are and will continue to be confronted with numerous real and imminent threats to their rights over their lifetime. These threats include loss of life or disruption of family life due to extreme heat, extreme drought, extreme precipitation, other extreme weather, rising sea levels and severe flood risks for large parts of the country.[136] The courts accepted that sufficiently real and imminent risks exist based on relevant predictions of climate change impacts for Europe, as supported by the IPCC, various other UN agencies and reports and the COPs of the UNFCCC. Moreover, such impacts were not contested by the Dutch government.[137]

Importantly, the claims of the nearly 900 individuals were dropped from the case after the District court had dismissed them partially on practical grounds.[138] All in all, this means that there was never a need to prove that climate change is or may be affecting *any specific indi-*

---

[134]   District Court The Hague, *Urgenda and others v the Netherlands* (24 June 2015) ECLI:NL:RBDHA:2015:7145; Court of Appeal The Hague, *The Netherlands v Stichting Urgenda* (9 October 2018) ECLI:NL:GHDHA:2018:2259; Supreme Court, *The Netherlands v Stichting Urgenda* (20 December 2019) ECLI:NL:HR:2019:2006.

[135]   Supreme Court (n. 134) paras 4.1–4.8, 5.4.2, 5.7.1–5.7.6, 6.3–6.6, 7.1. et seq.

[136]   Ibid. paras 4.1–4.2, 4.7. et seq.

[137]   Ibid. paras 4.1–4.2, 4.7, 5.6.2.

[138]   See District Court (n. 134) para. 4.109.

vidual person. It has been argued in the literature that the particular features of the case might prevent replication of similar lawsuits before the ECHR, or in other legal systems.[139]

States have individual obligations to mitigate GHGs under the ECHR

In terms of the merits, the *Urgenda* judgments stand out for several reasons.[140] First, the courts determined that the Netherlands bears an *individual legal obligation* to reduce GHGs in a sufficiently safe way as part of the *collective effort* of implementing the UNFCCC (and Paris Agreement). Second, and importantly, this binding obligation may not exist on the basis of the UNFCCC or Paris Agreement directly or solely, but may be construed on the basis of States' positive obligations under Articles 2 and 8 ECHR, as read *in light of their wider international law background*. In short, a binding minimum GHG reduction effort for the Netherlands may not ensue (only) from UN, EU or national climate law or policy, but (also) *from its positive obligations under Articles 2 and 8 ECHR*.[141]

The Supreme Court decided that it could establish a minimum GHG reduction obligation for the Netherlands under Articles 2 and 8 ECHR due to the 'settled' case-law of the European Court of Human Rights on environmental protection (see Section 11.3.2) and the fact that the European Court itself is accustomed to interpreting ECHR rights in light of their 'wider international law background' or any clear shared 'European consensus' or 'common ground'.[142] The 'common ground' approach essentially entails that the Supreme Court determined whether any sufficiently clear and widespread international consensus exists regarding the threats of dangerous climate change to human rights, and what it would take to truly credibly maintain global safe temperature limits and GHG concentrations.[143] In appraising such consensus, the Court looked at States' legal–political support for the UNFCCC, the Paris Agreement and its various principles of CBRD-RC and precaution, relevant UNFCCC COP decisions, the no-harm rule in international environmental law and international law on state responsibility, along with widely supported and cited IPCC reports – which were also not contested by the Dutch State.[144] All these sources together left the Supreme Court with limited doubt that there is a sufficiently clear and high degree of international consensus regarding the dangers of climate change, and what it takes to combat it. Specifically, the Court pointed to a consensus within the UNFCCC and the EU – not contested by the Netherlands either –

---

[139] See e.g. Suryapratim Roy, '*Urgenda II* and Its Discontents' (2019) 2 *Common Market Law Review* 133–4.

[140] See e.g. ibid; and C.W. Backes and G.A van der Veen, 'Urgenda: The Final Judgment of the Dutch Supreme Court' (2020) 17 *Journal for European Environmental & Planning Law* 307–21; Marjan Peeters, '*Urgenda Foundation and 886 Individuals v The State of the Netherlands*: The Dilemma of More Ambitious Greenhouse Gas Reduction Action by EU Member States' (2016) 25 *RECIEL* 123–9.

[141] Marlies Hesselman, 'Domestic Climate Litigation's Turn to Human Rights and International Climate Law' in M. Fitzmaurice, D.M. Ong and P. Merkouris, *Research Handbook on International Environmental Law* (Edward Elgar Publishing, 2021). See e.g. Supreme Court (n. 131) para. 7.3.3.

[142] Supreme Court (n. 134) paras 5.1 et seq.

[143] See ibid. 5.6.2–5.6.4, et seq.

[144] Ibid. paras 2.3.2, 7.2.10–7.2.11.

that for developed countries listed on Annex I to the UNFCCC, a minimum reduction effort in the order of at least 25–40 per cent by 2020, 49–55 per cent by 2030 and 80–95 per cent by 2050 were necessary for any credible possibility of limiting warming to a maximum of 2°C.[145] Moreover, the Court further referred to consensus about the existence of dangerous 'global emissions reductions gaps', and the small window of opportunity available to maintain critical carbon budgets and temperature limits.[146]

As such, and keeping in mind the precautionary principle along with the serious and possibly irreversible impacts of global warming (e.g. 'tipping points'), the Court felt sufficiently comfortable to concretize the Netherlands' positive obligations under Articles 2 and 8 ECHR in this manner. It agreed with Urgenda that the effective protection of ECHR rights requires the Netherlands 'to do more' to help curb global dangerous climate change, and to *at least* reduce its emissions by 25 per cent as part of its own minimum fair share of work to tackle the problem.[147]

Urgenda and effort sharing within the EU

A particularly interesting detail of the *Urgenda* judgment, for this chapter, is that by affirming that the Netherlands – as an Annex I country – must pursue a minimum safe reduction target of least 25 per cent by 2020, the Dutch courts also (in)directly suggested that EU reduction targets of 20 per cent by 2020 or 40 per cent by 2030, are non-compliant with obligations to effectively contribute to human rights protection under the ECHR.

Indeed, the Dutch courts explicitly noted on this point that the Netherlands cannot hide behind targets agreed within the EU. The Netherlands has its *own legal obligation* to ensure the effective protection of human rights under the ECHR (even if that obligation exists as part of a wider global effort to prevent dangerous climate change under the UNFCCC). Specifically, the Netherlands failed to explain how or why EU targets for 2020, and the Netherlands' (even lower) target of 17 per cent as a result of EU effort sharing, could be 'considered responsible in an EU context'. The Supreme Court observed that until 2011 the Netherlands had actually committed to a higher target of 30 per cent, and the EU had conditionally committed to 30 per cent by 2020 under the Kyoto Protocol.[148] All this supports that higher targets were in fact preferable. In terms of the feasibility of those targets, the Netherlands also failed to satisfactorily explain, on any scientific, economic or other grounds, why reducing more ambitiously was impossible or unreasonable. On the contrary, the government actually admitted it would be possible for the Netherlands to go faster, and that this would likely be more cost-efficient for society in the long run, as well as safer and more precautionary.[149]

As a result of these Dutch judgments, there was debate as to whether it was appropriate for Dutch courts to suggest (either directly or indirectly) the unlawfulness of the EU's collective

---

[145]   Ibid. paras 7.2.1–7.2.11.

[146]   Ibid. para. 7.2.9.

[147]   Ibid. paras 2.1, 4.1–4.4. 5.1–5.8, 7.2.10.

[148]   Ibid. paras 2.3.2, 7.2.6, 7.3.3, 7.4.6, also referring to Appeals Court.

[149]   Ibid. paras 2.3.1, 7.4.6, 5.2.1, 5.3.4, 5.5.3; District Court (n. 134) para. 4.76.

2020 targets, or to impose a higher target on the Netherlands, which might 'affect the EU legal order'. In particular, the CJEU may need to be given an opportunity to deliver a preliminary ruling on the matter.[150]

The *Carvalho* judgment, however, it seems to support that the current Effort Sharing Regulation lays down 'minimum contributions' for EU Member States towards Union targets for the period 2021–30 rather than maximum emissions limits.[151] Even so, there may be concerns that any additional reductions realized by EU Member States could be off-set by allowing lower emissions in another State through the ETS system (the so-called waterbed effect).[152] The Dutch courts responded to this by simply reaffirming the Netherlands' *own* legal responsibilities to do its minimum fair share and by stating that it cannot be automatically assumed that other EU Member States will implement less far-reaching measures. On the contrary, several Western European states are far more ambitious than the Netherlands.[153]

All in all, it may be concluded from the *Urgenda* case that the 'multi-layered' European legal order carries forth various implications and challenges, in terms of its diverse sources for both human rights protection and for EU and international climate law. Based on the *Urgenda* reasoning, it would seem that at least *the ECHR* places binding individual duties upon its States parties to mitigate GHGs sufficiently ambitiously, that is, in a safe and effective manner for human rights, in line with the objectives of the Paris Agreement. In turn, this may mean that the EU's and all other EU Member States' targets could also be called into question, including based on corresponding CFREU rights. As was evident in the previous sections, fundamental rights have so far not been thoroughly assessed in the context of EU climate target setting and policy-making.

Of course, the EU has now increased its collective targets to a 55 per cent reduction by 2030 and climate neutrality by 2050. Both targets seem to be on the upper limits of the 40–55 per cent and 80–95 per cent reduction targets previously supported by the IPCC, which could suggest they are human rights-compliant. At the same time, it is worth noting that these IPCC pathways may be outdated, and were supported in light of a maximum warming target of 2°C and not *well below* 2°C, or even 1.5°C as now agreed in the Paris Agreement. From this perspective, some further discussion on and appraisal of whether and which European targets should truly be considered a minimum fair share towards effectively contributing to human rights protection is welcome.

---

[150]   See e.g. Roy (n. 139); Peeters (n. 140); Backes and van der Veen (n. 140); Surya Roy, 'Distributive Choices in Urgenda and EU Climate Law', in M. Roggenkamp and Catherine Banet (eds), *European Energy Law Report XI* (Intersentia 2017) 47–70.

[151]   Case T-330/18, paras 11, 27.

[152]   Case T-330/18, paras 11, 27.

[153]   Supreme Court (n. 134) para. 2.3.2.

11.6 CONCLUSION

International and European human rights law clearly applies to climate action. It stems from a wide range of different legal instruments and rights recognized by the EU and EU Member States, both within the context of the EU and as parties to non-EU international legal instruments for the protection of human rights.

The 'multi-layered' nature of both international and European climate law and human rights law raises many legal questions, not only in terms of how to apply or interpret human rights within the context of each legal regime, but also in light of their interplay. While in the EU the CFREU is by now the leading human rights instrument for assessing the compliance of EU laws with human rights, it applies only insofar as the EU or its Member States are indeed implementing EU law. We saw that even then, EU rights have their origin in – and must still be interpreted in line with – other instruments, such as the ECHR and international human rights treaties.

Unfortunately, the fundamental rights culture within the EU institutions is still developing. The application and interpretation of CFREU rights to the design and implementation of climate change law specifically is rather poor. It is nevertheless important that the EU and its Member States pay proper attention to the protection of human rights in climate change action. In particular, the emerging trend of climate litigation shows that inadequate attention to human rights, or simply poor climate action, may lead civil society to call for better rights-based protection and scrutiny in courts.

In light of the serious threats posed by climate change, and the increasingly small windows of opportunity for States to 'turn the tide' on dangerous climate change, more citizens and civil society organizations may turn to fundamental human rights for a remedy. Time will tell whether rights-based litigation will be successful, or will become slowly redundant in the case that States step up in formulating and implementing sufficiently ambitious, safe and rights-compliant climate targets.

CLASSROOM QUESTIONS

1. Which human rights are relevant to climate change and what are the most important human rights instruments for their protection in the EU?
2. How can advanced interpretations of States' human rights obligations in international law or Council of Europe law be relevant to the application and interpretation of the EU Charter for Fundamental Rights?
3. To what extent are strict (interpretations of) *locus standi* requirements an important bar to access to courts in European climate cases? Is this desirable?

SUGGESTED READING

Books

Bogojevic S and Rayfuse R (eds) *Environmental Rights in Europe and Beyond* (Hart Publishing 2018).

Kellerbauer M, Klamert M, Tomkin J (eds) *The EU Treaties and the Charter of Fundamental Rights: A Commentary* (Oxford University Press 2019).

Wewerinke-Singh M, *State Responsibility, Climate Change and Human Rights under International Law* (Hart Publishing 2018).

Articles and chapters

Backes CW and van der Veen GA, 'Urgenda: The Final Judgment of the Dutch Supreme Court' (2020) 17 *Journal for European Environmental & Planning Law* 307–21.

Butler IDJ, 'Ensuring Compliance with the Charter of Fundamental Rights in Legislative Drafting: The Practice of the European Commission' (2012) 37 *European Law Review* 397.

Winter G, 'Armando Carvalho and Others v EU: Invoking Human Rights and the Paris Agreement for Better Climate Protection Legislation' (2020) Winter G, 'Armando Carvalho and Others v EU: Invoking Human Rights and the Paris Agreement for Better Climate Protection Legislation' (2020) 9 *Transnational Environmental Law* 137–64.

Policy documents

UN Human Rights Council Resolution 10/4. 'Human rights and climate change' (25 March 2009) UN Doc. A/RES/10/4.

O. de Schutter, 'The Implementation of the Charter of Fundamental Rights in the EU Institutional Framework: Study for the AFCO Committee' (European Union, November 2016).

HRC, Report of the Special Rapporteur on human rights obligations relating to the enjoyment of a safe, clean, healthy and sustainable environment on a safe climate (15 July 2019) UN Doc. A/74/161.

PART IV
CONCLUSION

12
The past and possible future of EU climate law

12.1 INTRODUCTION

The previous chapters provided an introduction to the 'essence' of EU climate mitigation law. The EU mainly targets carbon dioxide (CO_2) emissions from burning fossil fuels, but also other greenhouse gases, such as nitrous oxide (N_2O) from acid production, perfluorocarbons (PFCs) from aluminium production and fluorinated gases (F-gases) used in refrigerators and air-conditioners. We have analysed the directives and regulations related to EU climate action, as well as their implementation, and we have reviewed current policy and academic debates on the climate mitigation targets and instruments of the EU. What can we conclude from our analysis and review of essential EU climate mitigation law? And in particular, have the principles of cost-effectiveness and solidarity played a prominent role in the design of EU climate mitigation law, as we hypothesized in our introduction to this book?

This chapter links the findings of the previous chapters and gives the reader some food for thought about the next steps that the EU may or perhaps even should take to further improve climate mitigation law. Section 12.2 reflects upon the general lessons that can be learned from the development of EU climate law. Section 12.3 outlines some specific lessons for its cost-effectiveness and solidarity. Section 12.4 provides some insights into the broader picture of EU climate regulation by addressing the overarching issues relating to energy network regulation, multi-level governance and human rights. Section 12.5 considers EU climate law's possible future. Section 12.6 concludes.

12.2 GENERAL LESSONS FROM THE PAST

Three general lessons emerge from the development of EU climate mitigation law:

- Accelerating ambition;
- Growing complexity;
- Contested consistency.

12.2.1 Accelerating Ambition

For several decades, climate science has called upon governments throughout the world to strengthen their climate targets and instruments to avoid dangerous global temperature rise. We observe that the EU is increasingly responding to this call. The EU started off slowly. Back in the 1990s, it committed to reducing its greenhouse gas emissions by 8 per cent in the period 2008–12 below 1990 levels. Although this was more than the average reduction of 5 per cent promised by industrialized countries under the Kyoto Protocol (1997), the EU ended up surpassing its own mitigation target by reducing emissions by almost 12 per cent by 2012. A stronger reduction target of 20 per cent followed for 2020. Again the EU ended up outperforming its target by reducing emissions by around 24 per cent. For 2030 the EU initially aimed for a reduction of 40 per cent, under the Paris Agreement (2015), in order to achieve a reduction of 60 per cent in 2040 and ultimately 80–95 per cent in 2050. However, only a few years later the EU tightened its 2030 target from 40 per cent to a reduction goal of at least 55 per cent, under the European Green Deal (2019). This culminated in the proposal for a 'European Climate Law', a Regulation developed in early 2020 by the European Commission aiming for climate-neutrality of the EU by 2050 at the latest.[1] We observe not only ever deeper emission reduction targets, but also an increasing number of climate-related directives and regulations in the EU (from 25 in 2012 to 37 in 2020).[2] The Union's climate ambitions are clearly accelerating. Greenhouse gas emission reductions were foreseen from the start, back in the 1990s, but in the past couple of years the EU has replaced its linear mitigation path with an exponential increase of its climate ambition level. The EU's narrative has changed accordingly, from climate 'package' and 'framework' to climate 'law'. This sends a robust signal to companies and governments that the EU is serious about phasing out fossil fuels, which helps to reduce investment uncertainty in zero-carbon technologies and stimulates the international community to take comparable climate measures.

12.2.2 Growing Complexity

With the expansion and deepening of EU climate mitigation law, its complexity has increased dramatically. As implementation problems arose, additional measures were taken to address them. An example is the EU emissions trading system (EU ETS). The political issue of windfall profits incurred by companies with free emission allowances, for instance, led to a shift towards allowance auctioning. An Auction Regulation was thus added to the EU ETS, with auctions supervised by multiple agents (including an auction monitor, the European Commission and several national authorities) on the basis of various related directives targeting financial instruments (MiFID) and market manipulation (MAD). The amended and later revised EU ETS Directive introduced detailed and complicated rules on free allocation

---

[1]   Proposal for a Regulation of the European Parliament and of the Council establishing the framework for achieving climate neutrality and amending Regulation (EU) 2018/1999 (European Climate Law), COM/2020/80 final.

[2]   Climate Change Laws of the World (database): https://climate-laws.org/geographies/european-union.

for production growth, capacity expansion and installation closure for producers in industries exposed to international competition. A related issue was the reduced international competitiveness of large industrial energy consumers, such as steel companies, who saw their electricity bills increase because power companies passed carbon costs to them under the EU ETS. As a result of successful industry lobbying, state aid rules were updated to partly compensate them. Another problem was the huge allowance surplus, mainly caused by the economic crisis of 2008, which resulted in a market stability reserve (MSR) to take in and partly cancel some surplus allowances by controlling the allowance auction volumes. There was also value-added tax (VAT) fraud in the EU ETS which led to an amendment of the VAT Directive, while a new Registries Regulation was adopted to combat allowance phishing by computer hackers. The list goes on and on – and this is just in relation to the example of emissions trading. More or less similar complexity issues also arose in some other areas of EU climate law, such as renewable energy and energy efficiency legislation. Consider for instance the detailed Governance Regulation added in 2018 to moderate the non-binding character of Member States' individual contributions to the Union's renewable energy and energy efficiency targets. Even experienced climate law researchers say they find it challenging nowadays to oversee and fully understand the increased complexities of EU climate law.[3]

12.2.3 Contested Consistency

As a result of the increased complexity of EU legislation for climate action, partly caused by the political desire to regulate ever more aspects of climate change against the background of industry lobbying, some policymakers and scholars are challenging the consistency of EU climate mitigation law. Some perceive a 'clash of the climate laws': a partly inefficient policy interaction between the various climate targets and instruments in Europe. For instance, the directives for renewable energy and energy efficiency help to lower emissions but also lower demand for allowances under the EU ETS Directive, thereby undermining the allowance price. The MSR and its allowance cancellation mechanism should help prevent this, but there is still academic debate about the extent to which it does so. To the extent that it does not, the consequences would be postponed abatement and a weakening of the incentive to invest in renewable energy and energy efficiency. Others, however, stress that, for instance, the Renewable Energy Directive stimulates innovation of zero-carbon technologies and helps to repair some of the regulatory imperfections in the EU ETS. There are many other types of alleged inconsistencies in EU climate mitigation law. For instance, Member States employ support schemes to encourage investment in renewable energy production, but legal questions have arisen over their compatibility with TFEU provisions on the free movement of goods and state aid. Another example is carbon capture and storage (CCS), which is encouraged with subsidies, while offshore CCS is partly discouraged because the transport of captured CO_2 by ship (instead of by pipeline) to an offshore storage reservoir does not allow the transferring operator to subtract the transferred CO_2 from its yearly emissions. Consistency of EU climate law is also increasingly contested before the European Court of Justice. A continuous process

---

3   M. Peeters (2019), 'EU Climate Law: Largely Uncharted Legal Territory', *Climate Law* 9(1/2): 137–47.

of policy reconsiderations and legal rulings will help to create more clarity and unity in the EU's increasingly ambitious and complex legal framework for climate action.

12.3 SPECIFIC LESSONS FOR COST-EFFECTIVENESS AND SOLIDARITY

The principles of cost-effectiveness and solidarity indeed play a prominent role in EU climate mitigation law, as we hypothesized in the Introduction to this book. Two specific lessons emerge from the past of EU climate regulation:

- Improved but imperfect cost-effectiveness;
- Partial but sustained solidarity.

12.3.1 Improved but Imperfect Cost-effectiveness

Various laws and regulations are meant to improve the cost-effectiveness of EU climate law. Consider the possibility for companies covered by the EU ETS to trade emission allowances (EUAs), the limited transfers of annual emission allocations (AEAs) from one Member State to another under the Effort Sharing Regulation (ESR), the cooperation mechanisms that Member States have at their disposal to jointly achieve their targets for increasing renewable energy consumption, or the regulation that allows quotas for bulk hydrofluorocarbons (HFCs) to be transferred to another producer or importer in the EU. However, there are also regulatory developments that have reduced the cost-effectiveness of EU climate law, for instance due to industry lobbying and additional political demands. Examples are the financial compensation offered by taxpayers for the increased electricity bills of energy-intensive industries under the EU ETS; the uncertain financial securities required in relation to CCS, which raise costs for project developers; or the addition of a domestic carbon price floor in a number of Member States. Some of those interventions help to boost societal or industry acceptance of climate policy, but they also imply regulatory complexities and inconsistencies that prevent a full realization of the cost-effectiveness potential of the various (market-based) instruments of EU climate law. Cost-effectiveness thus plays an increasingly important role in EU climate law, although some inefficiencies remain.

12.3.2 Partial but Sustained Solidarity

Solidarity between Member States is sustained in most but not all areas of EU climate law, but this solidarity is only partial. Solidarity is clearly present in the differentiation of green-house gas emission reduction targets: Member States with a relatively high GDP per capita must achieve higher emission reductions than Member States with a relatively low GDP per capita. Solidarity can also be found in the design of the instruments of EU climate law. For instance, a small but significant part of the allowances to be auctioned under the EU ETS is redistributed from Western to Eastern European Member States. Another case in point is the Energy Efficiency Directive, which takes into account the higher costs of increasing energy efficiency in more economically advanced Member States. But there are also areas of EU climate law where solidarity is still underdeveloped. The regulation of fluorinated gases, for instance, does not contain any elements of solidarity. The 2018 Renewable Energy Directive requires Member States to collectively ensure the 32 per cent renewable energy consumption target by 2030. However, the contributions by Member States to this binding EU target are non-binding. The implication is that the Commission can only make recommendations to the Member States on how to improve their renewables policies. This move away from binding targets at Member State level will actually require a higher level of solidarity between EU countries to reach the Union's overall renewables target. The same issue of non-binding targets at Member State level applies to the Energy Efficiency Directive. The CCS Directive, another example, neither prescribes a common EU approach for the selection and develop-ment of CO_2 transport and storage infrastructure nor elaborates upon cross-border transport and storage of CO_2. This makes it more difficult for Member States without substantial CO_2 storage capacities to deploy CCS technologies. The absence of solidarity in the latter case may also have a detrimental effect on the cost-effectiveness of climate action in the long term. In most cases, however, solidarity does not compromise the cost-effectiveness of EU climate law as such, since it merely changes the intra-EU distribution (so not the efficiency) of the efforts required to combat climate change. Solidarity then takes some of the burdens of EU climate law away from the poorer Member States and places them on the shoulders of the richer ones, which is crucial for political acceptance in Member States with a relatively low GDP. Solidarity is certainly an element of EU climate law, as a crucial political lubricant, but it also appears to have its limits as a principle of European cooperation.

12.4 THE BROADER PICTURE OF EU CLIMATE REGULATION

Three lessons emerge from some of the so-called overarching issues of EU climate regulation:

- Path-dependent energy network regulation;
- Reinforced multi-level governance;
- Emerging human rights culture.

12.4.1 Path-dependent Energy Network Regulation

An important overarching issue for EU climate mitigation law involves the transport of renewable energy and thus, in particular, the networks connecting energy production and final consumption. As energy networks are key in facilitating the production and use of renewable energy, they play an indirect role in combating climate change which is, moreover, characterized by path-dependence. Path-dependence means that current and future actions may be constrained by earlier choices.[4] This clearly applies to energy network regulation: the legal framework and the qualification of types of networks arise from a period in which the energy system was based on a centralised approach where energy was fed into the upper level of the grid and extracted by consumers at the lower level of the grid. Due to the increasing number of renewable energy production facilities connected to the lower levels of the grid, this approach is changing. As a result, the law needs to be amended to facilitate a decentralised and bottom-up approach to network regulation. Apart from facilitating the transport of energy from renewable energy sources via existing networks, the decarbonisation process involves the development of a new type of networks, such as offshore cables connecting wind parks to shore and possibly networks facilitating local energy communities. These 'new' networks require a legal qualification and an assessment as regards the type of network management. Similar challenges will arise with regard to 'green hydrogen'. So far Member States have often taken the lead in these new developments and amended national laws to facilitate the development of new networks necessary for the decarbonisation process. A more goal-setting and coordinating EU approach governing the decarbonisation of networks may be a prerequisite to meet the net-zero greenhouse gas emissions target for 2050.

12.4.2 Reinforced Multi-level Governance

Another overarching issue in relation to EU climate regulation is multi-level governance between international climate agreements, EU climate policy, climate action by individual EU Member States and climate measures taken at sub-national and corporate levels. Multi-level governance is not just inherently complicated and problematic, but is also an unavoidable political reality that not only reinforces itself but also reinforces the role of the EU in shaping and administering climate mitigation law. In particular, where markets play a considerable role in shaping climate governance, the EU is a predominant player. International climate negotiations resulted in the Paris Agreement but the nationally determined contributions around the world are insufficient to limit global warming to 1.5 degrees Celsius, while EU Member States' implementation of various directives related to climate action suffers from delays and imperfections. EU climate policy stands in between and struggles with the growing complexities, contested consistencies and elements of inefficiency as outlined above. Moreover, implementing EU climate policy requires action primarily at the level of EU Member States, while the companies involved exert influence over and experience the effects of EU climate law at all three superordinate levels (international, EU and national). However, both climate law and energy market integration in the EU stimulate cost-effectiveness and solidarity between the Member States, for instance resulting in opening up incentive schemes for low-carbon technologies to companies across the EU. The involvement of the industry level and the role of markets in the Union's climate policy in general, as well as the need to take into account the international level, actually reinforces the role of the EU in a multi-level governance scheme that has been largely shaped by general EU law.

---

[4]   E. Woerdman (2004), 'Path-Dependent Climate Policy: The History and Future of Emissions Trading in Europe', *European Environment* 14(5): 261–75.

12.4.3 Emerging Human Rights Culture

A third overarching issue is the limited yet increasing role of human rights in EU climate law. Climate change may have an impact on health and a clean environment and could thus negatively affect nearly all human rights. The EU and its Member States are legally bound by their international and European human rights obligations when taking climate action, but law-makers in Europe often fail to adequately take human rights into account when designing and applying climate laws. This has triggered some human rights-based climate litigation, including the notorious Dutch case of *Urgenda v the Netherlands*, but applicants may face difficulties in claiming that they are personally affected by climate change. The EU Charter on Fundamental Rights (CFREU) is the leading human rights instrument for assessing compliance of EU climate law with human rights, but it applies only insofar as the EU or its Member States are actually implementing EU law. The fundamental rights culture within the EU is still developing and the application and interpretation of CFREU rights to climate law is limited. With mounting evidence of increasing human-induced climate damage, we expect to see more citizens and civil society organisations turning to fundamental human rights for a remedy.

12.5 EU CLIMATE LAW'S POSSIBLE FUTURE

Will the future of EU climate law present a straight path or a bumpy road towards carbon-neutrality in 2050? Based on the findings of this book, we argue that EU lawmakers should strive to (a) strengthen the effectiveness, (b) reduce the complexity and (c) increase the consistency of climate mitigation law. In particular, it would be desirable to (d) further improve its cost-effectiveness and (e) expand the solidarity between Member States in the context of EU climate mitigation law. However, it is not easy to attain these goals, for the following three reasons.

First, attaining these goals is not easy because there is no academic or political consensus on how big the aforementioned problems actually are, nor any greater detail on what the desired situation would look like. For instance, how serious is the 'clash of the climate laws' and, thus, how inconsistent is EU climate law as it currently stands? Is a 2030 greenhouse gas emission reduction target of 55 per cent, as foreseen by the European Council, sufficient to contribute to the global 1.5 degrees Celsius goal of the Paris Agreement? Should the EU adopt a carbon price floor in its ETS to stimulate zero-carbon technologies or would that make the 2050 net-zero emissions target unnecessarily expensive? Different studies appear to provide different answers.

Second, attaining these goals is not easy because there is no academic or political consensus on which (combination of) instrument(s) would be most suitable to realise these goals. For instance, should the EU strengthen its climate mitigation law by expanding the EU ETS to more sectors of the economy and by restricting the application of the directives for renewable energy and energy efficiency, or would that significantly raise complexity and undermine the Union's energy security? Should the EU increase its subsidies for CCS or should each EU Member State totally abandon the idea of storing CO_2 underground? Should a 'two-speed climate policy Europe' be allowed to emerge, where only the economically more advanced Member States impose domestic carbon taxes on sectors already regulated under the EU ETS, or would this undermine the consistency and solidarity within EU climate law?

Third – and this also becomes clear from the examples given above – attaining the afore-mentioned goals is not easy because satisfying one goal may come at the expense of reaching another. For instance, should the EU stimulate cost-effectiveness by linking its EU ETS to emissions trading schemes in countries with allowance price ceilings or with weaker levels of climate law enforcement, thus potentially undermining effectiveness? Should EU climate mit-igation law be strengthened by making the renewables and energy efficiency targets of the EU binding for individual Member States or would that infringe upon principles of cost-effective-ness and subsidiarity?

Due to these unavoidable trade-offs and uncertainties, the future of EU climate law is prob-ably one in which issues of complexity and consistency remain. We expect these problems to be added to, not only by continued lobbying by industries and NGOs but also by newly emerging political desires, as has been the case in the past. EU climate mitigation law, with its multiple directives and regulations, is firmly established in EU law and is still evolving. Due to path-dependence, which means that future actions will to some extent be constrained by earlier choices, it will not be easy to rebuild EU climate mitigation law or to change its course. But change is not impossible – and it is under way. An example is the requirement for each Member State to establish an integrated national energy and climate plan (NECP) for the period 2021–30, which stimulates coordination within and cooperation between EU countries. Another case in point is the emergence of energy communities in which end-consumers can be actively engaged in the production and supply of electricity as well as in grid management.

The path-dependence of EU climate mitigation law also presents an advantage. For many years now, the EU has been seriously dedicated to its greenhouse gas targets, such as those for 2020, 2030 and 2050, with mostly persuasive monitoring and enforcement mechanisms in place. Therefore, we do expect the Union to further strengthen the effectiveness of EU climate mitigation law and take measures accordingly. Justified or not, the EU has always perceived itself as the world's leader in climate action and we have found no signals that the EU intends to abandon this path. Although the effectiveness of EU climate mitigation law faces several challenges, such as the emission allowance surplus under the ETS or the non-binding nature of its renewables and energy efficiency targets at Member State level, the EU surpassed its greenhouse gas emission reduction targets for 2008–12 and for 2020. The 2030 target is likely to be met, and possibly even surpassed again, as a result of technical progress that enables companies to save energy and produce power and heat at increasingly lower costs.[5]

Moreover, the EU's sincere desire to expand climate action across the globe in a fair and cost-effective way is likely to lead to limited or partial linking. The EU ETS may be coupled with a few other emissions trading systems, probably only in industrialised countries with well-developed administrative infrastructures, and even then there are many legal, economic

---

5   E. Woerdman (2019), *Energietransitie en klimaatbeleid: tussen marktwerking en overregulering*, inaugural lecture [in Dutch] https://papers.ssrn.com/sol3/papers.cfm?abstract_id=3361527.

and political hurdles to overcome. Although this is difficult to predict, bottom-up climate policy via linking is therefore likely to be fragmented (with only a few countries linking their carbon trading schemes), just as top-down climate policy used to be (with only a limited number of countries accepting absolute emission caps and implementing emissions trading systems in the first place). In such an international context, it will also be exciting to see what will become of the Carbon Border Adjustment Mechanism (CBAM): an import levy at the borders of the EU to reduce the risk of carbon leakage, as proposed by the European Commission in its 2019 Green Deal. This would not only help to ensure that the prices of imports more accurately reflect their carbon content, but is also likely to lead to international political turmoil and legal discussions about compatibility with World Trade Organization (WTO) rules.

The biggest challenge to the effectiveness of EU climate law is probably its international context. Where the EU managed to reduce its greenhouse gas emissions by 24 per cent between 1990 and 2020, in spite of all its regulatory imperfections, outside Europe those emissions will continue to increase for years to come, mainly due to the absence of yearly declining absolute emission ceilings in China, India and the United States. As a result, despite the expected developments in the EU, we are heading for a global temperature rise of at least 3 degrees Celsius by the end of this century. To keep the temperature rise well below 2 degrees and preferably to 1.5 degrees Celsius, as internationally agreed upon in the Paris Agreement, it would be necessary for global greenhouse gas emissions to have peaked in 2020. However, with countries' current climate plans (Nationally Determined Contributions), this peak will not even be reached in 2030.[6] In spite of EU climate law, the earth is therefore likely to have warmed by around 3 degrees Celsius at the end of this century – with all the severe climatic consequences that this entails.

Having presented our thoughts on EU climate law's possible future, it is also possible to take a more practical view on improving the implementation of EU climate law as it now stands. The insights of the chapters in this book lead to some suggestions. For instance, the provisions for production growth, new entrants and installation closures under the EU ETS should be reconsidered, because free allowances to cover extra emissions trigger overinvestment in production capacity and because losing allowances upon closure provides a disincentive for closing old plants. Furthermore, the EU should elaborate the rules for cross-border CCS regulation and develop a common EU approach for the selection and development of CO_2 transport and storage infrastructure. Moreover, the EU needs to break out from path-dependent energy network regulation by providing further guidance at EU level in order to promote the decarbonisation of the energy networks. The need to switch from a top-down to a bottom-up regulatory approach has a clear impact on the development and use of energy networks, especially given the new role of small customers in energy production and supply. The decarbonisation process leads to the development of new types of energy network, both at the upper level of the grid (such as offshore park-to-shore cables) and at the lower level (including smart

---

6   UNEP (2018), Emissions Gap Report 2018, Nairobi: United Nations Environment Programme, p.21. See also S. van Renssen (2018), 'The Inconvenient Truth of Failed Climate Policies', *Nature Climate Change* 8: 355–8.

networks facilitating energy communities). Although the level of EU guidance differs per type of network project, a common EU approach should be aimed at, which thus should not be limited to those projects that have a cross-border effect.

12.6 CONCLUSION

After having examined all Union legislation related to climate action, we wonder: what is the 'essence' of EU climate mitigation law? We realise that there is probably no single correct answer. Nevertheless, we are inclined to say that its 'essence' is an increasingly ambitious legal framework of partly overlapping targets and instruments that together succeed in reducing greenhouse gas emissions in the EU, with some degree of cost-effectiveness and solidarity between its Member States.

The EU aims to be a global leader in climate action. EU climate mitigation law expanded and deepened in the past and will continue to do so in the future. Cost-effectiveness and solidarity are recurrent themes in the design of EU climate mitigation law, as previously noted, and they will continue to be key elements of any changes or additions to this dynamic area of EU law. That is not to say that EU climate mitigation law is fully cost-effective: some inefficiencies remain, for instance due to cost-raising subsidies for CCS or the numerous exceptions from carbon cost internalization under the EU ETS for industries exposed to international competition. Solidarity is and will be necessary to obtain political acceptance of climate action in Member States with lower levels of economic welfare, for instance by redistributing some allowances to them under the EU ETS or by submitting their non-ETS sectors to a lower level of emission mitigation compared to wealthier Member States.

The complexity of EU climate law has increased dramatically over the years. Its consistency is sometimes contested due to policy interactions, not only between the various climate targets and instruments themselves (including those for emissions trading, renewable energy and energy efficiency) but also with other related policy goals, such as protecting the competitiveness of European industry and safeguarding security of energy supply. Furthermore, there are some limitations to the effectiveness of EU climate mitigation law, for instance due to the emission allowance surplus under the EU ETS or the non-binding nature of the energy efficiency targets at Member State level. Fortunately, several mechanisms have been adopted or are proposed to strengthen the effectiveness of EU climate law, such as the Market Stability Reserve, the Governance Regulation and the increasing role of the courts, but their impacts on cost-effectiveness are debatable. EU climate law will most likely continue to be complex and contested, as well as ambitious and persistent, for many years to come.

INDEX

Printed and bound by CPI Group (UK) Ltd, Croydon, CR0 4YY

27/10/2024

14580414-0001